D0311903

HALF LIFE

ALSO BY FRANK CLOSE

HALF LIFE

The Divided Life of
Bruno Pontecorvo, Physicist or Spy

———————

FRANK CLOSE

ONEWORLD

A Oneworld Book

Published in Great Britain and Australia by Oneworld Publications, 2015
Originally published in the United States of America by Basic Books, a member
of the Perseus Books Group, 2015

ISBN: 978-1-78074-581-7
Ebook ISBN: 978-1-78074-582-4

Designed by Trish Wilkinson
Printed and bound in the UK by CPI Group (UK) Ltd, Croydon, CR0 4YY

Oneworld Publications
10 Bloomsbury Street
London WC1B 3SR
England

To Abingdonians, present and past

Contents

SECOND HALF

AFTERLIFE

Preface

"DID MI5 GET BACK TO YOU AFTER I FORWARDED THEM YOUR LETTER?"
When I started to research the life of Bruno Pontecorvo, the nuclear physicist who disappeared through the Iron Curtain at the height of the Cold War in 1950, I didn't anticipate receiving such an inquiry, let alone replying in the affirmative. Nevertheless, my correspondence with the British intelligence agency led me to solve a sixty-year-old enigma: Why did Pontecorvo flee so suddenly, just a few months after the conviction of his colleague, atomic spy Klaus Fuchs? The obvious answer—that Pontecorvo was "the second deadliest spy in history," as the US Congress later described him—has hung around for decades, but no proof that he passed atomic secrets to the Soviets has ever been presented, nor has there been any suggestion of the information he might have disclosed. Contrary to popular wisdom, neither the FBI nor MI5 ever located evidence against him. So if Bruno Pontecorvo was a spy, he was most successful. Pontecorvo, a communist who had managed to evade detection and join the Manhattan Project, always insisted that he fled for idealistic reasons, having felt persecuted following Fuchs's arrest.

Bruno Pontecorvo's passage through the Iron Curtain split his life into two almost-equal halves. This chronological split defined his scientific life: great insights at the end of the first half were frustrated by his move to the Soviet Union and may have cost him his share of a Nobel Prize. His personality was also divided into two complementary halves. On one hand there was Bruno Pontecorvo, the extroverted, highly visible, brilliant scientist, and on the other was his alter ego: Bruno Maximovitch, the enigmatic, shadowy figure who was secretly committed to the communist dream.

There are already two excellent books that provide extensive assessments of Bruno Pontecorvo: *The Pontecorvo Affair*, by Simone Turchetti, and *Il lungo freddo*, an Italian text by Miriam Mafai. Turchetti focuses on the first half of Pontecorvo's life, the political implications of his defection, and how the British government in particular downplayed his significance at the time of his disappearance. I have profited on many occasions from discussions with Turchetti, not least in evaluating some of the new facts that have come to light during my own investigation. Mafai's book is Bruno's life story as he would wish it to appear, based on a series of interviews with Pontecorvo late in his life.

Half Life takes a different approach. I am myself a physicist, and so I focused initially on Bruno Pontecorvo's life as a scientist. Klaus Fuchs, Alan Nunn May, and other players in the atomic spy saga were quality scientists, but are known only because of their role in the passing of secrets; Pontecorvo is unique in that he could merit a biography for his scientific contributions alone. The fact that his name has long been associated with those of proven atomic spies simply adds to his interest. Thus I also sought to understand his value to the USSR once he arrived there, to assess what information he could have transmitted to the Soviets before 1950, and to filter truth from myth with regard to his real agenda. I do not examine in any detail the interactions between MI5, the FBI, and their respective governments, mainly because Turchetti in his book, and Timothy Gibbs in his Cambridge University PhD thesis, have already done so. Nor do I offer any sociopolitical commentary on his political beliefs or his reactions to the profound changes he experienced during the dissolution of the USSR; Mafai has covered this, although her personal communist perspective mingles with that of Pontecorvo, and it is not always clear whether her views or his are on display. In order to make the scientific concepts digestible, I have avoided technicalities in several places. Readers who want a more in-depth study of Pontecorvo's work and its context can find it in the article "Bruno Pontecorvo: From Slow Neutrons to Oscillating Neutrinos" by Luisa Bonolis.

Frank Close
Abingdon, March 10, 2014

Prologue: Midway on Life's Journey
1950: The Gathering Storm

NEW YEAR'S DAY 1950: THE FULCRUM OF THE TWENTIETH CENTURY. When the century began, no one knew that the atomic nucleus existed, let alone that it was the custodian of huge reserves of energy. By the century's end, humanity had learned to live with the possibility of a thermonuclear holocaust. As 1950 dawned, however, less than five years had passed since the atomic bombs at Hiroshima and Nagasaki had ended World War II, and society was only beginning to realize the awful implications.

In a historic English market town near Oxford, one of the fathers of the atomic age celebrated the New Year with his family. Bruno Pontecorvo was thirty-six years old. Sixteen years earlier, as a student of physics, he had contributed to a discovery that would herald a new world of nuclear reactors and atomic weapons. That breakthrough would determine his destiny. By 1950, he had earned a reputation as one of the world's leading nuclear physicists, had recently published two papers that would lead to Nobel Prizes, and was being courted by physics institutions in both Europe and North America. This brilliant Italian scientist appeared to have an idyllic life. He lived comfortably in a pleasant home near the River Thames. He had an attractive Swedish wife and three young sons.

All seemed perfect, carefree. But Bruno Pontecorvo had a secret.

For more than ten years he had been a member of the Communist Party. At first glance, this might hardly seem to merit comment. Many intellectuals who had grown up in the 1930s and witnessed the vicious effects of fascism had chosen to ally themselves with the communist movement. By 1950, however, anticommunist hysteria was growing in the West and many lives were being ruined. For Bruno it was imperative that his communist links remain secret. During World War II his work had related to the atomic bomb, and now he was again engaged in secret work, at Harwell in the heart of England, where the United Kingdom was building the first nuclear reactor in Europe.

As it happened, the British and American intelligence agencies were already interested in Dr. Pontecorvo, and during the next few months their files on him would grow rapidly. Before February 1950, when his colleague Klaus Fuchs was arrested for passing atomic secrets to the Soviet Union, Bruno Pontecorvo's communist beliefs did not hinder his work or his life in general. Everyone involved in classified work had a security file; Pontecorvo was but one among many. But hysteria grew after Fuchs's arrest and conviction. Events were about to move out of Pontecorvo's control, leading ultimately to his midlife crisis.

GUY LIDDELL, DEPUTY DIRECTOR GENERAL OF THE BRITISH SECURITY service, MI5, sat down before his diary. A new year meant a new book, its blank pages soon to be filled with an insider's personal record of international affairs. On New Year's Day 1950, the implications of atomic weapons were at the top of his agenda.[1]

The scientists who had built these weapons were regarded as heroes. They had managed to unleash the immense forces contained within the atomic nucleus of a rare form of the element uranium, and also of a newly synthesized element, plutonium. It is hard today to fully appreciate the cataclysmic impact these developments had on the international scientific community. The war against fascism had been won, but with a Faustian pact: victory came with the release of the atomic genie. The explosions at Hiroshima and Nagasaki had shocked the world, but scientists already knew that even more devastating weapons were feasible. The fact that a major industrial city could be flattened by a single atomic bomb was bad enough; the next stage of nuclear technology, involving thermonuclear or "hydrogen" bombs, would have the potential to destroy life on earth.

The Western Allies briefly thought they had the power to rule the world, as they were the exclusive owners of these terrifying new weapons.

The United Kingdom, especially, was proud that its scientists had first conceived of the new technology, and then played leading roles in its development. However, the illusion of Western omnipotence was shattered irrevocably in 1949. Liddell began his diary, "The event of [1949] has been the explosion of an atomic bomb in Russia, which has thrown everyone's calculations out of date." Although the USSR had been an ally in the war against the Nazis, it had not been party to the Manhattan Project, which built the atomic bomb. As the West's relationship with the USSR grew increasingly tense in the years after the war, this fact provided some solace. Between 1945 and 1950, however, there were disquieting clues that some of those heroic Western scientists had been passing secret information about the weapon to Moscow. The Soviet Union had survived the war with large military reserves, capable of threatening American dominance. If Soviet espionage managed to neutralize the West's trump card (exclusive possession of the atomic bomb), the USSR would be a formidable enemy.

The first hint of this duplicity had come as early as the fall of 1945. That was when Western intelligence agencies learned that British physicist Alan Nunn May had taken samples crucial to the atomic bomb project from his laboratory in Canada and passed them to the Soviet Union. By the start of 1950, Western counterespionage had discovered the treachery of Klaus Fuchs as well.

Fuchs had passed substantial information to the Soviets, both during the war (when he was working on the atomic bomb at Los Alamos) and later, following his move to Harwell. Indeed, he transmitted enough high-quality data to the USSR to threaten the balance of world power. Liddell concluded his diary entry for the first day of 1950 by writing, "It is clear that by 1957 the Russians should have sufficient atomic bombs to blot this country out entirely."

He meant it. Two bombs had reduced the major Japanese cities of Hiroshima and Nagasaki to rubble. Given that a city could be destroyed by just a single atomic bomb, the Soviet Union could decimate the nerve centres of Great Britain with little more than a handful. If that wasn't enough to worry him, thermonuclear hydrogen bombs were already being developed in the United States and, we now know, in the USSR. In 1957, when British prime minister Harold Macmillan asked his science adviser, Sir William Penney, how many H-bombs would render the United Kingdom useless,

Penney (a gentle, peaceful man, who was no Dr. Strangelove) replied, "Five! Or let's say eight to be on the safe side."[2]

The atomic spies had given the USSR a fast track to this Promethean technology. Instead of a Western monopoly on nuclear weapons, the world now headed toward an unstable balance of mutually assured destruction.

In February 1950, Klaus Fuchs was arrested in London. The interrogation of Fuchs soon led to the arrest of Harry Gold, his courier in the United States, where Fuchs had worked (and spied) during the war. Gold's arrest and confession led the FBI to a Soviet spy ring, which included David Greenglass as well as Julius and Ethel Rosenberg, who were destined for execution by electric chair. In the US, Senator Joseph McCarthy started a witch hunt with his claim to have a list of 205 communists working at the heart of the American government. Today we know that this was fantasy, but in the ensuing hysteria it became risky for Americans to express views that were even slightly left-of-centre. One chemist who worked at the University of Wisconsin, in McCarthy's home state, later recalled, "We would not talk about anything if a third person might be listening."[3]

Until 1950, Bruno Pontecorvo had successfully hidden his life as a communist, but now his well-kept secret was threatened. A sister and a brother were also communists, as was a cousin, Emilio Sereni, who worked for the Italian government. It would be easy for Western intelligence agencies to discover this, should they choose to investigate. Bruno felt certain that they would do so. For, with the exposure of Klaus Fuchs, his colleague at Harwell, lightning had struck twice in his vicinity: four years earlier, Bruno Pontecorvo had been working in Canada alongside Alan Nunn May.

The British security services interviewed Pontecorvo in March, and again in April. His security clearance was withdrawn, and the authorities prepared to transfer him to a university, away from classified work at Harwell. In the feverish atmosphere of the times, Pontecorvo's communist associations were enough to foster suspicions that he too had passed secrets to the USSR. The intelligence agencies had no proof, but Pontecorvo was in the dark as to the contents of MI5's files. Klaus Fuchs and Alan Nunn May were prosecuted because they had confessed, and offered the intelligence agencies critical information that would condemn them. The atomic spy Ted Hall, by contrast, admitted nothing, was never arrested, and only came to public attention decades later. In Pontecorvo's case, there was one crucial question: Who would blink first in the game of cat and mouse? Suddenly, Pontecorvo disappeared, along with his wife and their three sons—Antonio (age five), Tito (age six), and Gil (the eldest at twelve). They went on holiday to Italy, flew to Stockholm and Helsinki, and then disappeared completely, only resurfacing five years later—in the USSR.

THE BRITISH GOVERNMENT DIDN'T HAVE A CLUE WHAT HAD HAPPENED. Bruno's brother Guido, who lived in Glasgow, didn't either. As of September 1, 1950, Bruno Pontecorvo had officially vanished. Even several years later, Pontecorvo's former teacher and mentor Enrico Fermi remained ignorant of his whereabouts, as Fermi's biography, written by his wife, Laura, in 1954, makes clear:

> Over three years have now passed since the Pontecorvos' disappearance. No word has been heard from them. Nobody has seen them. Their relatives deny knowing anything about them. Enrico and I have come to accept that Bruno and his family have probably passed to the other side of the Iron Curtain.
>
> The British Government has made no charge against Bruno. If anything at all has been found in England that could be construed as evidence against him, the existence of this evidence has never been revealed. And all this happened in the twentieth century![4]

I read Laura Fermi's book in the 1960s, when I was a physics student at Oxford. Her tale of the unsolved mystery of Bruno Pontecorvo was the first time I had heard his name. So it was a shock when, a few weeks later, I saw a new article in the scientific journal *Physics Letters* written by none other than Bruno Pontecorvo. His professional address was given as the Joint Institute for Nuclear Research (JINR), in Dubna, near Moscow. "Did anyone realize?" I wondered. My interest in Pontecorvo began at that moment.

Fortunately, the head of my group at Oxford was Rudolf Peierls, a man whose knowledge of the atomic bomb and the spy sagas it had spawned was second to none. In 1940, Peierls had calculated that an atomic explosion would require no more than a few kilograms of a rare form of uranium, a calculation that would prove crucial to the Allies' Manhattan Project, which culminated in the explosions over Japan. He went on to play a central role in the project at Los Alamos, where he had worked actively with Klaus Fuchs and been his closest friend. After the war, they both returned to the UK, and when Fuchs was exposed as an atomic spy, Peierls came under suspicion himself. Peierls had also known Pontecorvo, and told me that Pontecorvo's defection had been hardly less of a shock to him than Fuchs's exposure. Many colleagues dismissed the idea that Pontecorvo, extroverted and superficially naive, could have been a spy. However, in Peierls's opinion Fuchs had also given no hints of his secret life, so one could never be sure.

But there was one thing Peierls did know: Pontecorvo's presence in the USSR had been revealed in 1955. The Soviets, for reasons best known to them, had kept his defection a secret for five years, and then suddenly revealed him to the world's media. As shocking as it was, the disclosure had actually explained very little; questions as to why he had defected so suddenly, whether he had been coerced, and whether he had anything to hide remained unresolved.

Moreover, even after Pontecorvo's location had been revealed, little had been seen of him. His research reports were published in Russian journals, their English translations appearing only months later. These papers were like the tip of an iceberg, the visible sign of Pontecorvo's professional existence, while his life, and the circumstances that took him to the USSR, remained out of sight.

In 1973 I went to Eastern Europe to attend a specialized physics school, where I met Russian scientists for the first time. Pontecorvo had been listed as one of the lecturers, but he never materialized, and one of his research collaborators stood in for him. One evening, my confidence boosted by vodka, I asked my Russian colleagues about Pontecorvo but came away little wiser. Perhaps my companions were more expert with vodka than I. Pontecorvo the scientist was easy to learn about, but the man was an enigma. *Charismatic, extrovert,* and *life and soul of the party* were the headline descriptions, sentiments that I have heard repeated subsequently by numerous other colleagues. However, as to what lay behind this exterior, I learned little.

One of the Russians I got to know during this time was Alexei Sissakian. Alexei, who decades later would become the director of Dubna, was then, like me, a young theoretician. The tale of Bruno Pontecorvo fascinated him too. Alexei told me that he had heard Pontecorvo's name "while still a schoolboy. Its unusually ardent ring surprised me. It was always surrounded by an aura of mystery and legend. Very little was written about him. Schoolchildren and students of my generation knew little about him. We only knew that there was a 'secret' professor at Dubna, who for ideological reasons had decided to transfer with his family to the USSR. . . . I think we shall never succeed in understanding the mystery of his transfer to the Soviet Union."[5]

Two years after that encounter, I was working in England, at a laboratory adjacent to the one Pontecorvo had fled. What's more, I discovered that I was living five minutes from Pontecorvo's former home in Abingdon-on-Thames, and was working with some of his former colleagues. Some senior members of the Abingdon tennis club remembered him: he had been a champion, always neatly dressed in white. His son would ride a bicycle around the courts while he played.[6]

MY INTEREST IN PONTECORVO THE PHYSICIST WAS REAWAKENED IN 2006 following the death of Ray Davis, an American physicist who had won a Nobel Prize at the age of eighty-seven by building on one of Pontecorvo's ideas. Pontecorvo had died in 1993, and so missed out on a share of the prize. As I researched Davis's life, first for an obituary and then for a book about him, I discovered the extent of Pontecorvo's own brilliant contributions to physics.

Pontecorvo devoted much of his later career to the study of the enigmatic neutrino, a subatomic particle produced in nuclear reactors and in stars. His work inspired a new branch of science: neutrino astronomy. Pontecorvo's lack of recognition by the Nobel committees was no fault of theirs; rather, it resulted from a combination of bad luck and Pontecorvo's choice to live in the USSR. The vagaries of Soviet politics prevented him from performing critical experiments. His theoretical ideas were secreted for years in Russian journals, unknown in the West. Thus, instead of being one of the most famous scientists of the twentieth century, Pontecorvo is largely unknown, except as one of the "traitors" who leaked the secrets of the atomic bomb.[7]

The question of whether he was in fact a spy has been a cause célèbre for more than half a century. KGB agents have named him as one, but he himself never confirmed it. The intelligence agencies of the United Kingdom, the United States, and Canada all vetted him. None found any conclusive evidence. On more than one occasion they cleared him for classified work.

Decryptions of Soviet ciphers show that there were several spies operating in the West during the time in question, each identified by a nom de guerre, such as Elli, Kelly, and Moor, whose true identities have never been unambiguously established. For over sixty years there has been speculation as to whether Bruno Pontecorvo was one of them. Following his disappearance in 1950, the Western media claimed that he had been on the verge of being exposed as a spy, like Fuchs and Nunn May before him, and so had jumped ship. The British government, misled by their security services, attempted damage control by portraying Pontecorvo as a scientist who had never worked on the atomic bomb and, by implication, had no worthwhile secrets to give to the USSR. Meanwhile, British intelligence began a forensic investigation into Pontecorvo's disappearance, led by Ronnie Reed, the head of counterespionage against the USSR.[8]

Although Central Casting might have chosen the dashing Bruno Pontecorvo for the role of James Bond, the real-world intelligence officer could have passed for a bank clerk. Three years younger than his quarry, slightly built, with large ears, a prominent nose, and a wispy mustache, Reed was unlikely to strike fear into a suspect through his physical presence. Nonetheless, this former electronics engineer had monitored communications between secret agents during World War II, and in the war's aftermath had proved adept at identifying fleeing Nazis who had disguised themselves as civilians. Though Reed was no scientist, he shared Bruno's talent for methodical investigation, as well as his persistence, insightfulness, and healthy skepticism. He would need to uncover the mechanics of the Pontecorvos' flight, but this was less important than understanding the reasons. Reed prepared to build a complete picture of his prey: his expertise, his colleagues, and his politics. To do this, he began by researching Bruno Pontecorvo's history.

LIKE REED, I BEGAN MY WORK BY EXAMINING PONTECORVO'S BACKground. Sixty years have passed since Reed's investigation, so I had several advantages. We now know what became of Pontecorvo, how the Soviets regarded him, and how he, in turn, regarded them. Also, being a nuclear physicist myself, I could assess Pontecorvo's scientific value—to the international community of physics throughout his life, and to the USSR in

1950. In any case, one thing was clear to me from the start: if he was a spy, he paid a huge personal price, greater even than the price paid by Klaus Fuchs and Alan Nunn May, who each spent but a short period in jail. The Soviets placed severe restrictions on Pontecorvo's freedom (similar to the constraints placed on on British traitors Guy Burgess, Donald Maclean, and Kim Philby, fellow émigrés to the USSR), and for years cut him off from all contact with his family in the West. They also placed restrictions on his scientific research. Nevertheless, my knowledge of nuclear physics had for some time led me to suspect that Pontecorvo's significance to the USSR, during the final years of Stalin's tyranny, was probably far greater than has been generally recognized. After 1950, the USSR was desperate to build nuclear reactors, as well as other equipment that would enable them to develop an arsenal of atomic weapons, and hydrogen bombs in particular. Fuchs had passed atomic secrets to the Soviets before 1950, as had other spies, but his expertise was not available to them during Stalin's final years. Pontecorvo, however, was in the USSR, where his knowledge of nuclear science could be tapped by both the scientific community and the government.

This possibility crystallized my personal quest to resolve the lingering questions surrounding Bruno Pontecorvo's defection: Why did he go? What happened to him in the USSR? Did he claim, like Edith Piaf, "*Je ne regrette rien*"? In the 1980s I had discovered that his eldest son, Gil, who had been twelve years old at the time of his father's defection, was now a scientist based at Dubna, and part of a team doing experiments in Geneva, at CERN, the European Organization for Nuclear Research. I too had worked at CERN, but we never overlapped. Recently, by chance, I found myself on a CERN committee that periodically reviewed the team's progress. In 2011, after half a lifetime, Gil and I finally made contact.

"You want to see Volga River?"

Thanks to the marvels of Skype, Gil Pontecorvo's face filled the screen of my laptop. Then he turned his camera around, so that it showed the view through the window of his apartment: the Volga, four thousand miles away from my living room in Abingdon, and from the house he had left sixty years before.

At last I could learn firsthand what had happened.

FIRST HALF

"**Midway on our life's journey**, I found myself in dark woods, the right road lost."

—*Dante's Inferno*

ONE

FROM PISA TO ROME

MOST OF THE SCIENTISTS WHO WORKED ON THE MANHATTAN PROJECT grew up in the 1930s, in an era when fascism was on the rise. Large numbers of intellectuals rejected such tyranny; many chose to follow the red banner of communism instead. Bruno Pontecorvo was not unusual in this regard. The events that would lead to his singular role in the history of the Cold War stemmed from experiences during his youth and early manhood, and flowered as a result of the influential people he came in contact with. Their seeds lay in his family history.

The Pontecorvos of Pisa were a wealthy and intellectually gifted Jewish family. In the nineteenth century, Pellegrino Pontecorvo introduced the spinning jenny to the Italian textile industry. His son Massimo, Bruno's father, expanded the business, eventually owning three textile factories, which employed well over a thousand people.[1]

Bruno hardly knew his grandfather, as he was only five years old when Pellegrino died. Pellegrino nevertheless established the mould within which his children, and later their children, were formed. He was active in the international Jewish community, and in the 1880s rescued Jews fleeing the pogroms in Russia. He inspired the family's liberal ethics, which became more radical and explicitly antifascist following Mussolini's rise to power. Bruno and several of his relatives joined the Communist Party in the 1930s.

Pellegrino's funeral in 1918 was a big affair. The Russian Revolution of the previous year had inspired unrest among workers throughout Europe, not least in Italy. Even though many Italian tradesmen were threatening to rise against their bosses, the community's respect for Pellegrino was such that "labourers and industrialists alike" came en masse to his funeral to celebrate his life.[2] Indeed, Pellegrino was held in such high esteem that he was given the title of *Cavaliere del lavoro*, similar to a knighthood in the United Kingdom, in recognition of his dedication to labourers' rights.

That same year marked a sea change in global politics. World War I ended; the November Revolution overthrew the German Empire; Italy was in turmoil due to high unemployment and social conflict. The Bolsheviks had taken power in Russia, and there was a real possibility of revolution in Italy too. Into the mess goose-stepped Benito Mussolini.

In Pisa there was an antifascist demonstration in which several of Massimo's workers were involved. The local fascist leader, Guido Buffarini Guidi, came to the factory and ordered Massimo to reveal the names of the participants, and the ringleader in particular. Massimo refused. Buffarini Guidi challenged him to a duel.[3] Fortunately the duel never took place, but Massimo's workers always remembered the support their boss had given them. Bruno's sister Anna recalled that when one of them saw her father in the street many years later, the man threw his arms around Massimo's shoulders and exclaimed that it was "like seeing the Lord resurrected."[4]

It was into such a family, with antifascism at its heart, that Bruno Pontecorvo was born on August 22, 1913. Bruno was the fourth of eight children—three girls and five boys. Those were days of rigid gender roles: the girls were educated in the liberal arts; the boys were encouraged toward science and technical matters. The most intelligent of the children, in their parents' opinion, was the eldest, Guido, born in 1907. He emigrated to the United Kingdom in 1938 as part of the Jewish exodus from fascism. There, he became a distinguished geneticist, and a fellow of the Royal Society. Paolo, "the most serious," was born in 1909. In 1938 he moved to the United States, where he worked on radar and microwaves during World War II. The eldest of the three sisters, Giuliana, born in 1911, was "the most cultured." She became a journalist and prominent communist.

Bruno was followed by brother Gillo in 1919, sisters Laura and Anna in 1921 and 1924, and finally, in 1926, the youngest brother, Giovanni. The children's French governess, Mlle Gaveron, said there would be no need for her to spend time in purgatory "as she had been there already looking after the children—except for Bruno, who was heaven."[5]

IMAGE 1.1. Bruno Pontecorvo as a child. (COURTESY GIL PONTECORVO; PONTECORVO FAMILY ARCHIVES.)

Each child was talented, so much so that Massimo and his wife, Maria, did not regard Bruno as particularly intelligent in comparison to his siblings. Years later, Bruno remembered that his parents described him as "the most gentle but the most limited." They also said that his eyes showed him to be "sweet but not intelligent," an opinion that left him with a shy disposition and an "inferiority complex that haunted me for the rest of my life."[6]

Bruno inherited a natural aptitude for sports. Friends recall his love for alpine skiing, underwater swimming, and, above all, tennis. Throughout his life, Bruno would recount how, at age sixteen, he had been included on Italy's national junior tennis team and been invited to attend a training camp in France. His parents refused to allow him to go, as they regarded the activity to be a distraction from serious study and wanted him to spend his time preparing for college. The disappointment of a young boy came across in the tale, even after nearly half a century had passed. His parents consoled him. They assured Bruno that his achievements in physics were also first-rate, and that with suitable dedication he could achieve great things there too. Bruno acquiesced—sort of: "Yes, but I would also like to be the Italian tennis champion."[7]

Bruno's mother, Maria, had grown up in a highly cultured family. Her father, Arrigo Maroni, had been the director of a hospital in Milan, enjoyed the opera at La Scala, and was well known in Milanese society. Her religious background was Protestant.[8] Massimo Pontecorvo, however, was still a traditional Jew when Bruno was born. After Pellegrino's death, Massimo continued to lead the family rituals, but attitudes toward religion in the home were changing. The younger members took part in the ceremonies, but they did so halfheartedly. Their mother was Christian, and the children were not actively Jewish. A young brother—probably Giovanni—even asked one of his older sisters about circumcision, only to be informed that she didn't know what it was.[9] There were no bar mitzvahs in the Pontecorvo family, no bris rituals, no burials in Jewish cemeteries, but nonetheless they "were Jewish enough for Mussolini."[10]

Indeed, the Pontecorvos' privileged and idyllic life began to unravel with the onset of Mussolini's anti-Semitic laws. The family dispersed. Guido had already settled in Britain, and in 1939, with the threat of war looming, Giovanni, Laura, and Anna, still teenagers, moved there too. The three siblings completed their education in Britain, becoming, respectively, an agriculturalist, nurse, and language teacher. Bruno's exodus in 1936, the first of his three great upheavals, had been more gradual.

Many Italian intellectuals—Jewish and Christian alike—believed strongly in the ideals of liberal socialism. Those Jews who foresaw the consequences of fascism from the start had adopted strong antifascist positions long before anti-Semitism became formalized. Bruno's cousin Emilio Sereni was especially prominent in this regard.[11] Sereni's mother, Alfonsina, was Bruno's father's sister. Emilio, born in 1907, was a powerful intellectual, with a strong personality of almost overpowering intensity. By the age of twenty he was reading Marxist classics avidly, and he soon married Xenichka Zilberberg, the daughter of two heroes of the Russian Revolution.[12] Sereni joined the Communist Party of Italy in 1927. In 1929, following in the tradition of his parents-in-law, Sereni, along with his colleague Manlio Rossi-Doria, founded an underground communist organization in Italy. The following year, the fascist police arrested Sereni and Rossi-Doria, and the Special Court of State Security, which the fascists had created to "defend the state," sentenced them to fifteen years in prison. Granted amnesty and freed in 1935, Sereni fled from Italy to Paris, where he became the cultural manager of the Communist Party of Italy, and the chief editor of *Lo stato operaio* (*The Workers' State*). It was during this period, in prewar Paris, two years after Bruno left Italy, that Emilio Sereni would begin to have a considerable influence over his younger cousin—an influence that would frame the course of Bruno's life.

PHYSICS IN ROME

It was far from obvious that Bruno would end up a great physicist. Initially he followed the same route as his older brother Paolo, and at age sixteen enrolled at the University of Pisa to study engineering. After two years, he was doing well but disliked technical drawing, so he quit engineering and, in 1931, decided to concentrate on physics.

As it happened, Bruno's childhood coincided with the emergence of atomic physics. He was born in the same year as the insight that every atom is like a miniature solar system, in which "planetary" electrons orbit a compact nucleus at the core. He was a student when physicists realized that an atom's ability to shed energy through radioactivity results from the instability of the nucleus and began to home in on the neutron, a still-hypothetical constituent of the nucleus. This is when his eldest brother, Guido, made a pivotal intervention.

Guido was insistent: "For physics you must go to Rome."[13] Enrico Fermi was there, building a huge reputation. In 1926 Fermi had been appointed, at just twenty-five years of age, to a professorship at the University of Rome, funded by Orso Corbino, an influential Sicilian. At the time, nuclear physics was an exciting new field. Quantum theory was being used to build mathematical models of the properties of the nucleus, but experimentally it remained virgin territory. Fermi decided that the best way to revitalize Italian physics was to understand the atomic nucleus, in terms of both constituents and construction, and the relationships between the nuclei of different elements. With Corbino's support, Fermi established a laboratory in the physics department on the Via Panisperna, in Rione Monti, a few minutes' walk from the main railway station; to help in the endeavour, he gathered a team of young experimental scientists—a group that became known as the "Via Panisperna Boys."

Guido's insistence that Bruno go to Rome stemmed from his friendship with one of the Via Panisperna Boys, Franco Rasetti. He and Guido had been friends for years and had explored the Alps together as hiking companions. At that time, Bruno was a child, patronizingly known as "the cub." Rasetti paid him little attention. Years later, when Bruno presented himself to Rasetti, announcing that he wished to complete his studies in Rome, Rasetti teased him: "Just out of your diapers and you want to become a physicist!"[14]

Although he was confident and spoke with ease—and was, in the words of Laura Fermi, "uncommonly good looking"—Bruno had a tendency to blush at the least provocation. In response to Rasetti's joke about his youth, Bruno gave one of his familiar blushes, but Rasetti—well aware of the intellectual strength of the Pontecorvo family—encouraged Fermi to take a look at him.

Fermi gave him an informal exam. Years later, Bruno claimed that he showed only "mediocre knowledge." Fermi explained to him that there were two categories of physicists: theoreticians and experimentalists. He then added: "If a theorist does not have exceptional ability, his work does not make sense. As for experimental physics, there exists the possibility of useful work, even if the person has only average intelligence."[15]

Fermi was infamously slow to praise and blunt in his criticism. It's unclear if Fermi was delicately giving his opinion, so as to guide Bruno toward experiment, or simply providing idiosyncratic commentary. In any case, in 1931 Bruno Pontecorvo entered the third year of physics at the University of Rome. This meant he had the good fortune to be studying physics in the annus mirabilis of 1932, when the atomic nucleus was discovered to have a labyrinthine structure of it own.[16] By 1934 Bruno was ready to take part in genuine research as a member of Fermi's team, right at the dawning of a new science: nuclear physics. At age twenty-one, he was destined to be at the epicentre of one of the greatest and most far-reaching discoveries of the twentieth century.

THE PREHISTORY OF NUCLEAR PHYSICS

At the end of the nineteenth century, atoms were believed to be the fundamental seeds of all matter. The standard model of that time asserted that all atoms of the same element were identical, that different elements consisted of different types of atoms, and that compounds formed from atoms of the constituent elements.[17] Much of this remains true today. However, the scientists of yesteryear also believed that atoms were indestructible and impenetrable objects, like miniature billiard balls. This is not the case.

In 1911, working in Manchester, Ernest Rutherford discovered that an atom is mostly empty space, with a massive, dense kernel at its centre carrying a positive electric charge—a kernel that he called the nucleus. In 1913, the year of Bruno's birth, Rutherford's colleague, Danish theorist Niels Bohr, proposed that atoms are held together by the electrical attraction of opposite charges. In this model, negatively charged electrons orbit the positively charged nucleus.

At that stage, no one knew what an atomic nucleus consisted of. By the time Bruno started school, Rutherford had shown that the nucleus of a hydrogen atom is the simplest of all, consisting of a single positively charged particle, which he called a proton. Rutherford had deduced that the proton was fundamental to the nuclei of all atomic elements. As a student, Bruno would have learned that atomic nuclei are lumps of positive charge, made up of protons, and that the more protons there are in the lump, the greater the charge. It is the amount of this positive charge that determines how many negatively charged electrons can be ensnared in the outer regions of the atom. The chemical elements are distinguished by the complexity of their atoms—hydrogen, the simplest, consists of a single electron encircling a single proton, while helium has two protons in its nucleus, carbon has six—onward to uranium with ninety-two. The chemical identity of an element is a result of its electrons, and chemical reactions occur when electrons move from one atom to another.

This simple picture first started to change in 1932, when James Chadwick of Cambridge University discovered a third basic seed of matter, the neutron. Neutrons are similar to protons but carry no electric charge; they cluster in atomic nuclei and add to the nuclear mass without changing the total charge. We now know that neutrons are an essential component of all atomic nuclei, except for that of hydrogen, which normally consists of just a single proton.

Every atom of the element uranium has ninety-two protons in its nucleus. The number of neutrons may vary, however. A rare form of uranium known as U-235 contains 143 neutrons, while the most common form, known as U-238, has 146. Adding the neutrons to the ninety-two protons in each case gives a total of 235 or 238 constituents, respectively. These varying forms are called isotopes, from the Greek *isos* and *topos*— meaning "the same place" (in the periodic table of atomic elements). Although all the isotopes of a particular element have the same chemical identity, the behaviour of their atomic nuclei can vary dramatically. Indeed, the neutron number is the key to extracting energy from the nucleus, either gradually in nuclear reactors or explosively in weapons. For example, U-235 forms the raw material for both nuclear power plants and atomic bombs.[18]

An atomic nucleus, then, is more than just a core: it is a new level of reality. Within its labyrinthine structures, powerful forces are at work, which are unfamiliar in the wider world. The presence of these forces is suggested by the otherwise paradoxical fact that nuclei exist. Why do the protons, which all have the same electric charge, not repel each other and cause the nucleus to disintegrate? The answer is that there is a strong attractive force that grips protons and neutrons when they are in contact with one another. Within the nucleus this strong attraction between a pair of protons is over a hundred times more powerful than the electrical repulsion.

There is a limit, however, to the number of protons that can coexist like this. For any individual proton, the attractive glue acts only between it and its immediate neighbours. The electrical disruption, however, acts across the entire volume of the group. In a large nucleus, the total amount of electrical repulsion can exceed the localized attraction, in which case the nucleus cannot survive. The neutron, being electrically neutral, helps counteract this disruption and stabilize the nucleus. Even so, many neutrons are needed to do this, especially in larger nuclei. Uranium, with ninety-two protons, is the largest stable example in practice, requiring 140 neutrons to stabilize its protons. Yet the counterbalance is so delicate that the slightest disturbance can split a uranium atom in two, in the phenomenon known as fission.

It turns out that the strong attractive force acts most efficiently when the constituent neutrons and protons pair off exactly. Thus U-238, which has an even number of neutrons, is more stable than U-235, which has one odd neutron without a partner.

Even the simplest element of all, hydrogen, has isotopes. A proton accompanied by one neutron forms a stable isotope, known as a deuteron, the nucleus of deuterium. Water (H_2O) contains hydrogen; the analogous molecule consisting of deuterium—D_2O—forms "heavy water." A proton accompanied by two neutrons forms the nucleus of tritium. Tritium is mildly unstable, however, with a half-life of about twelve years.[19] These are all the possible isotopes of hydrogen; "quadium" or "H4," in which a proton is joined by three neutrons, does not exist.

Why don't clusters of tens or hundreds of neutrons exist on their own, without any protons mixed in? In short, it is because neutrons are inherently unstable. A neutron is slightly heavier than a proton. Given Einstein's equivalence between mass and energy, this implies that a neutron has slightly more energy locked within it than a proton does. This extra energy leads to instability—so much so that an isolated neutron cannot survive for more than a few minutes on average, whereas a single proton can exist for eons, possibly even forever. As a general rule: if there are too many neutrons in a nucleus, the assembly becomes unstable.[20]

The result of all this is that only a limited number of stable isotopes exist, namely those where the number of neutrons is close to, or larger than, the number of protons. As we go further up the periodic table of the elements, the atomic nuclei become larger, and the number of excess neutrons expands as well.

It is difficult to predict in advance which isotopes will be stable and which will not. Much of the work on atomic weapons and nuclear reactors in the 1940s and 1950s would rely on experimental tests or rules of thumb. The different isotopes of uranium, plutonium, and hydrogen would become central players in this saga, all as a result of the discovery of the neutron. Indeed, the whole course of history was changed as a result of this discovery, causing the astrophysicist Hans Bethe to describe the years leading up to 1932 as the "prehistory of nuclear physics." Everything that came after was history.[21]

It was the perfect moment for an ambitious young scientist to start a career—a moment when he could investigate the mysteries of the atomic nucleus. And Enrico Fermi's group in Rome was ideal. Bruno Pontecorvo was about to become an expert in neutron physics.

TWO

SLOW NEUTRONS AND FAST REACTIONS

1934–1936

It was in 1934, just as Bruno was about to join the team, that Enrico Fermi's genius began to bear fruit. The circumstances that inspired him resulted from a setback in his attempt to explain one of the fundamental natural processes: a form of radioactivity known as beta decay.

Ernest Rutherford had identified three different varieties of radioactivity in 1899, and named them after the first three letters of the Greek alphabet: alpha, beta, and gamma. Alpha radiation consists of massive, positively charged particles, which emerge in a staccato burst when spontaneously emitted by substances such as radium. (Despite its name, the "alpha particle" is not a fundamental particle, since it is built from protons and neutrons—two of each, or four particles in all. However, the quartet is so commonly produced in radioactive decays that it was identified before its nuclear structure was recognized, and the name stuck.) Beta radiation consists of electrons, not those preexisting in the atom but ones created when an unstable isotope changes into a more durable form. Gamma radiation involves high-energy photons, with much shorter wavelengths than visible light. Although these three forms of radioactivity were known at the start of the twentieth century, no one understood how they arose, or what effects they had on the nucleus, for another two decades.

Beta decay was especially tantalizing, since energy seemed to be disappearing without a trace. As an explanation, Austrian theorist Wolfgang Pauli proposed in 1930 that the electron—the beta particle—is accompanied by an unseen, electrically neutral particle, which carries away the missing energy. This is the neutrino, literally the "little neutral."[1] Unfortunately, Pauli's hypothetical neutrino was so ghostly that he feared it might never be detected. The idea of the neutrino excited Fermi, however, who used it in a theory of beta decay in 1933.[2]

Fermi's inspiration came when he visualized a nucleus made from neutrons and protons. He realized that a neutron behaves much like a proton with its electric charge removed, and he guessed that the neutrino might be similarly related to the electron. He proposed that beta decay occurs when a neutron, within a nucleus, spontaneously changes into a proton, conservation of electric charge is maintained due to the appearance of an electron, and the overall energy balances due to the creation of a neutrino.

Today we know that Fermi was fundamentally correct. However, the editor of *Nature*, the journal to which Fermi submitted his paper, "Tentative Theory of Beta Rays," for publication, rejected it on the grounds that it contained "speculations too remote from reality to be of interest to the reader."[3] Fermi's paper was eventually published in another journal, but the arguments with the editor of *Nature* had exhausted Fermi to such an extent that he decided to switch from theory to experiments "for a short while."[4] In fact, this change of focus lasted for the rest of his life.

In January 1934, Fermi went on a skiing trip to the Alps. It was on his return that he saw the way forward, thanks to a discovery made in France by Irène Joliot-Curie (the daughter of Marie Curie) and her husband, Frédéric. The Joliot-Curies had been exploring the uncharted inner space of the atomic nucleus since at least 1930. Four years later, after a series of misadventures, they made a discovery that would inspire Fermi and his group—including its new member Bruno Pontecorvo.

This discovery was made possible through the investigation of radioactivity, which enabled scientists to unravel the deep structure of atomic nuclei. Radioactivity intrigued many physicists, but for Irène and Frédéric there was a special motivation. Irène's mother, Marie Curie, had discovered that the element radium is so highly radioactive that it is warm—one can literally feel it pour out energy spontaneously. This energy is carried off by alpha particles. A few grams of radium can therefore function as a practical source of large numbers of alpha particles, which are like atomic bullets, able to smash into atoms of other elements. In this respect, Irène and Frédéric were in a privileged position. Thanks to Marie Curie, their laboratory in Paris had access to more radium than anywhere else in the world. This inspired the Joliot-Curies to use this invaluable element as a source of alpha particles, which they used to bombard atoms of other elements. The result was a memorable series of experiments in the early 1930s. In one such experiment, the Joliot-Curies bombarded a sample of aluminium with alpha particles. A Geiger counter near the target sample started crackling when the irradiation began; when the barrage ended, the crackling continued, decreasing to half its original intensity after about three minutes.

This is what had happened: An aluminium nucleus consists of thirteen protons and fourteen neutrons. The addition of an alpha particle to the mix temporarily supplies two more protons and two more neutrons; however, the collision of the particles chips off a single neutron from the nucleus, leaving a cluster of 15 neutrons and 15 protons. This group of thirty is a radioactive isotope of phosphorus, called phosphorus-30. It decays with a half-life of three minutes, which explains the behaviour of the Joliot-Curies' Geiger counter.

This was revolutionary work. In 1933, Ernest Rutherford had famously remarked that anyone who believed in extracting energy from the atomic nucleus was talking "moonshine." If natural radioactivity had been the only possible kind, Rutherford would have been right. However, Frédéric and Irène had discovered that it is possible to alter the nucleus, and thereby induce radioactivity in otherwise inert material, such as ordinary aluminium. Their experiment showed that it is possible to liberate part of an atom's latent nuclear energy at will, potentially in amounts far exceeding anything known to chemistry.

The vista the Joliot-Curies revealed included a wealth of opportunities for medicine, science, and technology. Frédéric and Irène received a Nobel Prize for their discovery in 1935. Upon receiving the award, Frédéric presciently remarked that by modifying atoms this way it might be possible to "bring about transmutations of an explosive type." He went on: "If such transmutations do succeed in spreading in matter, the enormous liberation of useful energy can be imagined."[5] It was a chance observation by Bruno Pontecorvo that began the transformation of this idea from imagination to reality, and marked the start of a new age.

THE VIA PANISPERNA BOYS

Fermi's group of young researchers, based at the laboratory on Rome's Via Panisperna, had been working together for about a year when Bruno joined them. This team of brilliant individuals was the brainchild of Orso Corbino, the head of the physics department at the University of Rome. Combative and quick, the Sicilian Corbino was an astute politician with sound judgment, and he would tirelessly pursue any goal that enthused him. He saw Fermi's talent, hired him, and provided the funds to build a research team.

Part of what enabled Corbino to accomplish this was his membership in Mussolini's cabinet, despite never having joined the Fascist Party. Fermi, although barely in his thirties, had enough political acumen to appreciate the delicacy of the situation, realizing that the group's resources were ultimately a gift of the government. He therefore insisted that physics and politics be kept separate within his team. This meant that Bruno's first experience of scientific research was as an apolitical enterprise. Much would change later.

With Corbino's support, Fermi attracted a handful of talented young people to the group. The eldest, Franco Rasetti, born in 1901, was the same age as Fermi. He would burst into high-pitched cackles of laughter at the least provocation. The two whom Bruno was closest to throughout his life were Emilio Segrè and Edoardo Amaldi. Segrè, four years younger than Fermi and Rasetti, was the most serious of the group, cautious and not inclined to go along with the tomfoolery of some of his colleagues. Amaldi, two years younger than Segrè, with a cherubic face and a mass of brown hair, was the baby until Bruno's arrival. Fermi, with his infallible intellect, became known as the Pope; Corbino, holding the purse strings, was the Eternal Father; Rasetti, Fermi's deputy, was the Cardinal Vicar. Segrè was called the Basilisk, reflecting what the others perceived as his rather irritable character, and Amaldi was called the Child. The final member of this holy caucus was Giulio Trabacchi, who provided them with a source of neutrons, which became key to their research; Trabacchi was thus known as the Divine Providence. When Bruno arrived, five years younger than Amaldi, he became known as the Puppy.[6]

Bruno's first piece of research followed the discovery by Amaldi and Segrè that the spectra of the gaseous form of certain elements alter when other gases are present. Fermi theorized that this is because the electrons at the periphery of heavy elements are nearly free, move relatively slowly, and bounce off the surrounding atoms. This changes the energy of the electrons and, in turn, the spectrum of light they emit. To test Fermi's theory, Bruno repeated the experiment, and measured the spectrum of mercury vapour in the presence of various gases. The measurements were delicate; their analysis complicated. Based on this work, Bruno published the first paper of his life, at the age of twenty-one. Fermi must have been impressed, for in the summer of 1934 he co-opted the young experimenter onto his team.

BY THIS TIME, FERMI AND THE VIA PANISPERNA BOYS HAD BEEN working on induced radioactivity for six months—ever since Fermi had returned from his ski holiday and learned of the Joliot-Curies' discovery.[7] Fermi had decided this phenomenon was ripe for his team to investigate— not using alpha particles, as the Joliot-Curies had, but neutrons.

In hindsight, this is an obvious idea, but at the time it was radical.[8] The fact that others hadn't immediately tried it came down to logistics: free neutrons are very rare. To create beams of neutrons, you first have to bombard atoms of the element beryllium with alpha particles. Because most of these fail to hit the beryllium nuclei, the process generates only one neutron per 100,000 alpha particles. This seemed so wasteful that most laboratories dismissed the project, if they considered it at all.

Nonetheless, Fermi persevered with neutrons because they had one huge advantage over alphas: neutrons are electrically neutral. Because alphas are electrically charged, like the atomic nuclei they are invading, getting an alpha particle into a nucleus is like forcing the north poles of two magnets to touch. When alphas (like those used by the Joliot-Curies) enter the dense forest of atoms in a bulk target, they are rejected by the positive nuclei and ensnared by the negative electrons, usually within a fraction of a millimetre of the sample's surface; even a sheet of paper can absorb them. There is little chance of an alpha particle hitting an atomic nucleus in so short a journey. Neutrons, being electrically neutral, can enter a nucleus without this difficulty. On March 20, 1934, Fermi accomplished his goal, inducing radioactivity in aluminium by means of neutrons, before doing the same with fluorine. In each case the balance of neutrons and protons in the target atoms is delicate, and the invader disturbs it. The new grouping gives up some energy and attains equilibrium by readjusting the ratio of neutrons and protons, which it achieves by emitting an electron or a positron—the phenomenon of beta radioactivity.[9] Fermi announced his discovery in a letter to *La ricerca scientifica* on March 25, 1934: "Radioactivity induced by neutron bombardment."

Next, Fermi attacked heavier elements. Frédéric and Irène Joliot-Curie had successfully induced radioactivity only in elements that were relatively light, mainly because such elements had only a limited amount of charge with which to resist the alpha-particle invader. Fermi saw that neutrons had a huge advantage when it came to bombarding heavier atoms, so he decided to launch a systematic attack—firing neutrons at every element on the periodic table.

This would require a team effort, so Fermi co-opted Amaldi, Segrè, and Rasetti, as well as a young chemist named Oscar D'Agostino. By the summer of 1934, they had tested about sixty elements, and induced radioactivity in about forty of them. Some elements released more radioactivity than others—hydrogen gave none; fluorine a little; aluminium more. These qualitative differences were clearly real, but some means of quantifying the results was needed.

Fermi's team developed a standard scale based on silver, which had been in the middle of the qualitative range. By this stage the team had mastered the techniques, meaning that the continuing work of recording these measurements was straightforward, ideally suited to a novice. So the task of building the scale fell to Bruno Pontecorvo, working with Amaldi.

BRUNO EXPERIENCES A MIRACLE

The team's protocol called for samples of each element to be engineered into hollow cylinders, into which they placed the neutron source. To protect the surroundings from radiation, they placed the sample and the source inside a box of lead and left them, giving the neutrons time to activate the sample. After a while they removed the sample and measured its activity.

Eventually Bruno noticed something odd: the position of the sample within the box, and the box within the room, influenced its ultimate degree of radioactivity, as if some strange telepathy linked it to surrounding objects. Bruno recalled his astonishment later: "There were wooden tables in the laboratory which had miraculous properties. Silver irradiated on these tables became much more radioactive than when an identical sample was irradiated on the marble tabletops in the room."[10] Bruno and his partner described this phenomenon to Rasetti—the Cardinal Vicar—who thought it was nonsense. Although he knew that Bruno had precocious abilities, Rasetti considered his laboratory work "extremely clumsy" and feared that Pontecorvo's sloppiness had infected Amaldi. Hearing of their observations, he diplomatically suggested that their results were nothing more than evidence of "anomalies due to statistical error and inaccuracy of measurements."

Fermi agreed that "the results did not make sense at all," leaving Amaldi and Pontecorvo to suffer a terrible couple of weeks.[11] However, as Fermi was always open-minded about the surprises nature might contain, he decided to investigate the phenomenon for himself, despite his misgivings. He later recalled, "It occurred to me to see what would happen if I put a piece of lead in front of the source of neutrons"—that is, between the source and the silver.[12] He was preparing the lead on a lathe very carefully, when he noticed a piece of paraffin wax lying around. Then, "without any conscious reason," he left the lathe and decided to use the paraffin instead of the lead. He confirmed that the radioactivity of the silver was much higher than it had been without the paraffin. Perhaps his criticism of Amaldi and Pontecorvo had been unfair.

It was the morning of Saturday, October 20.[13] Amaldi, Rasetti, and Pontecorvo were in their offices. Fermi showed them his results, and then it was time for lunch.[14] What happened next would become part of the folklore of physics.

During lunch, Fermi continued to ruminate. What do paraffin and wood have that marble does not?[15] He visualized a neutron in flight, bumping into atoms in its surroundings and slowing down. A lightweight atom, such as hydrogen, would be especially good at reducing the neutron's speed. Hydrogen is present in water, which is found in wood but not marble. It is also present in paraffin. Could slowed-down neutrons be the key to the riddle? Then he saw the answer: whereas alpha particles have a positive charge and need high speed to penetrate the repulsive electric fields that surround a nucleus, neutrons don't need any such aid. For neutral neutrons, impervious to electrical impediment, the rule is: the slower, the better. Lumbering neutrons, slowed to the point that their motion is no more than thermal agitation, remain in the vicinity of the target atoms for longer than fast-moving ones, giving them a greater chance of being captured and activating the sample. Fermi had experienced an epiphany: slow neutrons are especially good at inducing nuclear reactions.[16]

This was a remarkably bold conclusion. Up to that time, the received wisdom had been that the harder you hit a nucleus, the more likely it is to fragment. If Fermi was correct, then this wisdom was wrong: nature is more subtle. In fact the radioactivity would become especially strong if there were some means of slowing the neutrons radically. His musings had already suggested a way to do this: use a substance containing plenty of hydrogen, such as water.

Hydrogen is the lightest element of all, its atomic nucleus consisting of a single proton. For our purposes, the key feature is that the proton has almost the same mass as a neutron. As can be seen in the analogy of two billiard balls colliding, it is when two particles of the same or similar masses collide that energy is most rapidly dissipated. Bounce a billiard ball against the edge of the massive table, and the ball bounces back at (almost) the same speed; in the case of a neutron, this is analogous to the neutron hitting a massive atom of lead and recoiling unslowed. However, if one billiard ball hits another ball, which was initially stationary, they both recoil, the first ball slowing in the process. As for billiard balls, so for neutrons and protons. It was the presence of hydrogen—each atom of which contains but a single proton—that slowed the neutrons most efficiently. The presence of hydrogen atoms in the wooden tabletop, and their absence in the marble, thus explained the difference in behaviour that Pontecorvo and Amaldi had noticed. The hydrogen in the paraffin explained Fermi's results too.

This conclusion had not been obvious. The place to test it, however, was. Senator Corbino, who had founded Fermi's laboratory, had a spacious apartment in the building, with access to a walled garden. Enclosed by the physics buildings and the church of San Lorenzo in Panisperna, it contained an almond tree, a classical water fountain, and a goldfish pond. The physicists rushed to Corbino's pond, armed with their neutron source and silver sample.[17] They put them underwater and watched expectantly. Corbino's goldfish continued to swim unperturbed while the scientists leaned over the edge of the pond, full of eager anticipation. That historic afternoon—October 22, 1934—they found the answer. The activity in the silver rose dramatically. That same evening, highly excited, they drafted a paper for publication in a scientific journal.

Beyond supplying the pond, Corbino had not been involved, but he was always interested in the work of his "Boys." Sensing their animation, he asked what was going on. Once he was told about the slow-neutron phenomenon, he became excited, and joined them in Amaldi's small apartment as they drafted a paper. Corbino was initially relaxed, but when they started to write a second paper he erupted.[18] He waved his hands and screamed, "Are you crazy?" This Sicilian man of the world had realized what the young scientists, living in an ivory tower, had not: their discovery could have industrial applications. Previously, the quantity of radioactive material that could be created using alpha particles or neutrons had been trifling. However, the slow-neutron technique could produce it a hundred times more abundantly, and the practical implications were tantalizing. "Take a patent before you give out more details on how to make radioactive substances," he urged.[19]

SATURDAY NIGHT AND SUNDAY MORNING

The story just related was told by Laura Fermi in her biography of her husband, Enrico. The book was a best-seller, which helped turn the tale into folklore, and then into received wisdom. The story was then retold by Edoardo Amaldi and repeated by many, including Enrico Fermi himself. However some details are wrong, and reveal the tricks of false memory.

Enrico Fermi's laboratory notebook shows that the first hint of the breakthrough came on Saturday, October 20, as stated above. However,

Laura Fermi, Edoardo Amaldi, and Emilio Segrè, who wrote later from memory, placed it on the same day that they drafted the paper: October 22. Fermi's logbook, which dates from the time in question, shows that two days elapsed between the epiphany and the paper. What really happened?[20]

Fermi's own record shows that he performed tests, with and without paraffin, on Saturday, October 20. His insight about water matured during lunchtime. However, one cannot immediately rush to a pond, dunk samples in it, and see them spontaneously burst into radioactive life. First you have to irradiate the samples with neutrons, underwater, for some considerable time.

There seems to have been a bucket of water in the laboratory, which a cleaner had left.[21] Fermi immersed samples of cesium and rubidium nitrate in this water, and irradiated them overnight, from Saturday night to Sunday morning. On Sunday, he measured the amount of induced activity in these two samples.

The results convinced him that he was on the right track, so he continued the exercise. He now prepared samples of sodium carbonate, lithium hydroxide, platinum, ruthenium, and strontium. Overnight, from Sunday to Monday, he irradiated them "in the water."[22] On Monday morning, October 22, he measured the amounts of induced radioactivity in each sample. He began with the sodium carbonate at 9:45 a.m., continued with lithium and platinum during the late morning, and completed the task with ruthenium and strontium after midday.

The two-day discrepancy with regard to the date is not important in itself, other than as proof that memory can be an unreliable guide. The story of Corbino's pond is so delightful that I hope it really happened. By the afternoon of the twenty-second, Fermi was satisfied that the samples had become more active underwater. If the cleaner had removed the water bucket, as in some versions of the story, it is plausible that the excited youngsters would make a student demonstration in the goldfish pond. In any case, the fact that the paper was drafted on the evening of the twenty-second is certain. This took place at Edoardo Amaldi's house. His son Ugo, who was then just a baby, recalls being told at several family gatherings that "I was asleep upstairs" on that fateful night, and also that the next day Ugo's nanny asked his mother whether the "signori the night before had been tipsy."[23]

IMAGE 2.1. The Via Panisperna Boys, from left to right: Oscar D'Agostino, Emilio Segrè, Edoardo Amaldi, Franco Rasetti, and Enrico Fermi. The photograph was taken by Bruno Pontecorvo. (COURTESY GIL PONTECORVO AND DEPARTMENT OF PHYSICS, SAPIENZA UNIVERSITY OF ROME.)

PAPERS AND PATENTS

Fermi's name appeared as the first author on the paper. This reflected his leadership in the discovery. His collaborators then appeared in alphabetical order: Amaldi, Pontecorvo, Rasetti and Segrè. Four days later, the discovery became the subject of a patent: "To increase the production of artificial radioactivity with neutron bombardment." The patent owners are the above quintet, along with chemist Oscar D'Agostino and Giulio Trabacchi, who had provided the neutron sources.

The scientists knew they had stumbled upon something with potentially immense importance. To record the moment they took a photograph, which has since become iconic. It shows the young men—Fermi, Rasetti, Amaldi, and Segrè—and D'Agostino the chemist. Years later, Edoardo Amaldi's son Ugo asked Bruno why he too was not in the famous picture. The answer: "I was on the other side of the camera." As the youngest member of the team, Bruno was given the responsibility of taking the photograph.[24]

Although Bruno was the last person to join the team, his role in the discovery had been honoured by his inclusion on the patent. On November 1, he received more formal recognition, receiving an appointment as a temporary assistant at the Royal Institute of Physics and the University of Rome.[25] On November 7, his significance was further highlighted when a second paper about slow neutrons was sent to *La ricerca scientifica* for publication. This one had just three authors: Fermi, Pontecorvo, and Rasetti. This was an outstanding achievement: Pontecorvo's name stood alone between those of the two senior professors on the team.

This paper provided experimental confirmation of Fermi's conjecture: it is indeed the presence of hydrogen that causes neutrons to slow. It also reported that, in addition to being activated, the targets absorb the slow neutrons. Furthermore, the team discovered that there is an enormous range in the ability of various substances to absorb slow neutrons. This would become important later in selecting materials for use in nuclear reactors.

Soon afterward, Pontecorvo performed a series of experiments using substances that contained no hydrogen. He measured how effectively they slowed neutrons, and published a paper as sole author in April 1935. In less than a year he had become an expert in a new field of huge importance.[26]

The key to nuclear power is to slow neutrons efficiently, and the most effective way to do so is to use either heavy water or graphite.[27] At the time, however, it didn't occur to Fermi's team that this could be the key to practical nuclear power—further discoveries would be needed before that route opened.

Even so, others were already anticipating the future. Hungarian physicist Leo Szilard believed that energy could be liberated from the atomic nucleus so abundantly and cheaply that an "industrial revolution could be expected."[28] Corbino remarked that nuclear physics could become a new "super-chemistry," producing more energy than conventional chemical reactions, with potential benefits for national electricity production. In 1934, however, these were little more than well-considered speculations.

Bruno Pontecorvo's first real steps into physics had led to a patent for a means of inducing radioactivity through the use of slow neutrons. Years later he recalled how the process was sold to the US Government, leading to payments for many years—to "everyone except me."[29] The US patent, which was filed on October 3, 1935, includes the assertion "To obtain radioactive substances in quantities of practical importance." Uranium is explicitly mentioned. The implications of this discovery, and the corresponding patents, were to prove far-reaching. They would affect both the world at large and Bruno Pontecorvo's personal destiny. He had been a midwife at the birth of the nuclear age.

NIELS BOHR EXPLAINS THE NUCLEUS

Despite their success, Fermi's team was still exploring in the dark. They had stumbled on a phenomenon—the efficacy of slow neutrons—and exploited it, but the breakthrough had given them no real understanding of what was going on deep in the atomic nucleus.

In Copenhagen, Niels Bohr was puzzled by the Italian team's discovery that slow neutrons affected the nuclei of some elements more than others. Years before, he had published his model of the atom, which treated electrons like planets orbiting a central nuclear sun. He turned now to the nature of the nucleus itself.

Given the miniscule diameter of an atomic nucleus, a speeding neutron would pass through one in a billionth of a trillionth of a second. To capture a neutron, the nucleus first has to stop it, which involves absorbing its kinetic energy somehow. Because overall energy must be conserved, this kinetic energy has to be transferred somewhere, and there was no obvious way of getting rid of it in such a short time span. Fermi's measurements unambiguously showed that the neutrons were captured. Bohr took it upon himself to find where the energy went.

One day in 1935, while listening to a seminar in Copenhagen describing these problems, Niels Bohr suddenly "sat still, his face completely dead."[30] Others in the audience at first thought that he was ill. Then he stood up from his seat and exclaimed, "Now I understand it!"

Bohr went to the board and explained his vision of the atomic nucleus. He saw it not as a single lump of charge acting like a lone particle, but as a tightly packed cluster of protons and neutrons, which touch one another. During a reaction the constituents are excited into a temporarily unstable compound state, which returns to a stable configuration once the reaction is over.

Bohr's picture explained how Fermi's neutrons were slowed, reduced in energy, and captured, all in a way that was consistent with what the Rome team had found. Like a cue ball in snooker hitting the reds, a neutron hitting a crowded nucleus gives up its energy to the nucleus's individual components, which recoil, bump into one another, and spread the impact around, sharing the energy among themselves. The nucleus becomes hot, and then cools down by radiating gamma rays, but no individual constituent member escapes. Having been the first to envision a picture of the atom, he had now come up with the model of the nucleus that is still, in essence, the foundation of modern nuclear theory.[31]

The discovery by the Via Panisperna Boys had inspired Bohr's explanation of the dynamics of atomic nuclei, and this, along with the breakthrough by the Joliot-Curies, opened up possibilities for mining the energy latent within the nucleus. By the mid-1930s, nuclear physics was fast becoming the frontier area of research worldwide. In the opinion of Maurice Goldhaber, one of the foremost Americans in the field, the leaders were Rutherford's group in Cambridge, followed by Fermi's group in Rome and the Joliot-Curies' group in Paris. Igor Kurchatov, in Leningrad, led the team that Goldhaber ranked fourth.[32]

Known as "the General" because he was a leader and liked to give orders, Igor Kurchatov was energetic, argumentative, and prone to expressive swearing. In 1932, at the age of twenty-nine, Kurchatov heard about the discovery of the neutron and the Joliot-Curies' breakthrough. Although he had been working on the electrical properties of materials, he abruptly changed course to nuclear physics, and when the Via Panisperna Boys discovered the slow-neutron phenomenon, Kurchatov immediately saw its importance and decided to specialize in neutron physics. Between July 1934 and February 1936, his team published seventeen papers on induced radioactivity, one of which particularly impressed Bruno Pontecorvo.[33]

Kurchatov was at the start of a stellar career. Within ten years, he would lead the Soviet efforts to develop nuclear weapons, and be recognized as the father of the Soviet atomic bomb. In the 1930s, the number of scientists working in the field of nuclear physics was still small, and its practitioners around the world were all known to one another. The possibility that its future would be full of secrecy, paranoia, and military applications was still undreamed of.

THE VIA PANISPERNA BOYS HAD BECOME ACKNOWLEDGED LEADERS of a new field of physics. Although their famous breakthrough involved little more than a bucket of water (or perhaps a goldfish pond), many attacks on nuclear structure required large machines. In Cambridge, during 1932, John Cockcroft and Ernest Walton had built a five-metre tower of capacitors, which they could charge to about 500,000 volts. This created powerful electric fields, enabling the Cambridge team to accelerate electrically charged particles, such as protons. When these high-energy protons hit the nuclei of atoms in a target, Cockcroft and Walton discovered that these nuclei were shattered. They had built what later became known as an "atom smasher."

In Berkeley, California, Ernest Lawrence built a machine that used a mix of electric and magnetic fields to guide charged particles around curves, speeding them up as the arc grew bigger. This invention, known as a cyclotron, gave birth to what is known today as high-energy physics. Although Rutherford was reluctant to embrace large-scale physics at Cambridge, elsewhere—most notably in Berkeley—a new age of particle accelerators was beginning. Those who didn't join this new adventure were in danger of being left behind. James Chadwick, discoverer of the neutron, was disappointed in Rutherford's reluctance and left Cambridge in 1935. He moved to Liverpool, where he built a cyclotron with help from Cockcroft.

Fermi and his team recognized the importance of this new strategy, but they were unable to get the financial support needed to build an accelerator. The team began to break up, partly due to this difficulty, and partly due to the growing threat of fascism. For Pontecorvo, young and ambitious, it was time to move on.

In 1935, Frédéric and Irène Joliot-Curie won the Nobel Prize for the work that had inspired Fermi and set Pontecorvo on his own research path. Whereas Fermi's Italian team had shone like a supernova, bursting into brilliance and then fading, the Joliot-Curies' lab in Paris was emerging as a steady star of nuclear physics. The couple began to attract foreigners to their lab. That same year, Pontecorvo won a scholarship from the Italian Ministry of National Education. Funded by this award, he moved to Paris in 1936 to work alongside the Joliot-Curies. Pontecorvo was certainly well placed within the scientific community: he was a member of one internationally famous team of nuclear researchers and about to join another.

THREE

PARIS AND POLITICS
1936–1940

BRUNO PONTECORVO'S CHILDHOOD, ADOLESCENCE, AND EARLY adulthood spanned an era bracketed by two world wars. He was born just before World War I started, was five when it ended, and had recently graduated from college when World War II began. It is a cliché to say that much had changed during that quarter century, but for Bruno Pontecorvo and his family this was cruelly true.

Although Italy was involved in World War I for only three years, it spent more money in that short time than it had during the previous half century, and nearly two million Italian citizens were killed or wounded. Having suffered such extreme costs, both financial and personal, the Italians expected some reward for their contribution to the victory. Such hopes were soon dashed. At the Paris Peace Conference that followed the war, the "Big Three"—the United States, the United Kingdom, and France—regarded Italy's delegation as minor players. This slap in the face generated great resentment. Italians viewed their government as weak; dissatisfaction festered. Unions were formed. Demonstrations, strikes, and militancy quickly followed.

Soon, Italy was in turmoil. During 1920, many factories were occupied. Industrial unrest spread rapidly, and at one point half a million workers were involved, spearheaded by the Italian socialist and communist parties.

Fascism too began its rise. Benito Mussolini, having been expelled by the socialists in 1914, formed the National Fascist Party. By 1922 he was prime minister, and by 1925 he was the self-styled "Il Duce"—the Leader.

The Pontecorvo family's reaction was typical of many intellectuals opposed to fascist rule, with its censorship, overweening propaganda, and (later) active anti-Semitism. In 1936, following Hitler's example in Germany, Mussolini enacted laws forbidding Jews from holding positions of authority, such as in universities, and limiting their right to work in a variety of ways. Anti-Semitism soon erupted into violent persecution.

At the time, Italy had large numbers of people who were technically Jewish but didn't actively practice the religion. The Pontecorvos fell squarely into this category. However, in such a vicious environment, you became Jewish whether you liked it or not. It was in this climate that the Pontecorvo family dispersed.

The rise of fascism also led to the breakup of Fermi's group in Rome. Fermi himself emigrated to the United States in 1938. By this time Bruno had already moved to Paris, where he'd joined the team of Frédéric and Irène Joliot-Curie. As it happened, Frédéric was an active communist, and Irène was a fellow traveller. Against the backdrop of the 1936–1939 Spanish Civil War, which caused thinking people around the world to declare their political allegiances, Bruno soon joined Europe's intellectual nexus in the fight against fascism.

YEARS LATER, AFTER BRUNO DISAPPEARED INTO THE SOVIET UNION, the British security services would identify the Paris years as the time when Bruno Pontecorvo had been "exposed to the virus of communism."[1] Their informant identified several communists present in Bruno's circle, including his cousin Emilio Sereni, as well as Frédéric Joliot and a certain Professor Langevin.

Paul Langevin was a physicist who had been Marie Curie's lover after the death of her husband in 1906. Two decades later, his influence pervaded the Joliot-Curies' laboratory. Indeed, he had been Frédéric's mentor, and it was through Langevin that Frédéric had gained his introduction to the Curie laboratory. As fascism threatened to engulf Europe, Langevin, like many intellectuals, had chosen communism. By the 1930s, he was one of the most influential people in France, and dreamed of setting up a workers' university in Paris, built according to Marxist ideals. When Langevin proposed this idea, both Frédéric and Irène offered to give lectures.

Langevin was also a foreign member of the Soviet Academy of Sciences. In 1933, when he invited Frédéric to join him for ten days of scientific meetings in Leningrad and Moscow, Frédéric was only too happy to come along. Langevin introduced him to many Soviet intellectuals. On that occasion, Irène was unwell and stayed in France, but she joined her husband on several visits to the Soviet Union later.

With the rise of Hitler and Mussolini, France had become a haven for left-wing intellectuals fleeing fascist persecution. Communists in France joined the strategic Popular Front, which included all the parties of the left and centre. Irène and Frédéric were strong supporters.

This vibrant cosmopolitan community awaited Bruno as he set out from Italy in February 1936. After an overnight train journey, during which he had to stand, leaning against a window, Bruno Pontecorvo arrived at the Gare de Lyon in Paris on Leap Day, February 29.[2]

Irène and Frédéric Joliot-Curie were at the height of their power and influence: they had won the Nobel Prize the previous December for their discovery of induced radioactivity; in the month of Bruno's arrival they attended the first Mendeleev Conference in Moscow, where Frédéric gave the opening address, discussing their breakthrough. Following the conference, the couple spent nearly a month in the Soviet Union, meeting many influential people, in both science and government. Meanwhile, in France, within weeks of Bruno's arrival, the Popular Front swept to power, led by Léon Blum. Blum immediately invited Irene to serve as undersecretary of state for scientific research.[3]

At twenty-three years of age, Bruno Pontecorvo could hardly have failed to be impressed.

LEFT WING ON THE LEFT BANK

The picturesque narrow streets of Paris's Latin Quarter, on the Left Bank of the River Seine, weave in and around the buildings of the Sorbonne and the Collège de France. A ten-minute walk south of the former Joliot-Curie laboratory is the Panthéon. Originally a church, the Panthéon is now a secular mausoleum, containing the remains of French "national heroes," including Pierre and Marie Curie, though not their daughter or son-in-law. Immediately in front of the edifice is the Place du Panthéon, where, among the cafés and offices, stands an eighteenth-century building: the Hôtel des Grands Hommes. Eighty years ago, when Bruno arrived in Paris, this grandly named residence was quite basic, even sleazy. Its main attraction for students and young researchers at the Collège de France was that the accommodations were cheap.

Bruno rented a room there. As it happened, the owner supplemented his income by renting out rooms by the hour during the day for assignations. Bruno discovered this when he arrived home one afternoon and encountered the novelist André Malraux in the corridor, along with a "very showy girl."[4]

The rooms were cheap, but even so the bed linens were changed once a week, and there was a sink in each room, though it only provided cold water. The toilet was in the corridor, and if you gave reasonable notice, and paid in advance, you could bathe in the communal bathroom. Such was student life in prewar Paris. During this period, Bruno met and befriended several people who would have a huge influence on his life, both by establishing him as one of the world's leading nuclear experts and by igniting his passionate belief in communist ideology.

In Germany, Nazi thugs were rampant. Anti-Semitic laws were expanded, an axis of alliance with Mussolini was formed, and the Berlin Olympics were exploited to promote Nazism and Aryan supremacy. In Spain, with its brewing civil war, the contest between fascism and socialism was about to erupt into violence. Meanwhile, in France, Léon Blum led the democratically elected Popular Front of socialists and communists. After the Popular Front's victory in May (but before Blum took office as prime minister in June), the workers' movement launched a general strike, which led to a series of agreements known as the "Magna Carta of French Labour." Bruno Pontecorvo thus arrived in Paris at a time of vibrant political activism. His background had prepared him for such an environment. In Italy, his friends had embraced the Italian antifascist movement, as well as the ideals of Antonio Gramsci, founder of the Communist Party of Italy.

Nonetheless, Paris opened his eyes to a whole new way of life. Workers mingled with students and ate with them in the canteen. Back home, Bruno's only contact with a manual worker had been when one of his father's employees periodically came to their home for discussions. However, Bruno later recalled, "I never ate at the same table" as the worker.[5]

In Rome, Fermi had always claimed that his research work left him uninterested in politics. Bruno had formed the impression that this was a general truth in science. However, at the university in Paris he was surrounded by political activism, quite unlike his previous experiences. Rome and Pisa began to seem rather provincial, while Paris seemed like the centre of the world.

Bruno had gone to Paris because he was impressed by the Joliot-Curies as scientists. When he visited Irène and Frédéric at their home, he was surprised by their intense discussions of politics. What's more, the majority of his colleagues were actively left-wing or communist. All the framework was in place for Bruno's political confirmation. The completion of his journey, from antifascism to a lifelong belief in communism, occurred through the influence of his cousin Emilio Sereni.[6]

Sereni had fled from the Italian fascist police in 1935, and immediately became immersed in the communist organization in France. When the cousins met again in Paris—Sereni now thirty years old, and Bruno just twenty-three—Sereni had a huge effect on his young relative. Sereni took Bruno along to political rallies, where they befriended communist intellectuals and party officials. Soon Bruno was attending meetings almost daily.[7]

In particular Bruno recalled joining Sereni and a group of Italian émigrés in the fall of 1936 at a large rally led by Maurice Thorez, head of the French Communist Party. Decades later, Bruno still had a vivid memory of the enthusiastic and excited crowd, who waved flags, raised their fists, and had red handkerchiefs or scarves wrapped around their necks.

Sereni inspired in Bruno a surge of enthusiasm for what was happening in the Soviet Union "where the proletariat were in power" and were constructing "the new man."[8] Bruno made no attempt to hide these sympathies, at least while he was in France. In fact, he inspired his older sister, Giuliana, and his younger siblings Laura and Gillo to convert to communism.[9] Years later, in the 1980s, Bruno confirmed, "I went over to politics when I went to Paris in 1936, the years of the Popular Front, and had the opportunity to meet with political emigrants such as Sereni, Luigi Longo [a leader of the Communist Party of Italy, who became its secretary in the 1960s], Giuseppe Dozza [later elected five times as the communist mayor of Bologna] . . . and others."[10]

Bruno Pontecorvo's life would confirm Ernest Hemingway's observation: "If you are lucky enough to have lived in Paris as a young man, then wherever you go for the rest of your life, it stays with you."[11]

MARIANNE

During the spring of 1936, as Bruno was becoming immersed in Parisian life, eighteen-year-old Marianne Nordblom was nearing the end of a yearlong correspondence course in Sweden. In May she graduated with a pass in shorthand, and distinction in typing. As for commercial correspondence, she passed in Swedish but was "not approved" in German, English, or French. At the end of the summer, on September 7, she spent the day packing suitcases. The following day she left her parents' home in Sandviken, took the train one hundred miles south to Stockholm, and boarded the SS *Burgundie*. Her final destination would be France, where she planned to work as a nanny, study the language, and have adventures.[12]

Her diary records that there was "lots of rough sea. Everyone was ill except for a few—including me." She arrived at the Gare de Lyon in Paris late on the afternoon of September 15, reached her lodgings at 37 rue d'Anjou, north of the Place de la Concorde, and spent the next day "unpacking."[13] Her visa would allow her to stay for up to two years.

She had been in Paris a fortnight when she was introduced to La Bohème—a dance club in Montparnasse frequented by students. She went there regularly. On the evening of Thursday, November 12, Bruno too was at La Bohème. Marianne noted the fateful encounter in her diary: "At La Bohème, met Bruno Pontecorvo."

Slender and fair-haired, with high cheekbones, Marianne had classic Nordic features, the blond counterpart to Bruno, with his dark Latin charms. The encounter seems to have been a *coup de foudre*, as her diary records that they met regularly. Their first date occurred just two evenings later, on Saturday the fourteenth, when she accompanied Bruno to a "big ball at the Cité Universitaire." Her diary includes the comment "great." The two were soon spending a lot of time together. Sometimes they went to see the Paris Opera Ballet or visited a museum, but dancing at La Bohème seems to have remained one of their favourite activities.[14]

In those days, few people owned a telephone, so invitations were sent in the form of brief letters. Bruno was very busy at the laboratory at the

IMAGE 3.1. Bruno Pontecorvo and Marianne
Nordblom in Paris, c. 1937. (COURTESY GIL
PONTECORVO; PONTECORVO FAMILY ARCHIVES.)

end of December, which made it difficult for him to book a table for a
planned dinner on New Year's Eve. The mail, however, was very efficient,
as demonstrated by the fact that Bruno wrote to Marianne at 11:00 p.m.
on December 30—to make arrangements for the thirty-first. His letter arrived in time and they successfully met, at 9:00 p.m. on New Year's Eve; in
accord with Bruno's instructions, they wore "evening dress."[15] They dined,
celebrated the arrival of the New Year, and then went to a club where they
played *pile ou face* until 2:00 a.m.[16]

Bruno was struck by Marianne's "slender grace, sweet temperament
and long careful silences" as well as her "childlike need for protection."[17]
Everyone who knew Bruno Pontecorvo remembers the charismatic

extrovert, who loved sports, games, and performing tricks such as riding a bicycle backwards or balancing a stick on his foot or nose. It would seem that Marianne's experience was the same; after a lunch together in February, she wrote, "Italian—he's stupid."

MEANWHILE, BRUNO HAD BECOME GOOD FRIENDS WITH THE JOLIOT-Curies. Frédéric and Irène sailed in the summer, skied in the winter, and researched together at the Radium Institute. Bruno's interest in sports and his charismatic personality made him a favourite with the couple. He was a frequent visitor to their home.

Liberal ethics ran deep for the Joliot-Curies. Irène's parents, Pierre and Marie Curie, decided not to patent the process for extracting and purifying radium because radium was an element that "belongs to all the people and [is] not meant to enrich any one person."[18] As a young woman, Irène Curie had accompanied Marie around the battlefields of World War I to X-ray wounded men. This helped forge her belief in pacifism.

Irène was often blunt and undiplomatic. A telling example can be found in a letter in which she declined an invitation to some tedious function. Her secretary had included the formal mantra "regret that I cannot attend," only for Irène to delete it. Like Edith Piaf, she declared, "*Je ne regrette rien.*"[19] Her husband, by contrast, was thoroughly outgoing and loved spending time with people. Tall and slim, with a distinguished, lean face and dark, brushed-back hair, Frédéric had socialism in his genes. His father, who was reasonably prosperous, had been exiled from France in 1871 because he had fought for the Paris Commune, the government of the workers' revolution. In 1880 he was granted amnesty and allowed to return.

Frédéric Joliot, born in 1900, was the sixth and youngest child of the family, and in his youth was so enthused by Pierre and Marie Curie that he had a photograph of them mounted on his bedroom wall. In 1924, he went to Marie's laboratory and asked for a position as her assistant. Marie assigned him to work with her austere daughter, Irène, who was just completing her doctorate.

At first, Frédéric was an outsider in France's hierarchy, which was very much linked by family ties. Some in the establishment felt he was selected by Irène as the "prince consort to the princess," his "coarse good looks" fitting that image.[20] But Joliot was a first-rate scientist and a good judge of ability. He correctly believed that he was better than many who sneered at him, and this made him bitter. Even after he and Irène won the Nobel Prize, his genius was only grudgingly recognized.

ISOMERISM

Bruno's scholarship would last for a year.[21] In 1937 he declined the chance to apply to the University of Rome for a tenured post. His reasons are not known, but the fascist situation in Italy surely did not help. By this stage, his relationship with Marianne was becoming serious, which may have also played a role, and he had begun to impress Frédéric Joliot-Curie. His work also attracted the attention of Igor Kurchatov. The link between Bruno Pontecorvo and Igor Kurchatov stemmed from a common interest in a strange nuclear phenomenon known as isomerism, which Kurchatov had recently brought to light in Moscow.[22]

In 1935 Igor Kurchatov was inspired by the work of the Via Panisperna Boys, and while checking the slow-neutron phenomenon for himself, he noticed something unusual: after a neutron of a given speed hits a nucleus, the radioactivity of the resulting isotope varies from one experiment to the next, even though it has the same number of neutrons and protons. With this exception, the new isotopes appear to be identical. This phenomenon—in which isotopes of the same mass give off different levels of radioactivity—became known as isomerism, from the Greek for "equal masses." Kurchatov discovered the first definitive example of an isomer when he bombarded bromine with neutrons in 1935. Within months, the number of isomers began to grow rapidly. The question was: What was happening?

Frédéric Joliot-Curie held Kurchatov's work on isomers in high regard. This respect was reciprocated: Kurchatov followed the work of the French group closely. When Bruno arrived at the Joliot-Curies' laboratory, he started to investigate Kurchatov's phenomenon himself. His success in this quest helped solidify his emerging reputation as a leading expert in the use of neutrons in nuclear physics.

Since the year of Bruno's birth, the atomic model of Niels Bohr had successfully posited that electrons in atoms cannot go wherever they please, but are restricted, like someone on a ladder who can only step on individual rungs.[23] When an electron drops from a rung with high energy to one that is lower down, the excess energy is carried away by a photon of light. The spectrum of these photons reveals the pattern of energy levels within the atom.

By the time Bruno started doing research, the fundamental explanation for this behaviour had been found in the equations of quantum mechanics. When Niels Bohr proposed his model of the nucleus, built from neutrons and protons in contact, many scientists wondered if quantum mechanics not only applied to the electrons in atoms but also determined the energy levels of the neutrons and protons in atomic nuclei. Bruno's experiments in Paris helped confirm that quantum mechanics does indeed control the atomic nucleus.

Nuclei in "excited" states, with one or more protons or neutrons on a high rung, give up energy by emitting photons of light, much like electrons do. The main difference between the case of electrons and the case of atomic nuclei is the nature of the radiated light. In the former the light may be in the visible spectrum, made up of photons with relatively low energy, whereas in the latter (the case of nuclei) the light consists of X-rays and gamma rays, whose photons have energies that are up to a million times greater.[24] Whereas light radiated by atomic electrons may be seen with the eye, the photons emitted from nuclei can only be detected with special instruments. Pontecorvo became an expert in this art.

If this were the whole story, it might not be particularly remarkable that nuclei can form excited states, similar to those of atomic electrons. Typically, a neutron might be captured by a nucleus and form an excited state, which then decays by emitting gamma rays, leaving a highly stable state. If this stable state were the lowest rung of the ladder—the ground state—all would be straightforward. However, there was an unexpected and tantalizing development: there seemed to be various different end-states (the isomers), all with the same number of protons and neutrons, and all highly stable, lasting for more than a day in some cases.

The addition of isomers to the rich variety of nuclear states created confusion, until quantum mechanics provided the explanation. Quantum mechanics had successfully described nature at atomic scales, 100,000 times smaller than the macroscopic world where Newton's classical mechanics rule. The nucleus is smaller than the atom by a similar factor, so it was a revelation to discover that quantum mechanics applies there too. For example, according to the theory, if a nucleus is rotating much more rapidly—has more angular momentum—than the ground state, it can have unusual stability. The origin of so many highly stable isomers was thus explained: they were nuclei with a large amount of angular momentum.

Bruno Pontecorvo entered this new field with enthusiasm. Driven by some theoretical ideas of his own, he worked mostly alone, receiving occasional advice from Frédéric Joliot-Curie. During the first half of 1937 he began what would become excellent pioneering work in the field.

1937

It's said that in years past, Parisians did not include August in their diaries; the city closed down and its residents departed for a summer holiday. In 1937, Marianne accompanied the family she worked for to Boissy L'Aillerie, a village north of Paris, to stay at a delightful hotel: L'Oiseau Bleu. With a park where the children could roam freely while Marianne looked after them, the location was ideal. Bruno, meanwhile, was in the Italian Alps, and sent Marianne a postcard of San Martino di Castrozza, a beautiful mountain resort. He briefly hoped that she might be able to join him there, but on September 1 he wrote to her at L'Oiseau Bleu to say that he would have to leave on the twelfth to attend a congress in Venice.

Bruno and Marianne were apart until the start of October. That same month, Bruno delivered his first public report on his work on isomers at a congress at Paris's Palais de la Découverte.

Meanwhile Frédéric, now thirty-seven years old, was appointed as a professor at the Collège de France, the most prestigious post in France. His eminence brought influence along with it. He persuaded the National Funds for Scientific Research to buy an old electrical plant in Ivry-sur-Seine, a suburb southeast of Paris, and convert it into a nuclear physics laboratory. His plan was for this lab to provide man-made radioactive elements for use in research.[25] He assembled a small but talented team, which included Bruno. Irène, meanwhile, remained with her own research team at the Radium Institute.

There were two members of Frédéric's group who would have a significant impact on Bruno's life and career: Lew Kowarski and Hans von

Halban. Kowarski was large, a veritable Russian bear who could have become a concert pianist had his fingers not grown too big for the keys.[26] In 1937 Kowarski was thirty years old, and lacked confidence, fearing that he was already too old to make a mark with his talent for electronic gadgets. In Kowarski's opinion, "Pontecorvo and von Halban were the two most outstanding personalities in the laboratory after Joliot."[27]

At the age of twenty-nine, Halban was already an established physicist. Bruno was still only twenty-four years old, a student, but, with his work on isomers now added to his earlier research in Rome, he was rapidly gaining an international reputation. Halban and Bruno shared a love of outdoor sports, especially mountain climbing and skiing. In terms of personality, however, they couldn't have been more different. Whereas Bruno was everyone's friend—the stereotypical warm, extroverted Italian—the arrogant Halban, who hailed from from Leipzig, was "intensely unpleasant, brutal with the weak," and carried himself "like a Prussian officer."[28]

Kowarski had developed an electronic gadget to measure radiation over a greater range of intensity than was previously possible. Joliot-Curie had no need for it in his research at the time, so in early 1938 Kowarski consulted Bruno, who, he knew, "could teach me a lot." Kowarski's impression was that Bruno wasn't exactly discouraging but nevertheless gave him the unspoken message, "Why should I be interested; it doesn't look promising and you're too old anyway." This seems unlike the Bruno of most people's memory, and it probably reveals more about Kowarski's self-image than about Bruno's opinion. However, Bruno did make an offer that had profound consequences: he introduced Kowarski to Halban.

Bruno's lack of interest in Kowarski's overtures might also provide a glimpse of the personal pressure that Bruno was under at the time. For Bruno, the carefree joy of 1937 was now replaced by worry. His financial situation was insecure, but more serious was the looming responsibility he now had, as Marianne was expecting their child later that year. Added to this was the emotional worry regarding her visa, which was due to expire soon afterward. On January 4, 1938, Marianne left her lodgings to live with Bruno, at the Hôtel des Grands Hommes, on the Place du Panthéon.[29]

On July 5, 1938, Marianne celebrated her twenty-first birthday with Bruno. On July 30, their first son, Gil, was born. Marianne's position was grim: she was unmarried, and in six weeks she would have to leave France.

It isn't clear why Bruno didn't marry Marianne at this time, since it would appear to have been the obvious means of enabling them to stay together with Gil in France. To make matters worse, Marianne's baby was apparently unwelcome in her Swedish family home.[30] Whatever the reasoning was behind this attitude, the reality was that Gil would be unable to come to Sweden with Marianne.

So Marianne and Bruno placed Gil in a residential nursery, Le Nid (The Nest), on the avenue de la Terrasse in Montmorency, a northern suburb of Paris. In September, when Gil was only a few weeks old, Marianne and Bruno made their way to her home in Sweden. En route they visited Holland and Denmark, where they stayed in youth hostels, and they spent the last week of September in Copenhagen, where Bruno attended a physics conference. Here he met one of his friends from Rome, the theoretical physicist Gian Carlo Wick. Years later, Wick recalled seeing Bruno, in the company of a "very pretty Swedish girl." He was surprised by the change in Bruno's political attitudes. He remembered that during their time in Rome Bruno had been uninterested in politics, but now "he was very keen about international events." Wick was especially struck by "the strength of his belief in the USSR."[31]

Bruno and Marianne at last reached Sandviken, her hometown in Sweden. Two years earlier, shortly after her nineteenth birthday, Marianne had left home, carefree and set on adventure. Now she returned to the small, conservative town, twenty-one years old, her Italian boyfriend in tow, their six-week-old son left behind in a French nursery. In 1938, being an unmarried mother was not accepted as easily as it is now.[32] Bruno returned to Paris alone, after a few days. This would be the only time he visited Marianne's family.

Later in her life, Marianne would suffer from depression and experience chronic periods of mental breakdown, which required treatment in a sanatorium. One can only imagine her distress in 1938, as a new mother subject to postpartum depression, her child a thousand miles away, across borders closed to her by the bureaucracy of visas.

AFTER LEAVING SANDVIKEN, BRUNO MEANDERED BACK TO PARIS. HIS first stop was Stockholm, where he attended another physics congress. On October 11, he went to dinner at the home of Professor Manne Siegbahn, the Swedish Nobel laureate; of the thirty guests, only Bruno and Gian Carlo Wick were not in evening dress.[33] The next day he met Lisa Meitner and Niels Bohr.

Meitner, who was Jewish, had escaped from Nazi Germany just two months earlier. She had been working in the same laboratory as Otto Hahn, with whom she had performed experiments that would lead to the discovery of nuclear fission in December. Bohr, having offered the world both his conceptual picture of the nuclear atom and his celebrated explanation of nuclear structure, had established himself as the foremost theorist in atomic physics. Bohr was based in Copenhagen, but visited Stockholm regularly. (It would be on one such visit early in 1939 that he would learn from Meitner about fission's awesome potential.) Bohr's presence in October 1938 was the excuse for another "grand dinner," as Bruno described it to Marianne in his second letter from Stockholm. He ended with a P.S.: "Don't fail to give my best wishes to your parents and to your brother."[34]

Bruno's journey back to Paris was difficult. He travelled via Cologne, Germany, and then spent five hours at the Belgian border because he didn't have a visa. He had to return to the Belgian consulate in Germany to obtain one. When he finally arrived at the Gare du Nord in Paris, he spent four hours with the police commissariat and paid 150 francs to amend his French visa due to a recent change in the law.[35] The next day he visited Gil at the nursery.

The following Saturday, he was invited to dinner at Halban's home. The invitation, which asked him to bring "the blond Scandinavian lady," suggests that Bruno had kept some distance between Marianne and his colleagues. On Sunday Bruno again visited the nursery in Montmorency, and wrote to Marianne about Gil's progress. During the ensuing months, Bruno's letters to Marianne would be written on the way back from Montmorency; they contained reports of visits to Gil, occasional allusions to progress in physics, and plans to obtain a visa for one or the other, allowing them to get together. Bruno's only mention of politics came in November, when he confessed his fear that "fascism will come to France." The correspondence is in line with the opinion of family friends who insist that Marianne had no interest in politics. There are, however, clues that all was not well with her. On more than one occasion Bruno alluded to Marianne's health: "I hope to find in your letter that you are not sick at all"; "are you ill?"; "what is the illness you have?"

During the rest of 1938 and much of 1939, Bruno remained in Paris, apart from Marianne. Bruno's brother Gillo—still a teenager—had stayed behind in Italy, where he "continued to pretend to be a playboy—granted with the defect of being Jewish."[36] Bruno sent Gillo tickets, allowing his brother to join him in Paris. It was during this period that Bruno introduced Gillo to communism. It was also the time of his most significant work on nuclear isomers.

Viewed with the experience and hindsight of eighty years, Pontecorvo's first entrance into nuclear isomerism was as a descriptive taxonomist, who collects species and classifies them, but does not offer any insights into their behaviour and evolution, let alone their DNA. He predicted that relatively stable nuclear isomers should exist,[37] and he found the first example when he bombarded cadmium with fast neutrons.[38] In 1938 and 1939, he performed two significant experiments. One helped establish the correlation of angular momentum with stability; the other measured nuclear energy levels. In due course, these results helped others develop the theory of nuclear structure.

One way to find out how much angular momentum a nucleus has is to detect the gamma rays when it eventually decays.[39] Several physicists had tried but failed to find the anticipated rays. This is where Pontecorvo made a notable contribution.

He suggested that electrons in the outer reaches of the atom capture the gamma rays. The energy of the gamma ray is passed to the electron, and the impact knocks the electron out of the atom. He proposed a test: look for electrons with a specific amount of energy—namely, that of the "lost" gamma ray. This is very different from the case of beta radioactivity, whose electrons emerge with a range of energies.

Bruno performed the experiment and found an example in an isotope of rhodium, where electrons always emerged with the same energy, as he had predicted.[40] This result proved the hypothesis that the original isomer had a large amount of angular momentum.

By 1939 Joliot-Curie's electrostatic accelerator at Ivry was ready.[41] One of Irène's assistants, André Lazard, had designed a Van de Graaff generator with her years before, and Frédéric had commissioned him to build the machine at Ivry. Bruno and Lazard now joined forces at Ivry, and discovered what Frédéric Joliot-Curie would call "nuclear phosphorescence."

Ordinary phosphorescence, in molecules and atoms, occurs when light is absorbed, stored, and later released gradually as a glow, visible in the dark. In these cases, the electrons in the atoms are kicked to higher rungs on the energy ladder, and then release this energy as they fall back to ground. Pontecorvo and Lazard found analogous phenomena in atomic nuclei. Instead of visible light, they used X rays—higher-energy photons. This raised neutrons and protons up the energy ladder, while the nucleus remained intact. If one of the high-energy rungs was unusually stable, the neutron or proton stored the energy for a time, and later shed it when it fell to a lower rung, slowly emitting an X-ray in the process.

The energies of these X-rays are like a bar code that reveals the energy states of the nucleus, analogous to the atomic spectra that can reveal the electronic structure of atoms.[42] Frédéric sent Bruno fulsome congratulations for this discovery, which gave him great joy. During his time in Rome, Bruno had felt that Fermi only really respected him for his prowess at tennis; Joliot-Curie's praise assured him that he had now proved himself in physics.[43]

The phenomenon of isomerism was important in establishing that a nucleus is a rich collection of constituents, which can move, orbit, and vibrate relative to one another.[44] It has applications in industry and medicine. The expertise that Pontecorvo gained in his studies of isomerism, in which he used neutrons and detected gamma rays, would prove invaluable throughout his career.

FISSION

Uranium nuclei are so large and fragile that a mere touch by a slow neutron is enough to split the pack: the phenomenon known as nuclear fission. This was so unexpected that when Otto Hahn and Fritz Strassmann discovered the phenomenon in Germany on December 17, 1938, they didn't realize what they had achieved. The discovery came when they irradiated uranium with slow neutrons, and identified the light element barium among the products. Up to that time nuclei had been modified subtly, by chipping off just one or two constituents, transmuting the target into an immediate neighbour on the periodic table. The appearance of barium, far removed from uranium, was bizarre. Hahn and Strassmann announced their results, and said little more about them. Only during 1939 were the full implications of their breakthrough understood.

The Germans' paper arrived in Paris on January 16, 1939. Joliot-Curie immediately understood what must have happened; Irène had seen a similar phenomenon the previous year, though she had not been confident enough to confront the criticism of her results, and had backed down. Now Frédéric suspected that she had been correct after all, and that the neutrons must have split the uranium in two. For the next few days the news was the hot topic of discussion throughout the group. Kowarski recalled that "nobody talked of anything else." Irène raged at having missed out on getting credit for the discovery, and told Frédéric, "What fools we have been," or probably used "a somewhat stronger word."[45]

Frédéric Joliot-Curie wrote to certain Soviet physicists about the phenomenon.[46] As a result, Igor Kurchatov immediately investigated whether any secondary neutrons were produced during the fission process, as did Joliot-Curie himself in Paris, and others in the United States. The critical question was whether more than one neutron was released for every neutron that caused the fission in the first place. If this was so, a self-sustaining reaction could occur.

A uranium nucleus contains more than 140 neutrons, more than enough to satisfy the needs of smaller nuclei, such as barium, krypton, or lanthanum, the likely debris. So it seemed plausible that, during the fission of uranium, some extra neutrons would also be liberated. Joliot-Curie immediately devised a simple experiment to detect these neutrons—and failed. The source of neutrons irradiating the uranium was so intense that his attempt to identify additional particles was like trying to detect a rain shower while standing beneath a waterfall.

Over the next few days Joliot-Curie designed a new experiment, one that would look for evidence of radioactivity in the debris when the neutrons hit uranium. Halban had left Paris to go skiing, and so missed the ensuing drama. Kowarski, however, was present.

Joliot-Curie had engineered two brass tubes, one of which was coated with uranium. He also had some Bakelite cylinders, which were larger than the tubes and could surround them like napkin rings. He had already verified that none of these tubes or rings was radioactive, even after they were irradiated with neutrons. Now all was ready for the experiment. First he placed a neutron source inside the brass tube that was free of uranium; then he placed the Bakelite ring around the tube. After a few minutes he removed the ring, took it to a Geiger counter, and verified that there was still no radioactivity. Next he repeated the exercise, but this time used the uranium-coated brass cylinder; as before, the tube was surrounded by the Bakelite ring. When he removed the ring on this occasion, and took it to the Geiger counter, it set the device clicking. This showed that radioactive fragments of uranium had adhered to the Bakelite, which proved that the uranium had been shattered.[47]

Joliot-Curie did this in the presence of Kowarski during the morning of January 26. He repeated the demonstration that afternoon before four witnesses, including Irène and Bruno Pontecorvo.[48] The next day Bruno wrote to Marianne, "Work at Ivry goes very well and if it continues will be very important."[49]

Halban returned from his ski holiday to find the laboratory in a state of excitement. He and Kowarski soon found a clever way to detect the liberated neutrons. The trick was to irradiate the uranium with slow neutrons, and use a detector sensitive only to fast ones. By this means, they could distinguish the liberated, fast neutrons from the slow ones emitted by the source. They began this experiment during the last week of February.[50]

Bruno was occupied with his experiments on nuclear phosphorescence, at the Ivry laboratory in the suburbs, so he did not take part in this fission experiment personally; nonetheless, he was deeply involved intellectually. Joliot-Curie was away at the ski resort of Val d'Isère. On March 3 he wrote to Kowarski to ensure that Bruno, who was due to join him at the resort along with the Joliot-Curies' daughter, Hélène, brought the latest news. Bruno duly arrived to report great progress. On March 27 he wrote to Marianne and mentioned physics for only the second time that year: "Physics goes very well," he wrote, underlining the words.

Fission might have been little more than a curiosity, except for two features. First, when the nucleus of a uranium atom splits, the total energy released is about a hundred times as much as that released by radioactivity, and up to a hundred million times the amounts in chemical reactions.

Here was the first hint of how to liberate nuclear energy on a larger scale than had hitherto appeared possible. In April, Halban and Kowarski finally demonstrated that the fission of a uranium nucleus liberates more than one neutron. The potential consequences of this discovery were nothing short of awesome. The possibility that these freed neutrons could initiate further fissions and produce a self-sustaining nuclear reaction was out in the open.

Suppose, for example, that when a single neutron splits the uranium nucleus into two chunks, two neutrons are liberated. There is a chance that these neutrons will hit two additional uranium atoms and repeat the fission.[51] If the same thing happens during this and subsequent collisions, there will now be four neutrons freed to make four fissions, leading to eight, sixteen, and so on—the number of neutrons doubles at each step. Thus it would only be necessary to irradiate uranium with a few neutrons to set off reactions that would continue spontaneously until all of the uranium was used up. This creates the potential for an immense release of energy.

As soon as this news arrived in the United Kingdom, scientists alerted the government to uranium's strategic importance. Enrico Fermi was in the United States; he too immediately realized fission's implications. In Germany there was a similarly intense response: all reference to atomic energy and uranium reactions was immediately censored in the German media.[52]

In France, Joliot-Curie sprang into action. On April 22, he announced that his team had established that a chain reaction was possible, and in the first week of May he applied for three patents. Two dealt with the potential application of fission to nuclear power, and the third, which was secret, related to explosives. On May 8, he went to Brussels to negotiate the acquisition of uranium stocks from the Belgian Congo, with a view to building a uranium bomb in the French Sahara.[53]

MEANWHILE WORLD EVENTS MADE THE LIKELIHOOD OF WAR MORE certain. Early in 1939, Bruno became seriously worried about the march of fascism, and its effect on his future. He had no idea when—or if—he and Marianne would have the chance to be together, and Gil's future also weighed ever larger on his conscience. In February he wrote Marianne a long letter, which in parts takes the form of a manifesto: "(a) If democracy survives in France <u>and</u> if I will be paid after to live with you and Gil, I shall stay in France. (b) If there is a war, and if that war is a democratic war against fascism, provoked by the axis powers, I will take part in that war. (c) If fascism comes to France, I shall go to the USA. <u>Write me if you find that *juste*.</u>"[54]

She replied, and agreed with his plan.

Bruno then followed up on his manifesto and encouraged Marianne to come "immediately" to Paris: "I don't want to influence you but if you decide to come I think it would be better as soon as possible, not only for the pleasure of you being here but also because we have many things to decide, for the little one and for us."[55] It seems that, at this stage, Gil was totally the responsibility of Bruno, although financial support came periodically from Sweden. Discussions with Frédéric Joliot-Curie had increased Bruno's conviction that France, and indeed Europe, was on a path to disaster, and that war was probably inevitable. If Gil remained in the nursery at Montmorency, Bruno would be tied to Paris. In April, therefore, Bruno signed a legal document, which would be deposited in a sealed envelope, to the effect that if anything happened to him, "Mademoiselle Marianne Nordblom . . . [will] decide at her discretion the fate of my son Gil Pontecorvo."[56]

During Marianne's absence, Bruno discussed Marxism with his cousin Emilio Sereni and, along with Gillo, deepened his links with communist groups. Gillo's son, Ludo, later described the situation: "My father [who believed strongly in communism] always said he felt that Bruno believed even stronger than him. Bruno was literally a big brother, five years older than my father—a similar age gap as between Emilio Sereni and Bruno."[57] Sereni influenced Bruno, and Bruno influenced Gillo. It wasn't just physicists who took note of Bruno's emerging intellect; Pontecorvo's involvement with communists was noticed in fascist Italy, where the local intelligence agency opened a file on the young scientist.[58] It is not known whether Swedish officials were aware of Bruno's communist activities; in any event, he soon discovered that he was persona non grata in that country.

In June 1939, Bruno applied to visit Stockholm and "possibly Sandviken" where "I have friends." He named Marianne as a reference for both

IMAGE 3.2. Bruno's unsuccessful application to visit Marianne in Sweden in 1939. (SVEN-OLAF EKMAN AND SWEDISH NATIONAL ARCHIVES.)

personal and financial guarantees, and stated that he had visited there previously. There was no mention in the application that they were the parents of a one-year-old boy. The records show that Stockholm police constable Berger Hassler phoned Marianne to verify the legitimacy of the application.[59] She confirmed everything and fully expected to see Bruno. But a few weeks later he learned that his application had been rejected.

Bruno now started encouraging Marianne ever more desperately to return to France with an appropriate visa, asking, "Why not come to Paris immediately?" He added that he didn't "want to influence her" but put strong pressure on her with the remark, "we have many things to decide," for "*le petit* [Gil] and for us."[60]

Marianne's reply seems to have mentioned that she was not well. This was clearly a recurring theme, as Bruno replied, "But what I would like to know absolutely is—what is the illness you have?" On August 7, when he left Paris for Zurich, he had still heard nothing from her. As soon as he was back in Paris, Bruno visited Gil, as always. He wrote to Marianne the next day, having heard nothing for several weeks. He reminded her again that they must "decide many things" for Gil. Two days later a card arrived from Marianne, saying that she would arrive in Paris the next week.

The following day, August 23, Stalin agreed to a nonaggression pact with fascism's high priest, Adolf Hitler. "A Victory for Peace" was the headline in *L'Humanité*, the organ of the French Communist Party. It reported that fascism had been "forced" to deal with "the very power that it has always declared to be its implacable enemy." The USSR was portrayed as a peacemaker, which had "imposed [the pact] on Mr. Hitler." In summary this was a "triumph of Stalinist politics." For a twenty-six-year-old living in the heady atmosphere of socialist Paris, the message of *L'Humanité* was overpowering. That same day, Bruno joined the Communist Party, "to prove his faith with Russia."[61]

He frequented party meetings, and took part in their passionate debates. At these meetings, all were free to speak, although "the leaders always had the last word."[62] The young, ardent physicist, who had attracted international attention with his scientific achievements, had already begun to be noticed by members of Comintern, the international communist organization. Marianne, meanwhile, was still living with her socially conservative family, far removed from such politics. She did not return to Paris until September 6, 1939, just days after the Nazis invaded Poland, and war in Europe threatened to cut Bruno off entirely. Her French visa now very correctly referred to "Mademoiselle Nordblom accompanied by Gil, born in Paris 30/7/38."

Years later, following the Pontecorvos' flight to the USSR, several reports in the conservative British media referred to Marianne as Bruno's unmarried lover, or mistress, which helped bolster their portrayal of the Pontecorvos as "immoral." In fact Bruno and Marianne married on January 9, 1940, which enabled them to obtain visas for North America when the Nazis invaded France in June.

DURING THOSE TUMULTUOUS MONTHS OF 1939, THE FISSION experiments of Bruno's French colleagues gave tantalizing hints that a chain reaction could be made. The scientists desperately wanted to be the first to achieve this, as the promise of unlimited energy was akin to finding the philosopher's stone. As five tons of uranium would be required for the attempt, the operation was moved from the university to the more spacious laboratory at Ivry.[63] Of course, the laboratory required financial support. During the summer of 1939, as war loomed, Kowarski and Halban performed demonstrations to impress the French Ministry of Supply, rather than pursue "scientifically impeccable proofs."[64]

In the fall of 1939 *Time* magazine was so impressed with the team's work that it included Joliot-Curie on its front cover. The team's actual scientific paper, which was also their last open publication before secrecy enveloped nuclear physics worldwide, was less impressive. Kowarski recalls that one of Fermi's colleagues once asked the great scientist, "What do you think of this paper by Joliot?" Fermi replied, "Not much." Kowarski himself admitted, "Fermi was quite right. Scientifically the paper was not very impressive [although] as a demonstration [it was]."[65]

As the team began their work at Ivry, ordinary water was their first choice as a moderator, but it was not actually the best option. When a neutron hits a hydrogen atom in a water molecule, it may bounce off and lose energy—which is the key to its speed being "moderated," thus raising its ability to cause fission. If that were the whole story, everything would be fine. However, the neutron may instead be captured by the hydrogen atom and lost, in which case the reaction dies. This latter effect is so probable that ordinary water kills rather than feeds the chain reaction. If instead of using ordinary water you use *heavy* water, however, neutrons are no longer captured. When a neutron encounters an atom of heavy water, it bounces off and slows, which is ideal.

One question that immediately struck scientists was this: If a chain reaction can indeed liberate energy explosively, why are the rocks around us, which contain uranium and are being hit continuously by cosmic rays, not liable to detonate spontaneously?

Niels Bohr's unique ability to visualize the labyrinth of an atomic nucleus gave him a key insight about fission: the quirks of nuclear structure imply that fission is more likely in an isotope with an odd number of constituents, such as U-235, than in one with an even number, such as U-238. Slow neutrons bounce off U-238 without causing it to fission. U-238 can also act as a blanket that covers any nearby U-235, making a succession of fissions rare, and the chance of a chain reaction negligible. Only if the neutrons encounter some of the rare isotope U-235 before exiting the uranium target will fission occur.[66]

Thus, the fission of natural rocks is rare because they contain so little U-235. While that is good news for our daily affairs, it makes it difficult to extract nuclear energy from raw uranium. To do so effectively one must first increase the amount of U-235 in the target, a process known as enrichment. This is difficult but, as would soon become clear, not impossible. Bohr published his insight in *Physical Review* in September 1939. It is ironic that the means that would help bring World War II to a close made its debut in the same week that the war in Europe began.

By January 1940 the French team had decided that heavy water was the best means to moderate the neutrons, as the first step toward unlocking nuclear energy from uranium. Frédéric Joliot-Curie alerted the French Minister of Supply that uranium could be the key to abundant energy or to a weapon of immense power. Also, he emphasized the special role that heavy water could play. The only European producer of heavy water in large amounts was the Norsk Hydro electric company in Norway; Frédéric explained that they would need "the whole of this stock."[67]

Then the team had a lucky break. On February 20, they learned that one of the minister's military contacts, Jacques Allier, had been a banker before being called up. The good fortune was that Allier's bank was the majority stakeholder in the Norwegian factory.

Allier, accompanied by some members of the French secret service, went to Norway and explained the situation. They stressed to the Norwegians that it was essential to "rescue" the entire stock of heavy water before the Germans invaded Norway. The Norwegian government feared a German invasion and was trying to appear neutral, even though its sympathies were transparent.

The French delegation was successful. In the second week of March 1940, the scientists received a telegram with the heading "absolute secrecy." It summoned them to a meeting where they learned that the entire stock of heavy water—some forty gallons—had arrived. At last they could plan experiments to determine the necessary conditions for a chain reaction.

Within weeks, the Germans invaded France and entered Paris. Kowarski and Halban set out on an odyssey, escaping to England with the heavy water to keep it out of Nazi hands. If Joliot-Curie had got his wish, Pontecorvo would have been with them, but the British authorities vetoed this plan. When British security viewed Frédéric's list of scientists, it gave the following assessment: "Dr PONTECORVO, a collaborator of Professor JOLIOT, is regarded as 'mildly' undesirable: might possibly be allowed to work if vital to the war effort, but even if working should be watched." There was no reason given for their description of Bruno as "undesirable."[68] In any event, Bruno had to make other arrangements.

The precious liquid would eventually end up in Canada, where, by a sequence of coincidences, Halban and Kowarski would be reunited with Pontecorvo. After the invasion, Joliot-Curie remained in Paris, where he became active in the French Resistance. Pontecorvo, being a Jew, was a target for the Nazis, and, as an Italian in France, he was an enemy alien. The options he'd laid out in his manifesto, which Marianne had agreed to, had now crystallized. Fascism had come to France; it was time to escape to the United States.

FOUR

THE FIRST ESCAPE
1940

ON THE MOONLIT NIGHT OF MAY 10, 1940, GERMAN TANKS BREACHED the Maginot Line; the Nazi army invaded France. The French tried to mount a defense along the Somme and Aisne Rivers in the north, but failed and withdrew to the Loire, south of Paris. Within two weeks the Allies had sacrificed northern France in service of a greater strategy: "He who fights and runs away lives to fight another day." A desperate evacuation of British troops from Dunkirk began on May 27, and was completed by June 4.

In April, while still in Paris, Halban and Kowarski had made a few preliminary measurements using the heavy water. But as events began to unfold, and the German army approached, urgency turned to panic. On May 16, Raoul Dautry, the Minister of Armaments, phoned Joliot-Curie and urged him to transfer his project out of Paris.[1] In the belief that the Loire would act as a southern limit to the German advance, they decided that Halban and an assistant would go to Clermont-Ferrand, about 260 miles south of Paris, and rent a villa, which could be used as an emergency laboratory. Kowarski and the heavy water would then join them there.

Hundreds of thousands of refugees began to flee south from Paris and its environs. In the midst of the mayhem, Bruno met with a tense and anxious Frédéric Joliot-Curie. Frédéric was a French patriot and decided to remain in the country. He urged Bruno to leave, however, adding, "It is best that you do so very soon."[2] Joliot-Curie understood the awful conundrum of Bruno's position. As an Italian, he was now technically an enemy of France; as a Jew, he was no ally of the Nazis.

At this juncture, Bruno experienced a piece of good fortune, thanks to Emilio Segrè, his old physics colleague. Segrè had gone to Berkeley on a summer visit in 1938. While he was there, Mussolini had passed the anti-Semitic laws, which barred Jews from university posts in Italy. Segrè was Jewish, and, having become excluded from work in Italy, he remained in Berkeley. One day in 1939 he attended a meeting of European émigrés living in California, where he met two other migrants—nuclear physicists who were prospecting for oil, and on the lookout for a neutron expert.[3] Segrè appeared to be manna from heaven. The physicists invited him to visit their laboratory in Tulsa, Oklahoma. In May 1940 he did so, but they were unable to tempt him to join them permanently; Tulsa was no competition for Berkeley, where Lawrence had built his cutting-edge cyclotron, and Segrè could watch the sun set behind the Golden Gate Bridge, across San Francisco Bay. Segrè turned their offer down, but, in doing so, he recommended Pontecorvo.[4]

Segrè wrote to Bruno, saying, "There is a good chance I can get you a job." Thus it came to pass that, in the spring of 1940, Pontecorvo had an offer of employment in the United States, just days before the Nazis occupied Paris.

Bruno had already escaped fascism once, when he left Italy in 1936. Now, with his family in tow, he did so again. Bruno's sister Giuliana, and her husband, Duccio Tabet, had also quit Italy, in 1938, and had rented a house in Toulouse.[5] They had told Bruno, Gillo, and their cousin Emilio Sereni that if ever the occasion arose, they could regard the house as their home. That moment had arrived.

On May 24, Marianne and Gil received papers of safe-conduct, which allowed them to make a single journey by train, or as passengers in a car, to Toulouse. The documents were valid for three days, starting on June 2.[6]

Marianne, loaded with baggage, and Gil, deeply upset but too young to understand the crisis, crammed onto the train to Toulouse. When they arrived there, Duccio Tabet immediately drafted a written declaration that Marianne and Gil were guests in his house, at 16 rue Edouard Baudrimont. The Toulouse police approved this arrangement on June 7.[7] Meanwhile, Bruno remained in Paris.

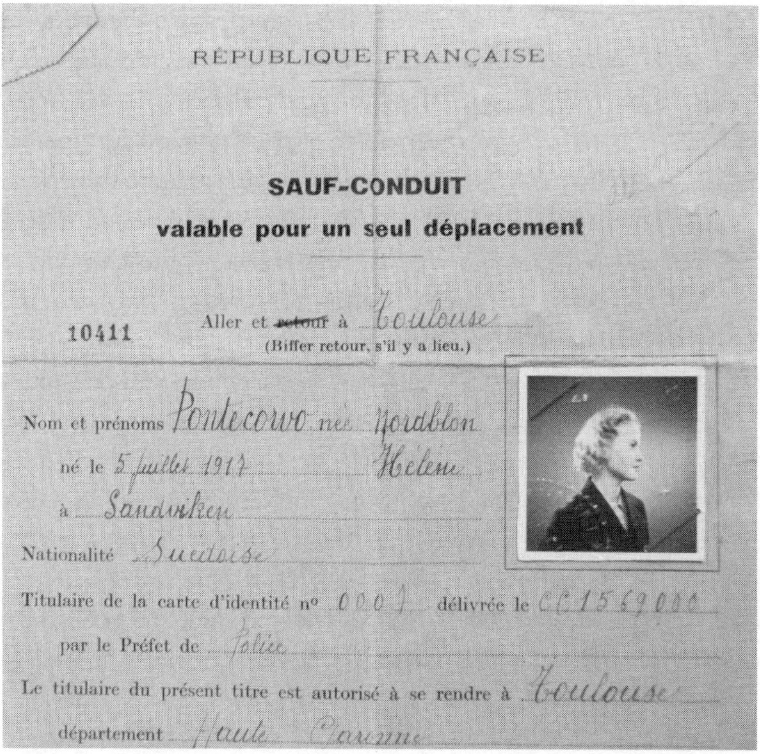

RÉPUBLIQUE FRANÇAISE

SAUF-CONDUIT

valable pour un seul déplacement

10411 Aller et ~~retour~~ à Toulouse

(Biffer retour, s'il y a lieu.)

Nom et prénoms Pontecorvo née Nordblom

né le 5 juillet 1917 Hélène

à Sandviken

Nationalité Suédoise

Titulaire de la carte d'identité nº 0001 délivrée le CC 1569000

par le Préfet de Police

Le titulaire du présent titre est autorisé à se rendre à Toulouse

département Haute Garonne

IMAGE 4.1. Safe-conduct document allowing Marianne and Gil to travel to Toulouse in 1940. (AUTHOR, CHURCHILL ARCHIVES CENTRE.)

THE START OF JUNE WAS HOT. THE SUN SHONE FROM A CLOUDLESS SKY.[8] Even the nights were warm; stars twinkled in the crystal clear air. Two miles above the city, a German pilot could see twelve boulevards radiating from the Arc de Triomphe, as if it were at the centre of a clock face. The Seine meandered through the suburbs, and separated in two to embrace the Île de la Cité at the city's heart, where stood Notre Dame, which had survived centuries of history and revolution. To the immediate south of the cathedral, the narrow streets of the Left Bank were laid out like a map. That is where Bruno Pontecorvo suddenly awoke, on the night of Monday, June 3, as the German aircraft released its bombs.

Also in Paris were Bruno's brother Gillo, their cousin Emilio Sereni, and Gillo's French girlfriend, Henrietta. Bruno's friend Salvador Luria, an Italian research colleague at the Radium Institute and a Sephardic Jew, was also worried about the Nazi advance.[9] Among others for whom that night would frame the rest of their lives was Irène Némirovsky, a Jewish writer. She depicted the events in her novel *Suite Française*. Some of her characters, such as the members of the Péricand family, started to evacuate immediately. Others, like the Michauds, hesitated until it was almost too late, even giving their apartment a final clean before they departed forever, naively dreaming that they would return someday. Her fictional refugees portrayed in microcosm the experiences of a million real ones, including the Pontecorvos, Sereni, and Luria.

Just before dawn on Tuesday, June 4, air-raid sirens sounded and the exodus began. During the final days of May, Kowarski had cleared out the laboratory he'd shared with Halban and prepared to move everything to Clermont-Ferrand, where he would join his colleague. In addition to the heavy water, he had to gather electronic amplifiers, radiation detectors, and a precious sample of radium and beryllium, which would provide their neutrons. The biggest task was moving several tons of lead bricks, used as radiation shields in the experiments.

By the fifth or sixth of June, Kowarski had loaded all of the lead, along with the rest of the equipment, onto several army trucks.[10] Then, accompanied by half a dozen soldiers, the entire caravan joined the tide of humanity headed south from Paris, away from the Nazi army. Kowarski's convoy reached Clermont-Ferrand on Friday the seventh; for security purposes he stored the heavy water in the nearby women's prison, in Riom. His wife and child managed to join him by train a few days later.

BY THAT WEEKEND THE MAIN STREETS OF PARIS WERE ALMOST EMPTY. Metallic shutters covered the windows of shops, abandoned by their owners, who had joined the throng of refugees. However, some still believed that the flight was the result of "hysterical rumours" circulated by "traitors."

Any doubts were removed on June 9, when the Germans assaulted the city. All major monuments were surrounded with sandbags; windows rattled, and "from the top of every monument birds rose into the sky" as the sounds of gunfire thundered across the rooftops.[11] "The Germans have crossed the Seine . . . even animals can sense the danger," declared one of Némirovsky's anxious refugees, whose husband still seemed in a state of denial. Gillo Pontecorvo vividly recalled that someone in the street shouted, "The Germans are at Pontoise"—about 15 miles northwest of the centre of Paris.[12]

The attack on the city and the rupture of the French defenses were so rapid that many were caught unprepared. For anyone with special reason to fear life under Nazi rule, the time had come to flee by any means possible: train, car, truck, or bicycle.

Frightened hordes descended on the railway stations. Their plans to escape were frustrated, however, when they found the platforms closed, guarded by soldiers. Would-be fugitives flooded the surrounding streets and offered taxi drivers small fortunes to take them out of the city. Money had ceased to have meaning; people, not possessions, were what mattered. But the taxis didn't have enough gas to reach even the relative sanctuary of Orléans, about 70 miles to the south. Those without train permits or gasoline, which was almost everyone, set off on foot. The roads south of Paris were clogged, as up to a million refugees dragged their luggage behind them, or humped it in wheelbarrows and carts.

These were the last days of freedom in Paris. As the crescent moon rose, the clear night sky was suddenly obscured. What at first appeared to be storm clouds was actually a smoke screen, put up deliberately to save the city from being bombed.[13] On Thursday the thirteenth, Paris was declared an open city, as the French government fled to Bordeaux. At this eleventh hour, Bruno, Gillo, Henrietta, and Emilio Sereni, along with Salvador Luria, joined the exodus—on bicycles. They were almost too late. The next day, June 14, the German army entered Paris.

AFTER SO MANY CLEAR, HOT DAYS, THE WEATHER BROKE ON JUNE 13. The rain showers may help explain the apparently casual nature of the group's departure; having cycled hardly more than a mile, they stopped to say farewell to friends on the rue Mouffetard.[14] These acquaintances could have been models for the Michaud family, who delayed until it was almost too late, or they might have been among those who were brave or foolish enough to remain and take their chances with the occupation. In any event, they insisted that the travellers stay a while and have something to eat.

Gillo, just twenty years old, and his girlfriend, Henrietta, seemed to regard the whole enterprise as a diversion, a frisson of excitement in the life of a self-confessed playboy. Their carefree attitude infected Bruno, who helped himself to two cream macaroons. Salvador Luria, a year older than Bruno and seemingly more aware of the seriousness of the situation, was shocked at his companions' casual attitude. They had wasted precious time and he urged them to leave. By nightfall they reached a guesthouse about forty-five miles south of the city limits.

The next morning, when Gillo opened the bedroom window, he discovered that the square below was full of German tanks. The previous day's carefree adventure was replaced by panic. He thought that further flight was pointless and that it would be better to return to Paris. However, Henrietta convinced him to continue the journey with the others. "Henrietta wanted to carry on south," he later explained. "And so we did."[15]

Whereas the day before they had been part of a steady exodus, now the streets were crowded and disorderly. People, desperate to escape, lined the roadside in the hope of finding some form of transportation. Thousands of cars littered the shoulders, broken down or out of gas. Streams of refugees and army vehicles were on the move. The main line of cars, their roofs piled high with suitcases, crowned with mattresses, looked like unstable piles of bric-a-brac on wheels. A procession of cars, vans, trucks, tractors, anything with four wheels, lumbered along in single file. The vehicles, which crept forward like snails, obstructed the torrent of pedestrians and horse-drawn wagons. In turn, the crush of refugees travelling on foot, meandering around and between the vehicles, blocked the roads and interfered with oncoming traffic.

Like a scene from a film noir, this mayhem formed the backdrop as Gillo and Henrietta, clad in shorts and carrying skis, cycled past on a tandem, along with Bruno, Sereni, and Luria. Someone shouted contemptuously, "Look at those bourgeois, going on holiday!"[16]

Up to that point their travels had indeed been like a holiday, in that they had stayed overnight at a hostel. Now everything abruptly changed. At Orléans and beyond, there wasn't a single bed free. People slept—or at least tried to—in their cars, in shop doorways, or even on the sidewalk, using suitcases as pillows. Sonia Tamara, a correspondent for the *New York Herald Tribune*, was in Orléans a day before the Pontecorvo party, and recalled that the scene near the railway station was "appalling." People lay on the floors inside, and filled the town square outside. Children were crying, and there was nothing to eat in the entire town.[17]

Refugees were everywhere. Some, like the Pontecorvos, had a specific destination in mind. Others just wanted to get away. No one knew where or when the Germans would appear to cut off their escape. The choice of the best route was a lottery. Many roads were closed, which forced refugees to take detours as soldiers and police made desperate attempts to let army vehicles through. Piles of stones and bricks blocked the outskirts of every village. In revolutionary tradition, peasants manned these makeshift barricades and examined the papers of the fugitives.

Yet in the midst of this mayhem the postal service appears to have kept functioning, at least in the south, away from Paris. For on June 14 Bruno wrote a postcard to Marianne, which reached her in Toulouse.[18] It would seem that by that evening the group had reached Beaugency, about ten miles south of Orléans and eighty from Paris, as it was there that he had time to write the postcard. It began, "*J'ai quitté Paris depuis deux jours en vélo.*" He continued, "I hope to get a train at Blois or some other place. If I can't I shall come to Toulouse by bike but I don't know when [I will arrive]. In Paris I didn't receive any letter from you but I understand the post is not operating [there] anymore. The Hotel Grands Hommes was closed as I was leaving. I can tell you more when I arrive."

On the morning of the fifteenth, the group set off once more toward Blois, thirty miles to the south. Their hopes of finding a train there were dashed, however, and by nightfall they had progressed to Vierzon, south of Blois. They had logged a total of about fifty miles for the day, and 130 in all from Paris. That evening, Bruno mailed the postcard to Marianne from Vierzon. There were still nearly three hundred miles to go.

Emilio Sereni decided to travel by trails and minor roads, which he felt would be more secure. The Pontecorvos decided to take their chances with the shortest, fastest route. They reached Toulouse in about ten days; Sereni took a few days longer. At some point Salvador Luria left his friends and joined a different group of refugees, eventually reaching Marseilles. He later made it to the United States, and in 1969 won the Nobel Prize in Physiology or Medicine for his work on the genetic structures and reproduction of viruses.[19]

FRANCE SURRENDERS

Halban, Kowarski, and the precious heavy water, meanwhile, had arrived safely in Clermont-Ferrand. However, by the time Bruno started his odyssey southward, the plans of his former colleagues had been thrown into disarray.

On June 16, the prime minister resigned, to be succeeded by Marshal Philippe Pétain, who negotiated an armistice. This was signed on June 22. In 1918, Germany had formally surrendered to its World War I enemies in a railway carriage. In 1940, Germany's revenge (and France's humiliation) was completed when the French surrendered to Hitler in the very same carriage, at the very same spot. Germany now occupied the north and west of France. The south, which included Toulouse and Marseilles, was nominally independent. Pétain took charge of this area.

On the day of the prime minister's resignation, Joliot-Curie had arrived in Clermont-Ferrand, in the independent south. He explained to Halban and Kowarski that, in his opinion, and that of government officials, France would remain divided for a period, but eventually the Germans would occupy the whole of the country. The plans for fission experiments at Clermont-Ferrand were therefore in jeopardy. Halban and Kowarski were ordered to go to Bordeaux at dawn the next day, June 17, along with the heavy water. Once there, they would receive further instructions.

They set off in two cars. In one was the Halban family; the other, a station wagon, contained the Kowarskis, along with twenty cans of heavy water piled in the back. They travelled roughly from west to east, crossing innumerable roads that spread southward from Paris. These roads, unsurprisingly, were full of refugees. It was dark when they reached Bordeaux and received their orders: "Proceed to England by ship."

Years later, Kowarski still recalled the drama and timing of these events. The Pétain government had been formed, and an armistice called for, but on this one day, June 17, the previous government was still concluding its business. As a last desperate act of defiance, the government "put its faith into these two departing magicians," with heavy water serving as their magic wand. It was clear to the pair that they were "carriers of a mission," entrusted with something of the greatest importance for the honour of France.[20]

Officials from the Ministry of Armament, assisted by highly ranked military officers, carried the scientists' household belongings and the cans of heavy water to the ship *Broompark*. By midnight Halban and Kowarski were on board, along with several other scientists. The few cabins had been assigned to women and children, so Halban and Kowarski had to fend for themselves. They eventually fell asleep on a heap of coal. The next day, Marshal Pétain ordered France to stop fighting, and surrendered to the Germans.

The heavy water had begun its own odyssey, first to England and then to North America. It arrived in England on June 21, just as Bruno Pontecorvo was reaching Toulouse and completing the first stage of his own adventure. Three years would pass before he would meet his colleagues again, across the ocean.

FROM VICHY FRANCE TO THE NEW WORLD

The south of France was the starting point for a great migration of refugees to North America. The Emergency Rescue Committee issued permits for immigration to America but only if the applicants satisfied very rigid criteria. Bruno, at least, had a job to go to. He now entered the corrupt lottery necessary to obtain the exit visas from the Vichy government. The character of Captain Louis Renault in the film *Casablanca* is a perfect reflection of Vichy officialdom: people who had France in their hearts, but were forced to work within the tight constraints of realpolitik. Their motives were mixed. Power over peoples' lives led to greed and corruption. Bruno Pontecorvo had to find his way through this labyrinth. The family planned to travel by ship from Lisbon, which required them to pass through Spain. So he also had to deal with the Spanish and Portuguese governments to obtain the necessary transit permits. Chaos and unpredictability reigned.

Bruno's attempt to leave Europe seems to have started on June 26 with a visit to the Portuguese consulate in Toulouse.[21] He then made a three-hundred-mile round trip to Bordeaux to visit the American consulate, on June 29, seeking permission to immigrate to the United States. This cost nine dollars per person.[22]

The following week was a rare period of relative calm before Bruno resumed crisscrossing France, accumulating further documents for his family's departure. On July 9, he was issued a visa in Perpignan—120 miles away from Toulouse, near the Spanish border—allowing them to travel across France "for the USA via Portugal and Spain." The visa was only valid until July 17; time was short. He stayed overnight in Perpignan. The next day, he visited the Spanish consulate, which issued more visas to the

Pontecorvos, "good for a single voyage from Marseilles via Spain en route to Portugal."

Having acquired Spanish entry and transit documents, Bruno next visited Marseilles—nearly 200 miles away—to obtain Portuguese documents, which he succeeded in doing on July 12: "Good for Portugal in transit to USA, valid for 30 days." Finally, with all these papers in place, official permission to undertake the actual journey and exit France seems to have been required; on July 17, the prefecture of the Haute Garonne department, in Toulouse, issued the last set of visas to Bruno's family, permitting them to leave France for the United States via Spain and Portugal.

Having come so far together, the band of refugees now went their separate ways. Gillo and Henrietta remained in southern France, as did Emilio Sereni and his family. Giuliana and Duccio joined Marianne, Bruno, and Gil en route to Portugal and the Americas.

At last they set off by train from Toulouse. They crossed the frontier into Spain at Portbou, on the Mediterranean coast, on July 19. They had travelled through France for half a day. It would be a further 500 miles before they reached Madrid, and five more days before they arrived in Portugal.

THE TRAIN FROM MADRID WAS PACKED SOLID WITH REFUGEES. THEIR luggage, stacked in the corridors, served as makeshift furniture for those who had not managed to grab seats. To add to the discomfort, it was the height of summer, very hot, and there was no water. Giuliana was pregnant, but stoically managed to deal with the hardships. Not so Marianne, who was also pregnant, and in pain.[23] She felt a little better when she was lying down, but lost consciousness several times. Bruno did not know what to do. For a long time she was able to sleep only by lying on the floor of the corridor.

On July 24 they reached the border with Portugal, and from there proceeded to Lisbon, the only port from which it was possible to reach the United States.[24] Thousands of exhausted and frightened migrants were pouring into the city. Many had abandoned comfortable homes, and sold their precious possessions in order to buy tickets, or to bribe officials in exchange for the necessary permits. And so it was for Bruno. After handing over yet more money, he obtained a permit for the group to reside in Lisbon for thirty days. They rented rooms in a small hotel.

After the privations of the journey, Lisbon was like heaven. Sunny and cooled by the breeze from the sea, the city also boasted coffee shops and the smells of pine trees, which added to the general ambience. The war was miles away, in all senses of the phrase. Marianne still had the "pallor of a sick child," however, and a few days before they were due to board the ship for the United States, she had a miscarriage.

She was in no state to travel, but they had come so far and had to continue. On the ninth of August, the two families set out together across the Atlantic Ocean, the final leg of their escape.[25] Their privations were not yet over, however. Although the weather was good, the voyage on the liner *Quanza* was awful. The Pontecorvos' cabin was low down in the ship, where the heat was insufferable. Marianne and Giuliana were seasick.[26]

To arrive in Manhattan after crossing the Atlantic by ship is, judging from my own experience, one of the great moments in world travel. After days of grey sea, with no distinguishing features to give a sense of distance or speed, the low-lying dunes of Long Island appear on the starboard, or right-hand, side. After another hour, the coast of the mainland becomes ever more prominent on the port side. You appear to be headed for disaster as the two landmasses of the North American continent and Long Island come together in front of you. Then, gradually, a narrow channel appears between Staten Island and Brooklyn. This is the strait known as the Narrows.

Since 1964 the Narrows have been spanned by a suspension bridge.[27] On a modern liner, such as the *Queen Mary 2,* the funnels of the ship seem certain to pierce the bridge above you. So it is with both relief and surprise that you rush beneath, and are suddenly within the confines of the continent. In 1940 there was no bridge, but the dramatic sense of arrival and relief as the Pontecorvos passed through the Narrows, from the Atlantic Ocean to the inland waters, would have been the same. The *Quanza* entered the Upper Bay, and turned to starboard, revealing the skyscrapers of Manhattan Island.

But first, for all travellers, and especially for refugees, there is one symbolic moment to savour: the sight of the Statue of Liberty. It was thus ironic that, on the day the Pontecorvos arrived, humid mist obscured this symbol of freedom. In any case, the ship reached New York on August 19; Bruno Pontecorvo was three days shy of his twenty-seventh birthday, which he spent with his brother Paolo at 503 West 121st Street. This was to be Bruno, Marianne, and Gil's temporary home until they moved to Tulsa.[28]

PARALLEL LIVES

When Bruno, Giuliana, and their families left France and headed for the United States, Gillo and Henrietta stayed behind. They moved to Saint-Tropez, where a strange assortment of individuals who were trying to escape the war had gathered. Gillo gave tennis lessons to the local bourgeoisie, and met exiled intellectuals such as the musician René Leibowitz, who taught him piano, harmony, and counterpoint.

Gillo later described this period, when war raged everywhere except this bizarre bubble, as "living outside history."[29] Then he found a way to reenter it. Having been converted to communism by Bruno, Gillo became a clandestine member of the Communist Party of Italy in 1942. While living in Saint-Tropez, he met Giorgio Amendola, an Italian communist who was secretly organizing opposition to the Mussolini regime. Amendola was desperate to find someone willing to go to Italy and reestablish contact with the antifascists and communists there. Previous agents who had made the attempt had been arrested just past the border. Gillo offered to try—and succeeded, using a variety of false identities.[30]

Soon he was making regular visits to Milan on courier and newsgathering missions. He worked on the party's underground newspaper, l'Unità, throughout the summer of 1943, while Milan suffered constant Allied bombardment. In 1944 Gillo was forced to go into hiding, but then went to Turin, where he began to organize young factory workers. After the liberation, he became the director of Pattuglia, a journal for communist and socialist youth, and then returned to Paris to become the Italian representative of the communist-backed World Federation of Democratic Youth.

Salvador Luria, as we saw earlier, reached Marseilles, and emigrated to the United States. Thus, our heroes survived, unlike some of Némirovsky's fictional ones, or indeed Némirovsky herself. Her chronicle of the flight from Paris mirrors the experiences of our group, but her own tale ended tragically: arrested by the Gestapo, she died at Auschwitz. Emilio Sereni, the final member of the group that fled Paris with Bruno, almost suffered a similar fate.

After fleeing Paris, Sereni was prominent in the communist partisan movement. Based in Nice, he encouraged Italians to resist Mussolini, and published a radical newspaper, *The World of the Soldier*. In June 1943 he was arrested, sent back to Italy, and sentenced to eighteen years in prison. He tried to escape, but was recaptured and sent to an SS camp in Turin, where he spent seven months, under threat of execution. In August 1944 he made a second escape attempt. This time he succeeded, and lived undercover for the few remaining months of the war.

After the war ended, Emilio Sereni—by now a hero as well as a prominent communist—took up politics full-time and became Minister of Public Works in the postwar government. At the Fifth Congress of the Italian Communist Party in 1945, he was elected as a member of the Central Committee and the Directorate. He soon became an influential member of Comintern, or the Communist International—the organization whose goal was to create an international Soviet republic. By 1950 Emilio Sereni would be a regular visitor to Eastern Europe, and well connected in Moscow.

FIVE

NEUTRONS FOR OIL AND WAR

1940–1941

BRUNO'S NEW CAREER AS AN OIL PROSPECTOR TOOK HIM AWAY FROM the frontiers of nuclear physics, which were now dominated by fission and the attempt to create a chain reaction. Meanwhile, Hans von Halban and Lew Kowarski had decamped to England with the precious heavy water, and begun experiments in Cambridge. The results of these tests appeared to show that a chain reaction might be possible in uranium when heavy water is used a moderator.

Even before Bruno's departure, the critical questions were obvious to physicists worldwide. In 1939 Niels Bohr had pointed out that using U-235 rather than U-238 was key, but how much uranium would be required to make a chain reaction in practice? And was uranium the only possible material that could be used? It's worth noting that if this quest was successful, it would highlight the irony of Hitler's persecution of the Jews: many brilliant Jewish scientists fled central Europe and became key players in the scientific war against the Nazis.

In England, at the University of Birmingham, two of these refugees, Otto Frisch and Rudolf Peierls, focused on the questions of whether a chain reaction could occur if fast neutrons hit a lump of pure U-235 and how much would be required to make a significant explosion.

In March 1940 the two young theoreticians worked out the equations and were astonished by the answer: "The energy liberated by a 5kg bomb would be equivalent to that of several thousand tons of dynamite."[1] The British, already at war with Germany, took immediate action: Frisch and Peierls's work was classified as top secret—so secret, in fact, that the two émigrés (technically "enemy aliens") who had made the discovery were barred from the official committee that first evaluated it.

The idea was not implemented efficiently, however, because of doubts that such a weapon could be built in time to influence the war, or even that it would work.[2] The enrichment of uranium to produce the necessary levels of U-235 would be a huge industrial enterprise. This inspired a new thought: Could there be other elements whose nuclei might fission easily? Scientists hypothesized that there existed unstable elements beyond uranium—now known as neptunium and plutonium—and that these could be likely candidates. However, no one knew whether they would be so in practice. Several grams of these exotic elements would be needed to find the answer, and because these "transuranium" elements do not occur naturally it would be necessary to make them.

Transuranium elements are occasionally produced, one atom at a time, when neutrons hit uranium. In June 1940, American physicists Edwin McMillan and Philip Abelson duly bombarded uranium, performed a chemical analysis of the sample, and identified the presence of some atoms of element 93—neptunium. They suggested that plutonium, element 94, might also be formed in such a way.[3] In Cambridge, Egon Bretscher and Norman Feather predicted that plutonium would have a strong capacity to fission, much like U-235. To test whether this was true in practice would require a lot of plutonium, which in turn would require tons of uranium, intense sources of neutrons, and an efficient means to moderate them. A nuclear reactor is the ideal machine for the task. Indeed, the production of these transuranium elements would become one of the major motivations behind the development of nuclear reactors.[4]

The project in the United Kingdom was already classified secret. When Abelson and McMillan's paper about neptunium and plutonium appeared in *Physical Review* that summer, available for anyone to see, scientists in both Britain and the US protested.[5] In Britain, a nation already at war, the dangers of advertising such a strategically important discovery were obvious. Although the United States was not yet involved in the war, influential refugees from fascism, such as Einstein and Fermi, were already in the country, and others, including Bruno Pontecorvo, were on their way. Thus, the Americans too became concerned at the potential implications of these discoveries. Starting in the summer of 1940, all research on fission in the West became a closely guarded secret, and no further papers on the subject were published in the open literature.

There were three main strategies in the quest for a chain reaction. One was to find some way to enrich uranium—that is, to increase the amount of the U-235 isotope. Another was to find some other fissile element—such as plutonium—and use that. In the USSR, Igor Kurchatov examined a third possibility: Could fast neutrons fission both U-235 and U-238 and initiate a chain reaction in natural uranium without the need for enrichment?

Early in 1940, at Kurchatov's suggestion, two junior colleagues— Georgii Flerov and Konstantin Petrzhak—used a range of sources that emit neutrons with different energies, put them inside a sphere of uranium, and measured how the neutrons flowed.[6] They found that fast neutrons create an insignificant amount of fission but also discovered something unexpected: fission appeared to occur sporadically in uranium without any neutron bombardment at all.

Their initial explanation was that cosmic rays from outer space were hitting the uranium atoms and splitting them apart. To guard against this possibility, they repeated the experiment underground in Moscow's Dinamo subway station, where the earth and rocks would shield the uranium sample. The spontaneous fission persisted. This proved that it was a real phenomenon, albeit rare. The team announced their findings at the Soviet Academy of Sciences in May 1940, and Kurchatov sent a short report to the American *Physical Review*, which published it in July.[7]

We now know that spontaneous fission occurs because even natural uranium is slightly unstable. Although the phenomenon is too rare to be a source of energy, it can nonetheless interfere with the delicate design of an atomic weapon. In a nutshell: to make a nuclear blast, fission must grow exponentially within a fraction of a second. Spontaneous fission, and fission induced by cosmic rays, can cause the nuclear energy to be released prematurely, before a chain reaction develops. Within a couple of years, this phenomenon would become one of the many problems to solve in designing a successful uranium bomb.

In the USSR—which was not yet at war—investigations continued. On August 29, 1940, Kurchatov, Flerov, and some colleagues drew up a plan for the "utilization of the energy of uranium fission in a chain reaction."[8] There was a lot of discussion about the plan in November at the Fifth All-Union Conference on Nuclear Physics, in Moscow. The need for large amounts of uranium was immediately obvious. How the USSR would find supplies of the scarce element, however, was not.

Two young Russian theoreticians, Yulii Khariton and Yakov Zel'dovich, now turned their attention to the same question that, unknown to them, Frisch and Peierls had already answered in March: How much pure U-235 was needed? In the fall of 1940 they found the same result as the British pair: a few kilograms of U-235 would do. The Soviet Union recognized the strategic importance of this discovery and classified their work as secret.[9] The Soviets already suspected that a secret nuclear physics project was under way in the West. Their hunch had developed during the summer, when Flerov and Petrzhak's paper appeared in *Physical Review*. Their discovery of spontaneous fission was of immense significance, and yet there was a complete lack of (visible) reaction to this in the United States.[10]

Until this time, every knowledgeable physicist could keep abreast of global developments in the field. There was no need for spies. The major secret in the West was the discovery made by Frisch and Peierls. Bruno knew nothing of this. The Soviets, however, thanks to Khariton and Zel'dovich had discovered it for themselves. The big unknown, of course, was what was going on in Germany. The potential of a chain reaction to fulfill the Promethean dream of creating unlimited amounts of atomic energy was common knowledge, and a year earlier, in April 1939, Germany had banned the export of uranium and started conducting experiments on fission. That was enough to spur the allies to develop nuclear technology.

It would be two years before Bruno rejoined the world of cutting-edge physics. During that time much would change, as nuclear physics became a tool of war, its research classified as secret. This secrecy applied not only to novel results but also to the nature of the quest itself.

BRUNO THE OIL PROSPECTOR

When the Pontecorvos arrived in the United States in August 1940, Enrico Fermi was based at Columbia University, in New York. One of the first things Bruno did upon arrival was visit the Fermis at their home in Leonia, New Jersey, two miles from Manhattan, across the Hudson River. Years later, Enrico's wife, Laura, remembered that this visit had disturbed her.

She recalled that Bruno had come alone. He explained that Marianne was "worn out" from the sea crossing and needed to rest. Laura had never met Marianne or Gil, and was disappointed that Bruno hadn't brought them along. She added that although Marianne's fatigue was understandable so soon after an "ocean journey on a boat crowded with European refugees," she was perturbed that Bruno declined her offer to visit Marianne, or to help her in any way.[11]

Laura Fermi wrote her account in 1953, at a time when Pontecorvo had disappeared behind the Iron Curtain, his whereabouts still unknown. Her account helped fuel the media myth that Marianne was a mysterious Svengali, silent because she had so many secrets to conceal. However, the real explanation of her behaviour is perhaps much simpler, and also more tragic. As we have seen, Marianne had been through a painful pregnancy and miscarried just days before crossing the Atlantic. Laura Fermi was presumably unaware of this. Marianne could also be painfully shy, at ease with friends but uncomfortable with strangers.

After a brief stay in New York at the home of Bruno's brother Paolo, the Pontecorvos set off on the final leg of their journey: 1,300 miles southwest to Tulsa, Oklahoma, the home of Well Surveys, where Bruno was to become an oil prospector.

BUILD A BETTER MOUSETRAP AND THE WORLD WILL BEAT A PATH TO your door.

In the 1930s, as the need for gasoline grew in the United States, the "world" meant the likes of Standard Oil, Texaco, and Phillips; the mousetrap, a means to locate precious new oil fields.

Well Surveys of Oklahoma was a company that specialized in oil prospecting; their big idea was that the natural radioactivity in rocks might reveal the geological formations where oil could be found. To realize this dream they needed an expert in nuclear physics and radioactivity, and, thanks to Emilio Segrè's recommendation, they hired Bruno Pontecorvo.[12]

Two of the scientists at Well Surveys—who were instrumental in hiring Bruno—were colourful characters whose backgrounds would later attract the attention of the nation's security agencies. Their names were Jakov (also known as Jake or Jack) Neufeld, a Polish émigré, and Serge Alexandrovich Scherbatskoy, a Russian who had been born in Turkey.

Neufeld had been born in 1906, learned nuclear physics at the University of Liège in Belgium, and then took a position at Cornell University. It was from there that he joined Seismic Surveys Corporation, a forerunner of Well Surveys. At the time, Scherbatskoy was the research director of the company.

Scherbatskoy had been born in 1908 in Constantinople, where his father, Alexander, was a diplomat at the Russian consulate. After the Russian Revolution of 1917, Alexander worked in the League of Nations offices in Berlin, and then moved to Paris. While the family was based in Germany, Serge studied engineering at Stuttgart. After the move to France, he enrolled at the Sorbonne, where he graduated in physics in 1926.

In 1929 he arrived in the United States, just before the stock market crashed. According to some accounts he had no passport, but in any event he was admitted, Serge Alexandrovich being reborn as Serge Alexander. He worked on telephones at Bell Labs, but he was laid off in 1932 due to the effects of the Depression, after which he took a variety of jobs in electronics. In 1936 he joined Seismic Services Corporation in Tulsa.

It was there that he met Neufeld. They used their combined experience in electronics and nuclear physics to design devices that detected gamma rays, which are emitted by radioactive atoms in rocks. Shales, which are harbingers of oil fields, contain uranium and thorium, which are radioactive. Sands and limestones, by contrast, are not. Their detectors could identify shale-rich oil fields more successfully than other techniques, which had failed or were at best inefficient.

Standard Oil was impressed with their research, and poured in money. In 1937, this led the pair to create Well Surveys, where Scherbatskoy was research director and Neufeld was at his side. Their plan was to find new ways of applying nuclear physics to prospecting.

Inspired by Frédéric and Irène Joliot-Curie's discovery of induced radioactivity, and the Via Panisperna Boys' demonstration that neutrons are especially efficient at activating it, Neufeld and Scherbatskoy decided to use this breakthrough to induce radioactivity in rocks. By this means, they believed that underground strata could be made to shine brightly—in terms of gamma rays—relative to the faint glimmer that is typical of natural radioactivity. This is what led them to hire Bruno Pontecorvo.

Bruno now set about developing the first industrial application of neutrons in locating oil-bearing rocks, a technique still in use more than half a century later. His invention paid off handsomely for his employers. In a 1990 interview he commented wryly, "I could have been a millionaire if I had patented my discovery. Instead I did not make anything, and now the patent belongs to the company I worked for. I do not have any practical sense."[13] Soon Bruno's invention would be used by the Manhattan Project to map the vast mineral resources in the Canadian wilderness, where it was secretly turned into a remarkably efficient way to find uranium. This would have profound implications for the development of nuclear energy, as well as for Cold War politics.

THERE ARE THREE PARTS TO THE CHALLENGE OF USING NEUTRONS to find oil. First, you must produce neutrons; then, having lowered the neutron source down a hole, you need to detect signals coming from the interaction between those neutrons and the underground minerals. Finally—the aim of the exercise—you need to decode the signals to learn about the nature of the strata.

For Bruno, making beams of neutrons was child's play. Typically he used radium, which produces a steady stream of alpha particles, and mixed it with beryllium, whose atoms emit neutrons when they're hit by alpha particles. In searching for oil-bearing rocks, the basic idea is that the neutrons interact with atoms in the vicinity of the borehole. Most bounce gently off other particles and pass through the rocks, never to be seen again, but some are reflected back toward the detector, or are absorbed by elements in the rocks. When neutrons are absorbed by a material, they convert the nuclei of the material's atoms into miniature transmitters. The end result is that information returns to the detector in the form of radiation, such as gamma rays, or the reflected neutrons themselves.

The whole apparatus that Bruno developed was contained in a thin cylinder, about two metres long and a mere ten centimetres in diameter, which could be lowered into deep boreholes. The cylinder contained his neutron source, an ionization chamber (similar to a Geiger counter), and an electronic amplifier for recording the faint signals created when radiation passed through the ionization chamber.

In Rome, Bruno had been party to the discovery that heavy nuclei tend to absorb slow neutrons.[14] This induces radioactivity in the nuclei, which may respond by emitting gamma rays. In this new application, the spectrum of these gamma rays could be used to identify isotopes of heavy elements present in the rocks. The data reports that were obtained from the neutron irradiation of rocks in prospective oil fields were therefore known as "neutron-gamma" logs.

Pontecorvo also developed an independent recording method—the "neutron-neutron" log—in which neutrons recoiling from the strata were detected. When a neutron hits a heavy element, the bulky atom stays where it is and the reflected neutron retains most of its energy. Light elements can also deflect neutrons, even reflecting them back to the source, but in this case the neutron has less energy than when it set out. Thus, if you could record the actual energies of the reflected neutrons, you would have a way (albeit a rough one) of differentiating between light and heavy elements.

This basic idea was simple, but there were challenges to overcome before it could be put into practice. For example, the source produced fast neutrons, whereas the detector was most efficient at recording slow ones. To resolve this issue, Pontecorvo embedded the detector in paraffin, which slowed the neutrons. Although this made his device more efficient, it created a problem of its own: to convert the number of counts in the detector into meaningful information about the neutron intensity, he needed to know what percentage of the original fast neutrons had slowed, and by how much. To find the answer, he performed a series of experiments in the lab. First he determined how his detector responded to a known source of neutrons, and then compared data from the field with these benchmarks.

He also needed to decide on the optimum amount of paraffin. A small amount would slow few, if any, of the neutrons to the low speeds necessitated by the detector. At the other extreme, a large amount of paraffin would absorb so many neutrons that very few would get through. Hence there must be some intermediate situation in which the intensity of slow neutrons is at a maximum. Only after these calibrations were done in the lab would it be possible to identify strata in the field.

During his time in Italy, Bruno had learned that the amount of radio-activity induced in heavy elements depends greatly on the speed of the incident neutrons. At a certain speed a neutron could have only a moderate effect, whereas if it were slightly faster or slower, the response could be huge. In 1936 Fermi and Edoardo Amaldi had measured the absorption and diffusion of neutrons in various materials, and plotted graphs—known as Amaldi-Fermi (or AF) curves—which in effect showed how the response varied with neutron speed.[15] AF curves were important in the practical interpretation of nuclear measurements, so Pontecorvo decided to build on that experience by seeing what happened when neutrons bombarded various strata containing a smorgasbord of minerals, and using the curves to help him interpret the results.

In doing this, he had to shield his detector from the source of neutrons; otherwise, radiation from the source itself would swamp the delicate signals from the surrounding rocks. This raised another question: Could he discriminate between the radiation coming from the rocks and the radiation coming directly from the source? He discovered that the source produced a steady level of radiation, while that from the rocks varied dramatically as the device passed from one stratum to another while being lowered down the hole. The amount of variation turned out to be sharp enough for him to identify the genuine signal. In June 1941 Bruno reported on the trials he had performed, and on the ability of the data to reveal the chemical composition and porosity of rocks.[16] His report showed that different rock strata give distinctive radioactive responses, providing remarkably clear differentiation between shale, limestone, and sandstone. What's more, this method of rock assessment was superior to the existing techniques, such as the gamma-ray device that Neufeld and Scherbatskoy had first used.

That same year, Bruno published a report in the *Oil and Gas Journal*, showing the superb correlations between signals in his detector and different varieties of rocks. He also identified a key asset of his device: "The strength of the neutron source can be made quite large, [so] the surveying speed which may be realized with [this] new method is very great." This report, in the open literature, included what amounted to an advertisement: "If favourable tests continue, the new process will be offered to the trade shortly."[17]

Next Bruno had to decide on the best neutron source. His use of paraffin to slow the neutrons limited the precision of his measurements. He began looking for sources of lower-energy neutrons, so that he could do away with the paraffin entirely.[18] This led him to seek out companies that provided radioactive materials such as polonium and actinium, to replace the radium-beryllium mixture he had used before. This occupied him throughout much of 1941.[19] By the end of the year Bruno was sure that neutrons were a remarkable new tool, which could reveal the transitions between layers of sandstone and shale, or between layers of sandstone and limestone.

The concept of "neutron" was entering the public awareness in unexpected ways. Periodically Bruno joined teams of engineers in the oil fields. During one of these visits, a truck suddenly stopped and its driver shouted to him, "Have you seen the neutrons?"[20] The trucker was looking for one of his colleagues, who had the team's exploration equipment, including one of Pontecorvo's neutron sources. For the driver, this source had become simply "the neutrons." Pontecorvo was quite humbled, and proud that "a particle dear to me, connected with my research with Fermi, had already entered the life of men, at least in the oil fields."[21] Bruno was establishing a whole new industry: one that used neutrons to locate the minerals that herald oil.

DURING THIS TIME, BRUNO, MARIANNE, AND GIL WERE HAPPILY settled in a typical American house in Tulsa. It was considerably more spacious than the student housing they'd occupied in Paris, with its communal bathroom.[22] In Tulsa they had a home to themselves, with bedrooms upstairs, a large living room, and a backyard. The local newspaper carried the story of the handsome, dark-haired Italian, his blond Swedish wife, and their curly-haired young son, who had escaped from the war in Europe. They were popular in the community, and planned to settle permanently in the United States.

Much changed, however, when the US entered the war in December 1941. President Roosevelt declared that "an invasion or predatory incursion [was] threatened on the United States" by Germany and Italy, and Bruno had once more become an enemy alien.[23] In this climate, Bruno decided to investigate the possibility of becoming a naturalized American citizen.

IMAGE 5.1. The Pontecorvo family in Tulsa, 1940. (COURTESY GIL PONTECORVO; PONTECORVO FAMILY ARCHIVES.)

SIX

EAST AND WEST
1941–1942

In June 1941, the Nazi army invaded the Soviet Union and headed toward Moscow. The Soviets viewed fission as a potential source of destructive power for a future war, but not relevant to the present one. The difficulty of extracting U-235 and controlling fission so as to guarantee an explosion seemed to be insurmountable challenges at the time. Pragmatism ruled: Why pursue fission, a long-term project of dubious success, when Moscow might fall within a few weeks? The Soviet work on fission ground to a halt.

Georgii Flerov, Kurchatov's protégé who had discovered spontaneous fission, thought this strategy was wrong. In his opinion, it would be a disaster to lose the race for the atomic bomb. He estimated that fast neutrons might initiate a chain reaction equivalent to 100,000 tons of TNT, without the need to increase the amount of U-235 through enrichment. He even designed a bomb, which, unknown to him, was quite similar to the one secretly envisioned by Frisch and Peierls in the UK. In December 1941 he sent a letter to Kurchatov, outlining his concerns. Unfortunately Kurchatov was ill with pneumonia and never replied.[1]

At this stage Russia and the United States were ostensibly allies, if only in the sense that the enemy of my enemy is my friend. In reality there was mutual suspicion. The US and UK kept their plans for an atomic weapon secret from their Soviet ally. However, this did not mean that the Soviets were unaware of it.

Klaus Fuchs was another refugee from fascism—a theoretical physicist who in 1941 was working with Rudolf Peierls in Birmingham. Following the German invasion of the USSR, Fuchs, who was a communist, decided that the Soviet Union had the right to know what the British were working on. Fuchs contacted the military attaché in the Soviet consulate in the summer of 1941 and told him about the British plans for separating U-235.[2] In addition to information obtained from spies such as Fuchs, there were also clues about the secret programmes to be found in the open literature, provided that one knew how to interpret them.

Early in 1942, Georgii Flerov was in the Soviet military. He had been removed from the relative luxury of physics and sent to fight at the front near Voronezh, about 300 miles south of Moscow. By this point, Voronezh State University had been evacuated, but the library shelves were full of the latest international journals. One day, when he had a few hours to spare, Flerov visited the library to read the latest American news on fission. He was astonished at what he found: there wasn't any.

Initially puzzled, Flerov flipped through the pages of the available Western journals and found papers on other areas of physics by a variety of authors, but on fission—nothing. That was only half of it: not only had papers on fission disappeared, but the leading nuclear physicists had also. None of the field's most prominent researchers—such as Fermi, Bethe, and Bohr—had published anything for several months.

Then Flerov realized the explanation: the papers were absent because American research on fission had become secret. This also explained the disappearance of the nuclear scientists: they were keeping silent as they worked on a nuclear weapon.

At least the Americans and the Soviets were on the same side in the war. What worried him was that there were also first-class scientists in Nazi Germany, who might have similar ideas. What's more, they also had access to large amounts of uranium and the means to enrich it. To sound the alarm, he wrote to Stalin in April 1942, urging him to call a meeting of the nation's leading nuclear scientists.[3] Flerov's intervention was backed up by information he wasn't aware of: the secrets obtained from Klaus Fuchs. Whether Stalin himself actually saw Flerov's letter is unknown, but Soviet scientists were eventually alerted.

By the spring of 1942 the Kremlin's attitude toward nuclear research was changing. One reason for this was that the immediate German threat to Moscow had been repulsed.[4] More to the point, Stalin's henchman Lavrenti Beria, who chaired the Special Committee on the Atomic Bomb, knew from Klaus Fuchs that work was under way in the UK to separate U-235. Thus Beria knew that the British considered the challenge of the atomic bomb to have been solved in principle. Its construction would now be an engineering project.

During 1942, the Germans regained the upper hand in their fight with the Red Army, overrunning Sevastopol and moving toward Stalingrad. This threat to the survival of the Soviet Union occupied most of Stalin's attention, but Beria's knowledge that a bomb was possible finally stimulated the establishment of a dedicated programme in the USSR. Although the information in Flerov's letter was probably old news, thanks to spies like Fuchs, the communication had an effect: it singled him out as an expert. In mid-July of 1942 Flerov was withdrawn from the southwestern front, summoned to Moscow, and put back into neutron research. In September, Kurchatov was brought on board. During October, while the Battle of Stalingrad was at its height, he was put in charge of the Soviet Union's quest for an atomic bomb.

MEANWHILE, IN THE UNITED KINGDOM, SCIENTISTS HAD BEEN investigating the practicalities of a nuclear weapon, building on the breakthrough made by Frisch and Peierls. By 1941, the British had the lead. They had investigated U-235 enrichment in experiments at Oxford, and were well advanced in the theoretical workings of a uranium bomb.

James Chadwick, who had started the saga by discovering the neutron, was now involved both in experiments and in organizing collaboration between the various teams in the United Kingdom. This dapper, hardworking experimentalist was now showing his brilliance as an administrator of science. Like many experts, including those in the Soviet Union, Chadwick initially thought that nothing would come of this work until after the war. Then, in the spring of 1941, British scientists found that it was possible to separate the fissile U-235 from naturally occurring uranium. What's more, the costs involved made this practical. Fuchs, who was party to this secret, duly passed the news on to the USSR. Chadwick's reaction was also dramatic, though more personal. Upon realizing that a nuclear bomb was not only possible but inevitable, he later recalled, "I had to take sleeping pills. It was the only remedy."[5]

Chadwick wrote a report, summarizing his conclusion that an atomic weapon was feasible, and by the end of 1941 Winston Churchill gave the top-secret project the go-ahead, under the bland code name "Tube Alloys." As before, the Soviets learned about this development from spies.[6]

Tube Alloys involved many scientists and engineers, but was managed by Wallace Akers, the research director of Imperial Chemical Industries (ICI), along with the company's senior administrator, Michael Perrin. By this point, Halban and Kowarski were in Cambridge, with the heavy water. Working with a team of physicists at Cambridge University, their goal was to create a chain reaction as a means of producing nuclear power. Although ICI was managing the enterprise on behalf of the war effort, the company saw excellent commercial possibility in the development of nuclear power, once the war was over.[7]

The heavy-water project now had real urgency. German interest in heavy water hinted that they too were on the hunt for nuclear technology. Although the United Kingdom was protected from enemy occupation by the English Channel, and had survived the aerial Battle of Britain in 1940, the possibility of a Nazi invasion remained real. Separating U-235, exploiting nuclear fission, and then building a bomb would be an industrial-scale enterprise, which would be vulnerable to enemy attack if based in England. Eventually, this undertaking would be subsumed within the Manhattan Project in the United States. In 1942, to safeguard the precious heavy-water project and, we may assume, to maintain control over its commercial potential, the operation was relocated to the relative safety of Canada. Thus began the Anglo-Canadian arm of Tube Alloys, based initially in Montreal and built around the senior personnel from Halban and Kowarski's Cambridge University team.

Halban's personality turned out to be a problem, both when forming the Canadian team and later when working with the Americans. He presented the group's work on heavy water in a self-aggrandizing manner. Kowarski resented Halban's habit of using their joint work for personal glory, and refused to go to Canada as second fiddle. So Kowarski remained in Cambridge.[8] Halban, meanwhile, visited the United States in 1942 to build the team, and to discuss nuclear strategy with Enrico Fermi.

FERMI'S PILE: 1942

When Fermi had first started researching the possibility of a chain re-action in the United States, he had used water as moderator. Like the French, he soon realized that this was not efficient because the water tends to capture neutrons from the beam. Heavy water was scarce in the United States, so Fermi decided to try graphite instead. Graphite contains carbon, which is light and slows neutrons efficiently.

During 1942, Fermi worked urgently to create a self-sustaining chain reaction using neutrons and uranium. To do so, he had to produce intense sources of neutrons, and assess which uranium compounds were most efficient. In addition he had to select suitable materials to moderate the neutrons—that is, to slow them to the optimal speeds that induce fission.

From his earlier work with Pontecorvo, Fermi knew which materials worked well in small-scale tests. Now he needed to attempt the same feat on a much larger scale, appropriate to a working reactor. In Chicago research into the properties of plutonium was under way, and, because one goal of a nuclear reactor would be to produce plutonium, Chicago became the logical place for Fermi to build the first reactor prototype.

The reactor he built there consisted of a stock of uranium oxide, in lumps about the size of tennis balls, embedded in solid blocks of graphite. The graphite blocks were stacked in a pile, which is the source for the colloquial description of this type of nuclear reactor as a "pile." Add a source of neutrons and you have the essence of a nuclear reactor—in theory. In practice the reactor includes a lot of additional hardware, such as metals and concrete, which form the floor and general infrastructure. When the neutrons hit these materials, they are scattered or absorbed, which influences their ability to cause fission in the heart of the reactor. So to construct a working reactor Fermi would need to know how neutrons react with a whole variety of elements, in complex mixtures and ores, not just with uranium. To obtain this knowledge would require an extensive series of tests, in which compounds were irradiated with neutrons to see how much energy is lost when they scatter, or if the neutrons are absorbed. While Fermi was in the early stages of planning this strategy, Bruno Pontecorvo paid him a visit.

Bruno had been outside this secret world for twenty months. However, although he was ignorant of the West's growing interest in nuclear reactors, he had unwittingly become an expert in the topics that are central to their design. Indeed, by 1942, when Fermi was beginning to build the pile, Bruno knew more than Fermi himself in some areas.

BRUNO MEETS WITH FERMI—AND HALBAN

Toward the end of 1941 Bruno was finding it increasingly difficult to obtain radioactive materials for his work. Radium was plentiful, but supplies of less common materials seemed to have dried up, as if someone was cornering the market. The reason, of course, is that the research work that would ultimately spawn the Manhattan Project was already using up most of the available resources. Bruno, a member of the general public, knew nothing of this, although he was aware that something was amiss. On November 17, 1941, he wrote a report about the shortage of radioactive sources and its effect on the development of new strategies for prospecting.[9] In April 1942 Bruno met with his mentor, Enrico Fermi, in hope of getting access to more supplies.

Fermi had spent the winter shuttling back and forth between New York and Chicago, in preparation for his reactor experiment. He too was an "enemy alien" and had to get permission every time he made the trip.[10] After April, Fermi moved to Chicago permanently but was still in New York when Bruno visited him. Bruno's colleague from Paris, Hans von Halban, and Czech physicist George Placzek were also present.

During the get-together, Bruno told the scientists about his work on neutron well-logging, including some of the technical details. Fermi was especially interested, making suggestions, showing deep knowledge of the subject, and asking questions. Bruno, who knew nothing of Fermi's secret project, was surprised. He assumed that his work had no obvious relevance outside the field of oil prospecting, yet Fermi showed a keen thirst for information. Of course, Fermi's interest stemmed from the reactor project. To build a successful reactor, Fermi would have to irradiate uranium with neutrons in the presence of graphite and other materials, so it was essential for him to understand as much as possible about how neutrons behave.

Bruno was unsuccessful in obtaining any essential materials from Fermi, but the meeting must have confirmed his suspicions that a major nuclear project was under way. Given Fermi's questions, the presence of Halban (who had demonstrated fission alongside Bruno in 1939), and Bruno's general knowledge of the state of nuclear physics before secrecy took over, it is inconceivable that he did not deduce what Fermi was up to. The details of the nuclear pile would have remained unknown to him, although it is clear that Fermi shared some of his own neutron data with Pontecorvo: in a report of this visit, written on April 15, Bruno commented that the data he had received from Fermi "had not been published, and cannot be published for a long time to come, because of their confidential character."[11]

Contrary to the spin propagated by the British government after Pontecorvo's defection in 1950, it seems he was aware of at least some front-line data in 1942, several months before Fermi completed the first nuclear reactor. Moreover, Pontecorvo now knew that neutrons had become central to wartime nuclear research.

Bruno was not alone in deducing what was going on. Others in the US had also put two and two together. Bruno's brother, Paolo, knew Harry Lipkin, who later became a distinguished nuclear theorist but at the time was specializing in microwaves. Working together, Paolo and Lipkin were testing a receiver designed to provide early warning of planes coming over the sea, and in 1943 the pair spent two weeks together in Maine's Acadia National Park. During this time they became friends. In Lipkin's opinion, Paolo had concluded that the Americans were working on the atomic bomb.[12] Paolo had come to this opinion because so many of Bruno's Italian nuclear-physicist friends, including Fermi, had suddenly disappeared, and their research publications had abruptly stopped. Furthermore, the only way to contact them was to send a letter to a PO box that gave no hint of its location. It was obvious that there was a supersecret project going on, and it was clear to Paolo and Harry what it must be.

Bruno Pontecorvo's visit to Enrico Fermi in April 1942 would lead him too to become part of that project. It was not Fermi, however, who recruited his former pupil; Bruno joined Halban's Anglo-Canadian team.

THE CANADIAN CONNECTION

In 1930, Gilbert LaBine, a Canadian mineral prospector, had discovered deposits of radium and uranium in the Northwest Territories of Canada.

At that time radium was highly valued, due to its benefits in cancer treatment, but uranium was considered useless. The world supply of radium was shared between the Canadian deposits and mines in the Belgian Congo. When the war started in Europe, markets for radium shrank and the Canadian mining operation struggled to make ends meet. Everything changed with the discovery of the chain reaction and uranium's potential as an enormous energy resource.

In 1942 LaBine's Eldorado Mine started to supply the US with uranium for Fermi's reactor. The company was taken over by the Canadian government in 1943.[13]

The agent for LaBine's Canadian mine was Boris Pregel, based in New York. Pregel, who came from the Ukraine, had moved to Paris after the October Revolution in 1917, and then escaped, like Bruno, when the Nazis invaded in 1940. Upon arriving in the US, Pregel set up the Canadian Radium and Uranium Corporation of New York, which made radioactive neutron sources and luminescent signs, as well as trading in Canadian ores.[14] This explains Pregel's interest in Bruno Pontecorvo, and toward the end of 1942 Pregel began to court him. His goal was to persuade Bruno to join his New York laboratory.[15]

As luck would have it, Pontecorvo was becoming increasingly frustrated at Well Surveys. His quest for radioactive materials, which were necessary for his work, was getting nowhere. He was about to accept Pregel's offer when fate intervened in the form of three former colleagues from his Paris days: Bertrand Goldschmidt, Hans von Halban, and Pierre Auger.

Goldschmidt (a chemist) and Auger (a physicist) were about to join Halban at the Anglo-Canadian reactor project in Montreal. Auger was to be the head of the physics division, while the head of theoretical physics would be George Placzek, who had been present when Pontecorvo visited Fermi's laboratory earlier that year. The pivotal piece of the puzzle, Goldschmidt was to be the section leader of the chemistry division. Goldschmidt is central to our story because he had been working at Pregel's New York laboratory, where Pontecorvo had been offered a position. In Goldschmidt's judgment, that job was beneath Pontecorvo's talent, and he recommended that Halban hire Bruno.

By the time the project was operating, about one hundred scientists and engineers were involved. In building his team, Halban, protective of his patent ambitions, had been careful to keep at arm's length anyone who might have claims of their own.[16] This was one reason for Kowarski's exclusion, and was also a bonus for Bruno. Bruno had the expertise, and had watched the heavy-water experiments in Paris, but had not been involved in the patents. Thus, as Halban knew Pontecorvo from their time in Paris, and Placzek had met him earlier that year, they arranged an interview in November 1942. On the basis of this interview, they invited Pontecorvo to join the Anglo-Canadian reactor team. The team's goal was to design a nuclear reactor whose fundamental ingredients would be uranium and heavy water, in contrast to Fermi's uranium-and-graphite reactor in Chicago. Ostensibly the Canadian reactor's purpose was to provide nuclear power in the postwar era; however, it would also produce plutonium and more exotic forms of uranium, which would later have applications for atomic weapons.

THE OPPORTUNITY TO MOVE TO CANADA CAME AT A GOOD TIME FOR Bruno, as life in the United States had begun to have some unwelcome consequences.

Being a foreigner in the United States during the war could involve unexpected hazards. One day Bruno was driving a truck full of geophysical instruments, and made some illegal manoeuvre on the road. Some police officers noticed the indiscretion, and set off in pursuit. Bruno stopped, but the police became suspicious when they noticed the array of unusual instruments in the truck, covered with Bakelite knobs and dials. When Bruno started to explain the purpose of the instruments in the van, they realized he wasn't American and exclaimed, "Enemy alien!" Bruno then attempted to retrieve his documents from his back pocket. In what he described as "like a scene from a movie," the police, who thought he was reaching for a gun, immediately immobilized him.[17] When they finally understood the nature of the situation, the police explained that he had risked being killed on the spot.

This incident ended happily, and left Bruno with a story to add to his collection. Another experience would end less happily, however—a visit by the FBI, whose consequences are reminiscent of chaos theory, in which the flapping of a butterfly's wings can cause a storm far across the globe.

On September 25, 1942, two FBI agents visited Bruno's home, in view of his status as "enemy alien."[18] Bruno was away for several days at the oil fields, so the agents spoke to Marianne. At the time this seemed of little consequence except for one thing: the agents noticed "25 or 30 books or pamphlets containing Communist literature" in the house. They asked Marianne whether Bruno was a communist. She replied that she "didn't understand and didn't know what a communist was." The agents noted that Marianne "spoke English well and appeared to understand all questions that had previously been asked." According to the agents she also said that Bruno "studied under Madame Curie while he was in France," which shows that there was a misunderstanding somewhere. Years later Bruno commented that in America he had felt an "intense anti-Sovietism" and that no one had seemed to share his "passion for the USSR."[19]

SEVEN

THE PILE AT CHALK RIVER

1943–1945

WHILE BRUNO HAD BEEN OUTSIDE THE BUBBLE OF SECRECY, MUCH had happened. The events that would eventually absorb him began in 1941, when British prime minister Winston Churchill made the decision to develop nuclear weapons, under the aegis of the Tube Alloys project—which by the time Bruno joined was about to be subsumed into the Manhattan Project.

The Manhattan Project itself included several components. At Oak Ridge, Tennessee, hundreds of acres of forest were cleared to make room for an immense gaseous diffusion plant, its purpose to separate the fissile U-235 from naturally occurring uranium. The actual design and construction of the atomic bomb occurred at Los Alamos, New Mexico. Among the nuclear physicists who moved from the UK to Los Alamos as part of Tube Alloys were Rudolf Peierls, who had first realized the bomb's capability, and his colleague Klaus Fuchs, who was already passing information to the Soviet Union.

Nuclear reactors were a key part of the Manhattan strategy, as they had the potential to make plutonium for a bomb. Led by the pugnacious General Leslie Groves, the project began in earnest on December 2, 1942, the day that Fermi's pile became a self-sustaining fission engine, which liberated energy at a steady rate. In the jargon, the pile became "critical." The development of nuclear reactors was the strand of the Tube Alloys project to which Bruno Pontecorvo would soon be co-opted.

The project was of course top secret, and its members were thoroughly vetted. The fact that Halban's operation was known as the Anglo-Canadian project is ironic as its genesis was in France, and its early members more French than English. The senior team consisted almost exclusively of émigrés from France and central Europe, who had been working together at Cambridge University with Halban and Kowarski. When Halban proposed that Bruno Pontecorvo join the project, Edward Appleton, the secretary of the British Department of Scientific and Industrial Research (DSIR), initially objected that he did not want to "add to the number of non-British nationals" working on this secret project.[1] In light of later events, this objection was ironic: the sole British member of the team borrowed from Cambridge was Alan Nunn May, later imprisoned for passing secrets to the Soviets. It is not clear how the Czech George Placzek and French Pierre Auger, who were already deeply involved in the project, felt about Appleton's insular attitude. In any event, they argued successfully that "the brilliant physicist Pontecorvo" should be included.[2]

Bruno had worked with neutrons for nearly a decade, and had been trained by Fermi in both experiment and theory. A further attraction was that Bruno would be able to consult his mentor in Chicago, to the obvious advantage of the Anglo-Canadian project. Bruno's work in Tulsa had taught him how neutrons behave in a range of minerals, which would be key issues in designing a working reactor. At the time, the judgment was that there was "no specialist of the same character [as Bruno Pontecorvo] available in North America."[3]

On November 4, 1942, the DSIR asked the British embassy in Washington, DC, to make "discreet enquiries [about Pontecorvo] before any direct approach to employ him." On November 30 the embassy's reply deemed him "quite satisfactory from point of view of security for employment by any British agency." It added, rather patronizingly, that Pontecorvo is "Italian by birth and Hebrew by race, but his record makes it quite clear that he is entirely in sympathy with the Allied Cause."[4]

On December 9, the British security authorities gave Bruno "an unusually enthusiastic report." The recommendation described him as follows:

"One of the ablest of the younger Nuclear Physicists and is acknowledged to be an expert on slow neutron physics—a subject on which he was doing research before the Atomic Energy project was started."[5]

BRUNO RESIGNED FROM THE OIL BUSINESS, AND MADE PREPARATIONS to move to Canada. On January 7, 1943, he travelled from Tulsa to Kansas to obtain a passport, returned home the next day, and then went to New York on the fifteenth to be formally appointed to Tube Alloys. He spent the rest of January in the city, where he was briefed about the project, and finally transferred to Montreal with his family on February 7.[6] He would work on secret nuclear physics programmes for the Allies and the British for the next seven years.

Unknown to the security authorities in Canada, a fortnight after Pontecorvo's appointment three letters were exchanged between the FBI and the British Security Coordination in Washington. The FBI was concerned about Pontecorvo's sympathies with communism. The origin of this concern was the visit by the two agents to his home in Tulsa a few months earlier, when they had quizzed Marianne in Bruno's absence. Although the letters were exchanged *after* Bruno's initial appointment to Tube Alloys, he was still only a probationer, and his final approval had yet to take place. Inexplicably, "by some organizational error,"[7] these letters were not available to the relevant authorities, and on March 3, 1943, his security clearance was approved.[8]

Canada would be strategically important for the Allies. There were vast deposits of uranium at Great Bear Lake in the Northwest Territories, and a uranium refinery at Port Hope, Ontario. Bruno's skill in locating minerals underground would be put to good use. The objective of the reactor programme was to complete the Nuclear Reactor X (NRX), the most powerful source of neutrons in the Western world, and a groundbreaking source of nuclear power. The Chalk River reactor in Canada would be the experimental facility from which Britain's postwar nuclear programme would grow, and it would also serve as a means to produce plutonium, as a "second string to the nuclear bow."[9] As a result, Chalk River would become a key target for Soviet agents.[10] Whereas overt communists, such as Frédéric Joliot-Curie, had been excluded from the Manhattan Project for security reasons, Bruno Pontecorvo had hidden his enthusiasm and slipped through the net.

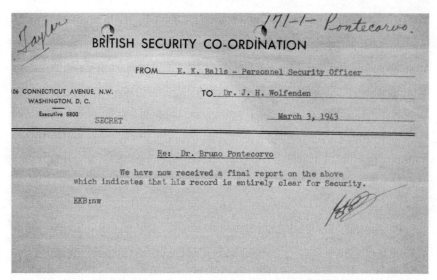

IMAGE 7.1. Bruno Pontecorvo's security clearance, issued March 3, 1943. (AUTHOR, THE NATIONAL ARCHIVES.)

ARRIVING IN MONTREAL

Bruno Pontecorvo's move from Oklahoma to Montreal was a shock. Having grown up in the warmth of Italy and lived for six years in France and Tulsa, where winters were mild, Montreal was brutally cold, dipping below zero even on the Fahrenheit scale.[11] Bruno's breath turned to icicles in front of his face, as the deep freeze continued for several weeks. There were a couple of days in February when the temperature peaked at the freezing point before plummeting once more, to –20 degrees Fahrenheit.

The Pontecorvos rented an apartment at the south end of Mount Royal, in a side street off Chemin de la Côte-des-Neiges, about two miles from the city centre. The family home was not large by North American standards, but was comfortable nonetheless, with a good view of Saint Joseph's Oratory, an imposing landmark that was reached by ascending a large number of steps. Bruno liked to entertain guests by taking them out on the balcony to watch the worshippers who mounted the stone staircase on their knees.

For Europeans, who constituted much of the membership of the Anglo-Canadian project, 1940s Quebec was a cultural backwater, dominated by religious divisions. The French Catholic province had backed the fascists in the Spanish Civil War, and was now supporting Marshall Pétain's collaboration with the Nazis in France. This contrasted with the local English-speaking community, who had favoured the Spanish Republicans and now supported the French Resistance. There was little science being practiced at the University of Montreal, and faculty membership was restricted to practicing members of the Roman Catholic Church. This astonished the newcomers, who found it hard to believe the immense control that the Church exercised. It censored books and films. It also banned drive-in movie theatres, which it regarded as dens of immorality.

On the plus side, downtown Montreal was cosmopolitan, with a large selection of restaurants: French, Chinese, and especially Jewish. This area was located some distance from the university laboratory, so although downtown was a popular dinner destination, midday lunch tended to consist of sandwiches eaten in the university common room. Over lunch, conversation thrived. The members of the team came from around the world, and had a wide range of backgrounds. Chatter initially centred on world affairs, but soon more formal lunch-hour discussions were organized on a range of issues.

It was at these gatherings that Bruno began to make his mark. They not only informed him about all aspects of the project and physics at large, but also made Bruno Pontecorvo known to everyone in the laboratory. In this way, he managed to cut through some of the bureaucracy. The formal pattern of work was divided between senior and junior members. Bruno was in the former class, even though he was barely thirty years old. The senior scientists tended to work closely with one another, whereas the junior staff was assigned specialized tasks, subject to the "need to know" principle, with little interaction between members of different groups. In the memory of one junior scientist: "Hierarchy prevailed. The atmosphere was more military than academic."[12]

And there, in a nutshell, is the tension, common among scientists throughout the war years. Science advances through free-ranging discussion, the sharing of half-baked ideas, from which unexpected synergies emerge. Military administrators such as General Groves were obsessed with security, however, and wanted to create firewalls to keep pockets of knowledge contained within small groups. For them, the ideal was that information should be compartmentalized, so that only a handful of bosses would have the complete picture. This was anathema for scientists, who were reared on skepticism and imbued, then as now, with a distrust of "the suits." Stories of how these rules were bypassed at Los Alamos are legion; in Canada the same was true. Scientists talk together; it's a fact of life. It's great for making progress, but a nightmare for security during wartime.

One member of the theoretical physics group recalled that Bruno had a broad interest in physics, science, and philosophy.[13] Although his primary role in the project was that of an experimentalist, theory remained close to his heart. He began to seek out the theoretical group's members in order to have discussions about physics that went beyond the project's immediate problems.

The theoretical team consisted mainly of young mathematicians with limited knowledge of nuclear physics. Bruno's interaction with them was a two-way street. His expertise in neutrons gave them the background they needed to apply theory to problems in nuclear physics; in return, these regular contacts brought his own talents in theoretical physics to life.

Several of these discussions touched on the fundamental ideas underlying radioactivity and nuclear transmutations. It was these talks that led Bruno to appreciate the singular role and importance of the still-hypothetical neutrino, the ghostly particle that had briefly interested his mentor Fermi years before, when Bruno was a student. After the war's end, the neutrino would become a lifelong interest for Bruno.

HALBANIA

Bruno's first task was to catch up with the progress that had been made on fission. In Paris he had been party to the experiments with heavy water, and he knew that his colleagues believed that each act of fission liberated two or three neutrons. Now he learned that, during his interregnum, Halban and Kowarski's experiments at Cambridge had convinced them that a chain reaction should be feasible.

Unlike Enrico Fermi, whose uranium-and-graphite pile had become critical on December 2, Halban's team had not established this fact. Yet the aggressive and arrogant Halban appeared to believe that his "proof," achieved in Paris and Cambridge, that a chain reaction could be produced using uranium and heavy water gave him a patent claim over all future work. The Americans were not impressed, questioned the French patents' validity, and even doubted the results on which they were based. In their opinion, the fact that Enrico Fermi (who distrusted Halban's results) had produced a chain reaction in practice moved the question far beyond mere theoretical possibility. Any collaboration between the British and the Americans would be under Fermi's scientific control, something that the autocratic and prickly Halban could not accept.

Thus Bruno found himself at a laboratory where stimulating research had yet to blossom. Even before his arrival, heated arguments had developed in the higher echelons of politics about the respective national interests of the US and UK.[14] The atomic bomb had been conceived in Europe and was being developed in America; who would control it postwar? For the US administration and General Groves, the answer was obvious: the United States, on whose soil the bomb would be built, and whose taxpayers were funding most of the work. The British, however, regarded the bomb as their invention. Their scientists had conceived the possibility, developed the early ideas, demonstrated that U-235 could be separated in principle, and were prominent in North America.[15]

Another goal of the Allied collaboration was the development of nuclear power, potentially the solution to the world's energy needs, and perhaps the most valuable prize of all. It did not sit well with the United States that ICI, who was managing Tube Alloys, had a clear commercial interest in developing nuclear power. The British, on the other hand, suspected that the official US agenda was hegemony in nuclear technology after the war. So, despite being united in fighting a common enemy, the two nations' divergent long-term political agendas created the need for firewalls, which held back sensitive data. Fermi's breakthrough, in December 1942, changed the politics considerably, and from that moment the American policy was to severely limit information exchange with the Anglo-Canadian project in Montreal.

The fact that the Montreal team was unhappy is no surprise, but there was also strong dissatisfaction among the American team of scientists with the policy of restricting access to data.[16] Ultimately, members of both teams found ways around this imposition.[17] In February 1943, for example, Pierre Auger—head of the Montreal experimental physics programme, and Bruno's immediate boss—visited Fermi's team in Chicago, along with chemist Bertrand Goldschmidt. The pair had been consultants at Fermi's laboratory and still retained their security passes, so they simply walked in the front door. They found the American scientists welcoming and helpful. Auger and Goldschmidt returned to Montreal the same evening, having persuaded the Americans to give Auger the basic details of the pile, and to give Goldschmidt two tubes, one containing a portion of fission products, the other containing four micrograms of plutonium.[18]

When a reactor is operating, the uranium inside is flooded with neutrons. Those that cause fission liberate energy, which is the goal, but many are captured and lost. The result of adding two neutrons to uranium is that, after a few days, natural radioactivity produces plutonium in the residue. By 1943, physicists knew that plutonium was a better fuel for chain reactions than U-235, and, by implication, a better fuel for weapons as well. Thus, although a reactor physicist might be driven by the goal of producing power for the good of society, and not wish to be party to weapons research, the reactor itself makes no such distinction. Plutonium is plutonium. To make use of plutonium, however, you have to remove it from the reactor by some chemical means, as tons of uranium lead to the creation of mere grams of plutonium per day. As a result of Goldschmidt's success in Chicago, the radiochemists in Montreal now had enough material to learn how to extract the element.

Bruno, meanwhile, set out to establish the facts about heavy water and fission. During his first weeks in Montreal, he had "many discussions on the Halban-Kowarski papers" with the head of the theoretical physics group, George Placzek, and with experimentalist Alan Nunn May. Nunn May, who boasted pince-nez spectacles and a small mustache, was a caricature of the 1940s egghead scientist, in marked contrast to the film star glamour of Bruno Pontecorvo. A nuclear physicist from Cambridge University, Nunn May was an expert in making precision measurements of nuclear processes.[19] He was also a communist, a fact that seems to have escaped notice when he was vetted for the Anglo-Canadian team, even though his views were well known to his colleagues in Canada.[20]

Halban as director was a mixed blessing. True, he had been intimately involved with fission from the outset, but many viewed him as having more ostentation than substance. His self-confident arrogance was pithily summarized by a remark in his obituary: "He moved easily from the laboratory, through the ministerial office, into the board room."[21] Halban's style and chutzpa were on full display at a meeting of the British and American teams in New York. On a Saturday morning he phoned his colleagues in Montreal, demanding that they take some documents from the Top Secret drawer and courier them immediately under diplomatic cover to New York. A call came back from Montreal, checking which papers he needed. "Any will do," Halban replied. "It doesn't matter which."[22] The purpose, in Nunn May's opinion, was simply to make grand theatre, to impress the Americans. The feeling among some members of the Anglo-Canadian team was that the purpose of the heavy-water work was less to further scientific knowledge than to bolster Halban's and ICI's commercial visions. Nunn May in particular felt that the scientists in Montreal were "pawns in a game between the British and Americans"[23] and living in a state of "Halbanian politics."[24]

SPENDING ONE'S TIME EVALUATING THE RELIABILITY OF HALBAN'S DATA on fission and chain reactions is a poor substitute for real experimental work. In 1943, the only immediate source of heavy water in Montreal was the batch that Halban and Kowarski had rescued, but these 185 kilograms were still in Cambridge in the United Kingdom. Even if this batch made it overseas, the team would still face problems: this limited amount of heavy water would be enough for the initial research process, but was utterly inadequate for a large-scale reactor, in which more than a ton of the liquid would be needed. The Anglo-Canadian project was in danger of being stillborn. Many found these early months frustrating, with Alan Nunn May later recalling, "The purpose of [going to Canada] seemed to have evaporated. [We] had no equipment, no laboratory, and no prospect of obtaining any materials, and the Americans were cut off [from us]."[25]

At last, in the spring of 1943, the heavy water arrived from Cambridge by air. Serious investigations could now begin. One of the first tasks was to repeat the Halban-Kowarski experiments with more sensitive detection equipment. The original findings—that fission released neutrons, capable of initiating a chain reaction—were confirmed, to great relief. Morale slowly declined, however, as the British and American authorities continued to argue over the control of the whole atomic project.

In August 1943, Prime Minister Winston Churchill and President Franklin D. Roosevelt met in Quebec. Part of the agreement they came to called for greater cooperation between the two teams. This raised the

Anglo-Canadians' hopes. They eagerly anticipated the arrival of sufficient uranium and heavy water to start their project in earnest. The Americans, however, were slow to respond to the agreement, and it was not until December that preparations began for the first conference between the two teams, held in Chicago in January 1944.

In the interim, an important change had occurred: Halban had quit as director and stepped down to head the physics department. The new director of the Anglo-Canadian project was John Cockcroft. Cockcroft was famous for developing the particle accelerator at Cambridge that first split the atomic nucleus in 1932, and by 1944 he had become an effective administrator of scientific research on radar. As the Soviet army pushed the Germans back, however, research on radar was deemed to be less urgent than the development of nuclear physics, which was seen as vital.[26] Cockcroft's arrival would transform the Anglo-Canadian project, as well as the fate of Bruno Pontecorvo.

SPYING FOR BRITAIN

Cockcroft soon gave the project a new impetus. When he arrived, the reactor was still on the drawing board, and the possibility of a serious accident occurring when it was completed convinced him that it should be constructed at least 100 miles from any major city. He found a suitable location in the forests about 130 miles west of Ottawa: the village of Chalk River. The nearby Ottawa River would provide water to cool the reactor.

One of Cockcroft's first actions was to encourage his team to coax information from the Americans. One result of the political constraints imposed by the nations' governments and militaries was that the Americans would give aid to the Canadian team in their pursuit of a heavy-water reactor, but not in relation to other technologies, such as the graphite reactor. Britain, with an eye to the future, saw graphite reactors as key to its postwar energy needs (as indeed was the case). Consequently, the British, who regarded the political restrictions as unreasonable, did their best to extract information about graphite reactors from the Americans. Thus was born the officially sanctioned strategy in which the governing authorities were bypassed, and the community of scientists shared data and ideas among themselves.

One example of such sharing occurred in March 1944, when Nunn May and a Canadian colleague, Ted Hincks, shipped equipment to Chicago for use in an experiment. In Chicago, the head of experiments at the pile was a man named Herbert Anderson. Nunn May talked with him.[27] Anderson agreed that General Groves's instructions were too restrictive, and were being applied overzealously. The Canadian team's engineers and physicists, frustrated by the difficulty of obtaining information from the Americans, had given Nunn May a "shopping list" of urgent questions, for which they wanted answers.[28] Nunn May was successful, as Anderson allowed him to read several reports that had not yet been released to Montreal. This open sharing of data became a regular occurrence during a series of visits.

Later in the war, after the atomic bomb tests, several leading scientists from Los Alamos passed through Canada en route to Europe. They were "debriefed" at Cockcroft's insistence. When Aage Bohr, the son of Niels and a future Nobel laureate, visited Canada, Cockcroft ordered Bruno Pontecorvo and Nunn May to "pump him dry."[29]

Since Nunn May was spying for the USSR during this time, and it has been alleged that Pontecorvo was also, it is ironic that Cockcroft in effect ordered them to spy on behalf of the UK. During 1944 and 1945, Nunn May was given access to Fermi's laboratory to perform several experiments. The possibility that Bruno Pontecorvo was passing information to the Soviets at this stage has been suggested, but never established. In any event, his intimate involvement with Nunn May would leave him compromised when the latter was exposed as a spy in 1945.

Years later, Pavel Sudoplatov, who was the deputy director of atomic espionage for the USSR during World War II, claimed that the Soviets obtained information from Pontecorvo, along with Bohr, Fermi, and others.[30] Many have taken this claim to mean that these famous scientists were active spies, which in the case of Bohr and Fermi is so bizarre that some have dismissed Sudoplatov as a fantasist. Although some of Sudoplatov's claims have inconsistencies, the accusation of "fantasist" might be due to a misunderstanding on the part of his critics: it is now clear that the Soviets did receive information, via Nunn May if not Pontecorvo, which *originated* with Fermi and other leading members of the American side of the Manhattan Project. It is also clear, as we saw earlier, that Fermi indiscreetly shared secret data with Pontecorvo in April 1942; while there is no evidence that this data would be of any interest to the Soviet Union, or even that Pontecorvo passed this information to anyone other than his colleagues at Well Surveys, the fact that Fermi was occasionally casual with information cannot be discounted.[31]

Of course, Pontecorvo had never tried very hard to hide his commitment to communism. His brother Gillo later said that Bruno had made no attempt to hide the fact that he was a communist in France or the United States, but that when he was invited to join the atomic project in Canada, he immediately became "nonpolitical." Gillo was convinced that this was a cover because whenever they met after the war, Bruno "was too well informed on communist literature and events."[32] However, even after the move to Canada, there were clues as to Pontecorvo's political affiliations. When Bruno and Marianne's second son was born in March 1944, they named him Tito Nils—*Nils* after Bruno's scientific hero, Niels Bohr, and Tito, as Bruno explained later, "in honour of the communist who had led the war of liberation in Yugoslavia."[33] Two years after Marianne had claimed to the FBI that she didn't know what a communist was, here was a public display of the family's sympathies.

1944

Bruno was part of the small group that visited Fermi's team in Chicago regularly during 1944 and 1945. To design a working reactor, one needs precise data on how neutrons interact with a variety of materials. In Montreal, there was no source of neutrons powerful enough to make good measurements. The Chicago reactor, however, produced more intense beams of neutrons, which could give clear answers. Obtaining these answers involved several collaborative visits and exchanges of information. Upon returning to Canada, the experts would report the new data to their colleagues, evaluate them, and plan new lines of attack. Bruno was the author of several of these reports.

A successful nuclear reactor needs to maximize the number of neutrons available as fission-creating bullets in the reactor's core. Therefore, one must avoid materials that absorb them. Neutrons can travel anywhere. They might induce radioactivity in unexpected ways and poison the reactor. In extremis, it is possible that this could cripple it fatally. In designing the reactor, it would be necessary to know the chance of neutrons being captured by all manner of elements—not only how they interacted with the fuel (the uranium and the heavy water or graphite) but also how they interacted with the concrete blocks and the potpourri of other materials in the assembly and surroundings. This was one of the reasons Fermi had been so interested in Pontecorvo's work with neutrons in 1942. The team was working in a new field, where hands-on, practical experience trumped any amount of theory. Thus Bruno's expertise in this area would now be invaluable in Canada.[34]

The first of Bruno Pontecorvo's visits to Chicago occurred in January 1944, when he made the journey with Nunn May, Auger, and other senior members of the team.[35] In line with the usual practice of compartmentalization, physicists such as Pontecorvo were only present for discussions of the physics programme; chemistry and engineering were discussed in separate groups.[36] This visit to Chicago kick-started the Canadian work on the nuclear reactor. Just nine people were involved in the discussions at the first two meetings, which evaluated detailed questions about the materials for the reactor and their optimal use.

The Canadian team visited Chicago on several occasions, and Bruno generally wrote the reports.[37] These reports confirm the significance of this work, and justify the British government's description of his role in both planning and research to have been "indeed of the greatest importance."[38] Bruno was in Chicago during the first week of March, and returned home in time for the birth of Tito Nils on March 20. Then he left for Chicago once again during the first week of April.

The physical principles of a chain reaction might be relatively straightforward, but the construction of a nuclear reactor is not. As the Canadian team forged ahead on the project, one problem after another arose.

In a heavy-water reactor, for example, is it better to mix some uranium compound with tons of heavy water to form a homogeneous sludge, or instead to immerse solid rods throughout the liquid? If the former, what is the optimal mixture? If the latter, what is the best matrix arrangement?

Some problems were relatively mundane but potentially serious nonetheless. One example is the fact that a working reactor produces energy, and gets hot. Solid uranium rods might expand or even buckle; how dangerous was this for a working reactor? Was it better to cool the rods by bubbling air or other gases through the heavy water, or by running ordinary water through pipes? Ordinary water could keep the uranium rods cool, but it might absorb some neutrons, slowing the reactor or stalling it completely. At their next meeting in May, the two teams discussed this issue. The knowledge gained in solving such problems took time to achieve, but was invaluable. Igor Kurchatov, the scientific head of the Soviet atomic bomb project, knew that if he could obtain solutions through agents in North America, a Soviet reactor programme could save months or even years of effort. Even so, having access to all the data and rules of best practice is no substitute for hands-on experience.[39]

One of many uncertainties the reactor physicists faced in 1944 was what happened when neutrons hit heavy water. Experiments by the team in Chalk River and the Americans in Chicago gave contradictory results. So Nunn May went to Chicago in the spring of 1944 to make use of the American team's intense beam of neutrons and repeat the experiment there.[40]

AT THE TIME, SCIENTISTS HAD ALREADY ESTABLISHED THAT FISSION IS most likely to happen in isotopes that contain an odd number of neutrons. Uranium-235 and plutonium-239 are well-known examples, but there was a possibility that another odd-numbered isotope of uranium, U-233, might be a suitable fuel too. Consequently, a third line of collaborative research focused on creating uranium-233 in the Chicago pile, one nucleus at a time, until there was enough to determine its properties.

Uranium-233 had two apparent advantages over its competitors. First, the chemists thought it would be easier to refine than plutonium. This was the scientific reason for pursuing this strategy, as understood by both the Anglo-Canadian and American teams. However, there was a covert political reason too. To breed U-233, neutrons are fired at atoms of the element thorium. This excited the British, as thorium was plentiful in India, which at the time was under British rule. The strategy was especially seductive because natural uranium, the seed for breeding plutonium, was in short supply. Thus, the British saw a potential niche market for thorium, which they'd be able to dominate, though they withheld this motivation from the Americans; on paper, the potential ease of refining U-233 was reason enough to be interested. With this scientific goal in mind, the Americans used their reactor to irradiate thorium and make a few milligrams of U-233, enough to compare it against U-235.

Nunn May went to Chicago to carry out these comparisons in September 1944. This too was in collaboration with Herbert Anderson, as part of the programme agreed to in January.[41] Thanks to Anderson's collegial indiscretion, Nunn May received a small sample of U-233—and he was fortunate to receive it. At the time, Chicago was the only source of U-233 in the world.

Cockcroft's diary records pithily that "Alan Nunn May returned to Montreal with very useful information." He returned with more than that, of course. Nunn May kept his sample of U-233 initially, but then passed it to a Soviet contact.

BRUNO THE URANIUM PROSPECTOR

Using electronic instruments to detect radiation and identify its source was Bruno's professional forte. This expertise now bore fruit, as his techniques were applied to the search for uranium, and other scarce minerals essential to the atomic project. Specifically, the technique was used to map areas of northern Canada.

In January 1944, Bruno once again travelled to the United States, this time to New York to see Gilbert LaBine, the Canadian mining entrepreneur we met in Chapter 6. LaBine's Eldorado Mine had already supplied the US with uranium for Fermi's pile, but to fill the needs of the Chalk River reactor and the overall Manhattan Project would require new sources of the element. As an outcome of the New York meeting, LaBine commissioned some of Bruno's former colleagues from Tulsa, including Serge Scherbatskoy, to determine the optimal means of locating further deposits by performing tests in areas of Canada known to have uranium. This occupied the group during the summer and fall of 1944. For five days in September, Bruno himself joined them at Port Radium, near Great Bear Lake in the Northwest Territories.

The team compared different ways of detecting uranium in pitchblende, a naturally occurring ore. They found that although ionization chambers and Geiger counters were equally likely to locate the mineral when the rocks were in outcrops, the ionization-chamber technique was better when the rocks were a foot or more beneath the surface. The group performed surveys using lightweight instruments in helicopters, as well as ground surveys in trucks, which carried more cumbersome apparatuses.

Bruno's favoured technology, ionization chambers, worked best for identifying the radioactivity of ordinary rocks and pebbles. The challenge in this case was to identify the signal of uranium amid the morass of other radioactivity.[42]

Bruno presented Cockcroft with an extensive summary of the group's findings, and then announced the conclusions in October, at a special meeting in New York. Senior members of the US military were present. This marriage of nuclear physics and geology proved so successful as a means of finding uranium that the Manhattan Project adopted it. This was another area where Pontecorvo's expertise would later have potentially huge importance for the USSR.

ZEEP AND THE END OF THE WAR

The uranium quest was Pontecorvo's primary endeavour during the summer and fall of 1944. Then, for much of the first half of 1945, he investigated whether the fission products of radium-226 and uranium-233 might contaminate the reactor.[43] Also, starting in June 1944, he was involved in a new project, which was the brainchild of Alan Nunn May: the Zero Energy Experimental Pile, or ZEEP.

Halban left Canada in 1944, and Nunn May took over as head of the physics division. He proposed that a small reactor—ZEEP—be built in order to test the soundness of the team's theoretical calculations on the optimal distribution of the uranium rods in the tank of heavy water for the NRX.[44] Big enough to be a reactor, but small enough that it produced negligible power and thus didn't need concrete shielding or a cooling plant, ZEEP could be used to test different configurations of rods.

In Nunn May's conception, ZEEP had a secondary benefit, not stated in the official documents. Although he only discussed the idea verbally, Nunn May felt it was imperative that, when the war ended, the Anglo-Canadian team be able to demonstrate a working reactor to politicians and journalists.[45] This was a problem because it was clear from the outset that it would take a long time to complete the NRX.

Lew Kowarski had remained in Cambridge since 1940, but with his bête noire, Halban, no longer in Canada, Kowarski now joined the project. He was given special responsibility for constructing ZEEP. Kowarski later recalled his first sight of a working reactor, when he visited Chicago in September 1944. To outward appearances the machine was unimpressive, merely a cube of painted concrete. Herbert Anderson took Kowarski over to the pile and said, "Touch it. It's warm."

This visit helped Kowarski decide on his plan for ZEEP. Pontecorvo briefed him on the basics of constructing a pile, knowledge that had emerged from the American team's work since December 1942. Kowarski recalled that Bruno did so "in a sort of lecture in a single afternoon, which was quite enough."[46]

Although ZEEP was important for the Anglo-Canadian project, back in the United States General Groves was skeptical. His goal was to complete the atomic bomb and win the war. On a visit to Chalk River, Groves met Kowarski and bluntly asked, "Are you the man who is building this damn fool unnecessary experimental reactor?" Kowarski confirmed that he was, to which Groves replied, "America gives most of the heavy water for it, and it's very very costly stuff. Make sure that you don't squander it."[47] Although stated brusquely, the point was well made: heavy water was precious.

ZEEP went critical on September 5, 1945, three days after Japan surrendered and brought an end to World War II. The team held a party to celebrate their success with ZEEP, and Nunn May made a speech. He declared that Kowarski should receive an accolade and, because Kowarski came from France, Nunn May proceeded to kiss him on both cheeks. The timing was ironic. Nunn May, who had been passing information to the Soviets, was about to be exposed. At the very moment when the celebrations began, Igor Gouzenko, a cipher clerk at the Soviet consulate in Ottawa, was feverishly filling his briefcase with a wealth of documents.[48] These papers contained information about a ring of spies working for the Soviet Union. Gouzenko intended them to be the down payment on his application for asylum in Canada. The information in Gouzenko's briefcase would lead to the arrest and conviction of Alan Nunn May six months later, in March 1946.[49]

By the end of the war, Pontecorvo was a much sought-after prize, the recipient of several job offers at universities in North America and Europe. To the surprise of several colleagues, in February 1946 he chose to join the Atomic Energy Research Establishment, the infant nuclear laboratory in

Harwell, England. Then, having made this unexpected choice, that same month he prevaricated and decided to stay in Canada. His stated reason was that he wanted to work on the NRX reactor, which was the focus of the project. From 1943 to 1945 he had devoted a huge amount of effort to its design, writing some twenty-five reactor-related reports. In 1945 and 1946 he developed sensitive neutron monitors for the initial start-up of the NRX, capable of confirming the transition from zero flux to the first feeble reactions. Because of this responsibility, Bruno was one of only four physicists allowed in the NRX control room at the start-up.[50]

The NRX became critical on July 21, 1947; by 1949 it was the most powerful neutron flux in the Western world, a status it held for several years. By that time, however, Pontecorvo had indeed moved to Harwell, to work on the British reactor programme. Why he chose this course is one of the questions that came up following his defection: Was it a personal choice or part of a clever strategy orchestrated by others? As we shall see, there were seemingly legitimate personal reasons for Pontecorvo's decision. On the other hand, its timing also fits uncannily with events involving Nunn May and Fuchs, and some regard this as being more than a coincidence.

EIGHT

PHYSICS IN THE OPEN
1945–1948

IN 1945, WITH PEACE RESTORED, BRUNO PONTECORVO HAD SOME TIME to devote to pure science.

By the 1940s the fundamental building blocks of atoms had been identified as electrons, protons, and neutrons. In addition there was the neutrino, theoretically predicted by Wolfgang Pauli, which was likened to an electron but without electric charge or mass.

Although the idea of neutrinos had been around for two decades, and Enrico Fermi had incorporated them into his 1934 theory of beta radioactivity, no one had been able to establish their existence experimentally. It was now that Bruno began what turned into a lifelong quest: to understand this ghostly and enigmatic particle.

THE GHOSTLY NEUTRINO

Today, neutrinos are the focus of investigations both in particle physics and in a new branch of science: neutrino astronomy. Pontecorvo's interest in the subject had begun while he was still a student, when he had learned about the theory of the neutrino and beta decay from his teacher and mentor, Enrico Fermi. According to Fermi's theory, a nucleus undergoing beta decay emits a neutrino; however, when a neutrino hits an atomic nucleus, this transmutation can happen in reverse: the neutrino becomes a negatively charged electron, while the nucleus increases its positive charge. In effect, the atom moves one place forward in the periodic table, becoming an entirely new element, all due to the action of the neutrino.

Thus, like H. G. Wells's Invisible Man jostling the crowd, a neutrino may reveal its presence by bumping into something. The problem was that, according to Fermi's theory, the chance of this happening was so small that it would be extremely difficult to detect. Most people believed it to be impossible.

In 1945 Bruno began to think about neutrinos a great deal. He was ideally placed to make a breakthrough. Having set Fermi on the road to nuclear power in 1934, Bruno was now designing a nuclear reactor. These experiences in nuclear technology inspired a thought: if Fermi's theory were correct, a uranium reactor should make over a million billion neutrinos each second.

In Pontecorvo's opinion, the goal of finding the neutrino was "not out of the question." Although there is almost no chance of capturing a neutrino, *almost* is not the same as *none*. An individual neutrino produced by beta decay may travel for several light-years without interruption, but when you have an intense source producing billions of neutrinos each second, one or two might occasionally get caught in the atomic net of a sophisticated detector. He believed that, with "modern experimental facilities" such as a nuclear reactor, it could be done.[1]

Bruno considered the requirements. He realized that, to capture elusive neutrinos, the target must be large, so its material would have to be cheap. A liquid was ideal, as its volume would be limited only by the size of the container. If the atoms resulting from neutrino capture were radioactive, their decays could be recorded by a suitable detector, giving proof of their transitory existence. The fact that they had existed at all would then be evidence that a neutrino had struck.

Cleaning fluid is cheap, and contains chlorine atoms in the form of carbon tetrachloride; sporadic hits by neutrinos will convert chlorine into argon, one atom at a time. Argon is chemically inert, and can be extracted simply by boiling the liquid. The argon atoms created by the neutrino collisions are mildly radioactive, and survive long enough for these metaphorical needles in a haystack to be extracted and identified. If you wait long enough, the amount of argon created will be large enough to be measured, and the neutrino's existence will be confirmed. This, in a nutshell, is the basic strategy for pinning down the neutrino's existence, as articulated by Bruno Pontecorvo in his publicly available report, written in 1946.

However, this report derives from an earlier document, written on May 21, 1945, which shows that Pontecorvo did not develop these ideas alone.[2] First, the original 1945 paper credits the idea of using carbon tetrachloride to Jules Guéron. Guéron, a squat Frenchman with a chubby face and high forehead, was a chemist at Chalk River. He was also an expert in radioactivity. It was Guéron, too, who suggested that the production of radioactive argon atoms could be key to the endeavour.

Second, the pursuit of the neutrino appears to have inspired wide interest at the Chalk River laboratory. Guéron's observation about radioactive argon inspired a further discussion between "the author and Dr. Frisch." This led Bruno to conclude that argon's advantage of being chemically inert made the carbon tetrachloride strategy "the most promising method according to Dr. Frisch and the writer."[3]

Bruno estimated that several cubic metres of cleaning fluid would give him a reasonable chance of winning the lottery if the fluid was located near a nuclear reactor. A small quantity of radioactive argon atoms would be produced, their number revealed by the amount of radiation they emitted. From this, one could deduce how many neutrinos had struck. Thus the idea that the vicinity of a nuclear reactor would be a good place to search for neutrinos was original to Bruno; many details, however, were not. And the actual discovery of the neutrino—in 1956, by two American physicists, Frederick Reines and Clyde Cowan—did not use this technique, as the chlorine-argon method was not actually appropriate for the emissions from a reactor. No one knew then what we do today: reactors produce antineutrinos—the antimatter counterparts of neutrinos—rather than neutrinos themselves.

Finally, we come to the idea of solar neutrinos, for which Bruno subsequently became famous. In the body of his original paper, Pontecorvo had focused exclusively on the opportunities offered by a nuclear reactor; there is no mention of solar neutrinos at all. He had concluded that a reactor with just a little more power than the one being designed in Canada might produce enough neutrinos to give success. At the end of the paper, however, after Bruno's signature, there is a footnote. This appendage is an afterthought, inspired by Maurice Pryce, a British theoretician at Chalk River.

Pryce pointed out to Bruno that if the sun is indeed powered by nuclear fusion, as astrophysicists theorized, it could be irradiating the earth with a neutrino flux of 10 billion neutrinos per square centimetre every second. Bruno Pontecorvo credits Maurice Pryce unambiguously with this suggestion: "Dr. Pryce pointed out to the author that the flux of neutrinos from the sun is quite considerable."

Thus, the father of the solar neutrino idea is Maurice Pryce. However, at the time, he and Bruno dismissed it because the intensity of solar neutrinos at the earth's surface would be "too low for an experiment of the type suggested." They estimated that a flux a million times stronger than this would be required for success.

Such was the sensitivity to anything "nuclear" that this paper was immediately classified as secret, and remained so for two decades.[4] On September 4, 1946, however, these attempts at secrecy became moot when Bruno gave a talk at a nuclear physics conference at Chalk River, which was subsequently published as a paper. It is this latter version that has entered the public record and become famous. As this document does not mention the source of the solar neutrino idea, or the roles of Guéron and Pryce, Pontecorvo has been credited with a string of ideas that actually originated elsewhere.[5]

IN 1948, AMERICAN CHEMIST RAY DAVIS JOINED BROOKHAVEN National Laboratory in Long Island, New York, a facility that specializes in the research of fundamental science. During the war Davis had become an expert on chemical explosives, and at its close he had joined the US Atomic Energy Commission to work on radiochemistry—the chemistry of radioactive materials. Davis's new boss at Brookhaven advised him to visit the library, read the research literature, and "choose a project that appeals to you."

His good fortune, which determined his life's work, was to find in the library an article about neutrinos.[6] Davis had heard mention of this hypothetical particle, but that was the sum of his knowledge. Here was a chance to learn something. As he began to read the article, he quickly discovered that no one else was much wiser. He had stumbled upon a field that was wide open, and rich in problems. Then his excitement grew. The paper briefly mentioned that in 1946 Bruno Pontecorvo had suggested a method of finding the neutrino using chlorine and radiochemistry. Davis realized that this was right up his alley.

For Davis, Bruno's proposed experiment seemed all too easy, and so it was. His first attempt at the Brookhaven reactor failed because the impacts of cosmic rays were so numerous that it was impossible to discern any faint signals of neutrinos coming from the reactor. In 1955 he built a large detector, using 4,000 litres of carbon tetrachloride, at the powerful reactor near the Savannah River in South Carolina; Davis dutifully shielded the detector from cosmic rays, but saw no sign of the neutrino. This eventually helped confirm that reactors produce antineutrinos, to which the chlorine-argon method is not sensitive, rather than neutrinos.[7]

Davis missed out on discovering the neutrino, but by 1956 the existence of the ephemeral particle had been confirmed, and the subtle difference between the processes at work in a nuclear reactor, which produce antineutrinos, and those in the sun, which produce neutrinos, were also understood. Thus, in 1959, Davis set out to use Bruno Pontecorvo's method for a slightly different application: looking for neutrinos that have come from the sun.

Forty years would elapse before the quest for "solar neutrinos" was completed to everyone's satisfaction, and Davis, at the age of eighty-seven, received his Nobel Prize having spent his career "doing just what I wanted and getting paid for it."

WHO ORDERED THAT?

While at Chalk River, Pontecorvo had an insight that now forms a cornerstone of the Standard Model of particle physics. Its genesis involved cosmic radiation.

During the 1930s and 1940s, cosmic rays—high-energy particles from outer space—revealed varieties of matter previously unknown to scientists. For nuclear physicists, interested in the basic particles from which matter is built, cosmic rays became the new frontier. Earth's upper atmosphere is being continuously bombarded by extraterrestrial radiation, including the nuclei of elements, many of which were produced during the explosion of distant stars. When these rays hit the upper atmosphere their energy dissipates as they disrupt atoms in the air, creating showers of less powerful subatomic particles. These finally reach the ground as a gentle "rain," which is interesting to scientists but poses no real hazard to humans.

Among the debris produced in collisions between the primary rays and the atmosphere were new forms of matter, previously unknown. The first example of antimatter—the positron, or antielectron—was discovered in cosmic rays as early as 1932. Other discoveries included the muon in 1937, as well as the so-called "strange" particles and the pion, both in 1947.

Electromagnetic forces give rise to radiation, such as light, which in quantum theory occurs in staccato bundles—the particles known as photons. In 1935, Japanese theorist Hideki Yukawa predicted that a similar phenomenon should arise from the strong nuclear force that holds atomic nuclei together; agitate an atomic nucleus violently and, under certain circumstances, it might radiate energy in the form of the particles later known as pions.

Within two years, cosmic rays had revealed novel particles that seemed to confirm Yukawa's theory. This discovery formed the frontier of fundamental physics as World War II began, and was still the frontier as physicists, including Bruno Pontecorvo, took up open science again in 1945. The cosmic rays seemed to have completed the fundamental understanding of atomic nuclei and particles.

However, scientists soon realized that they had caught the wrong suspect. Three young Italians, hiding from the Germans in Rome toward the end of the war, had set up a makeshift laboratory in a basement.[8] They used an array of Geiger counters to reveal the passage of cosmic rays; their hope was to measure the lifetime of Yukawa's novel particles. Yukawa's theory predicted that the negatively charged versions should be attracted by the positively charged nuclei of atoms, and thus would be captured and absorbed by the strong force before having a chance to decay. Positively charged particles, by contrast, should be repelled and then decay. The Italians' experiment succeeded, but with a surprise: they saw both negative and positive versions decay. This implied that the novel particle had no affinity for the atomic nucleus, and so could not be the pion that Yukawa had predicted. The eventual discovery of the pion in 1947 completed the understanding of the powerful forces of the nucleus, but left scientists with an enigma: What was this other particle that had been discovered in 1937?

To distinguish it from the pion, physicists gave it a name: *muon*. A name is a form of classification, and provides a sense of control perhaps, but is not an explanation. The discovery of the pion now thrust the question of the muon's identity to centre stage. This seems to have been the moment when Bruno shifted his focus from nuclear reactors to cosmic rays and the nature of the muon.

As we have seen, Bruno was familiar with the phenomenon of beta radioactivity, which causes the transmutation of elements in nuclear reactors, and, we now know, in stars. Nature also plays this sequence in reverse: it is possible for a proton in a nucleus to capture an electron. This is known as the "inverse beta process." Bruno noticed that an atomic nucleus captures a muon in a similar fashion, and took the brave step of assuming this similarity to be significant. His conclusion: the muon is a heavier version of the electron. "Who ordered that?" physicist Isadore Rabi famously exclaimed, and more than thirty years would pass before the beginnings of an answer emerged.

A first step toward answering Rabi's question was to find how muons are born. Experiments quickly established that the muon is produced when pions decay. Bruno proposed that this process is fundamentally analogous to ordinary beta radioactivity. He then turned his attention to how muons decay. Unlike electrons, which are stable, muons only live for about a millionth of a second before decaying into other particles. Bruno addressed the question of what the resulting particles might be. An electron is the only electrically charged particle that is lighter than the muon, and because the muon is electrically charged, the only way that the electric charge can survive when the muon decays is if an electron is created. To balance the total energy and momentum in the transmutation, one or more other particles must be created also. The most likely candidates were a single photon, some previously unknown massive particle, or a pair of neutrinos.[9] If a muon was simply a heavy form of an electron, the muon should be able to shed energy by radiating a photon, thereby converting to an electron and a photon. If the muon's relationship to the electron was more subtle, however, this decay would be very rare or absent. To decide between these alternatives, he devised an experimental test.

If a muon decayed into an electron and a photon, the latter pair would emerge back-to-back, with each carrying a specific amount of energy. If the decay produced an electron and two neutrinos, however, the energy could be shared among the particles in a variety of ways. So Bruno proposed a test: measure the energy of the electron in hundreds of examples of muon decay, and determine whether its energy is always the same (suggesting the presence of a photon), or whether it has a range of values (suggesting the alternative). He completed a paper that laid out this proposal in June 1947, and it was published later that year.[10]

Having proposed the idea, Bruno then began to set up the experiment, along with his colleague Ted Hincks. The basic idea was to have the electrons hit a sheet of material that would absorb them. The more energy an electron has, the more material it can pass through. So they measured the quantity of electrons that managed to penetrate the material, to see how this varied with its thickness. If the decay of muons always produced electrons with the same energy, as in the electron-photon scenario, the quantity of electrons getting through the absorber would suddenly drop when its thickness reached a critical level. However, the quantity would change gradually if the electrons' energies covered a range of values, as would be the case if each electron were accompanied by two neutrinos.

Their experimental results appeared in 1948.[11] Pontecorvo and Hincks were able to show that the muon does not decay into an electron and a single massive particle, but the question of whether the electron was accompanied by a photon or two neutrinos remained unresolved. In any event, Jack Steinberger, an American physicist working in Chicago, beat them to the finish line. Steinberger conducted a similar experiment, and published his findings in the same journal.[12] He was the first to show that the scenario in which a muon decayed into an electron and two lightweight particles (neutrinos) fitted best with the experimental results.[13] This would not be the last time that Steinberger beat Bruno to a crucial discovery.

THE UNIVERSAL WEAK FORCE

In the seventeenth century, Isaac Newton realized that the rise and fall of ocean tides, the orbits of the moon and planets, and the descent of apples to the ground were all manifestations of the universal force of gravity. In 1947, halfway through the twentieth century, Bruno Pontecorvo proposed that a variety of apparently diverse phenomena in atomic physics could be due to a universal "weak" force. These included the transmutation of the elements in beta decay, the production and decay of the muon, the instability of the pion, and the behaviour of "strange particles."

It would be disingenuous to push the comparison with Newton too far. In both cases, a variety of disparate phenomena were recognized to have a common fundamental origin. In the case of gravity, Newton both recognized its universal character and worked out the quantitative laws governing its behaviour. Pontecorvo made a qualitative insight regarding the existence of the weak force, and performed experiments that helped establish its reality. However, another quarter of a century would pass before the quantitative theory of the weak force was established by others. Nonetheless, this time span itself bears witness to Pontecorvo's foresight in recognizing the presence of the universal weak force, which choreographed the dance of these particles.

Today, the weak force is recognized as one of the four fundamental forces of nature, along with gravity, electromagnetism, and the strong nuclear force. The discovery of the universal weak force is one of the most significant scientific advances of the twentieth century.[14] Bruno also realized that this universality held the key to understanding how the electron, muon, and neutrino are related. Having confirmed that the muon is a sibling of the electron, he later applied this same idea to the neutrino, which he saw as having two varieties: one appears to be a sibling of the electron; the other of the muon. Bruno's pairing of these fundamental particles into distinct families was the seed for the modern Standard Model of particles and forces.

These insights regarding the weak force and the choreography of particles are of Nobel Prize quality. Bruno Pontecorvo would be involved in all of them.

BRUNO THE INSTRUMENT MAKER

Between Bruno's 1947 paper on the muon's genealogy, and his investigation of its decay, he travelled to Europe to see his parents, visit Harwell, and spend a few days with his brother Guido in Glasgow.

At this time, Bruno and his colleagues at Chalk River were developing a novel instrument, in preparation for the experiment on muon decay.

Today known as the proportional counter, this instrument measures the energy of radiation. If radiation has enough energy, it can knock electrons from atoms in an inert gas, leaving positively charged ions in its wake; low energy radiation, however, has no such effect. The liberated electron and the positively charged ion are known as an "ion pair." In a proportional counter, the radiation passes through a small chamber filled with an inert gas, creating ion pairs. In the process, the radiation loses energy until eventually it is unable to create further ions. Thus the number of ions is proportional to the energy of the initial radiation. These electrically charged ion pairs create a signal in a detector, whose magnitude reveals the energy of the original radiation. Thus a proportional counter is especially good at measuring the energy of radiation, a technique in which Bruno Pontecorvo became an expert.

Meanwhile, in the physics department of the University of Glasgow, Samuel Curran, an expert on Geiger counters, was collaborating with John Angus, a research student of nuclear physics. Their goal was to develop new ways of detecting beta particles and measuring their energy. Their interests were very similar to Bruno's. During his visit to Guido, Bruno called on the Glasgow physics department and met with Curran.

There is no record of whether Bruno was already aware of Curran's work, and was making the visit to compare notes, or whether he learned of the project only when they met. In any event, the Glasgow team apparently felt that they had given out more information than they had gained, as a dispute arose over the invention of their technique. The relevant university documents note that, during the course of the Glasgow team's work, "Curran had described the method to a member of the staff at Chalk River Laboratory, Canada."[15]

In the August 21, 1948, edition of the journal *Nature*, Curran and Angus included a brief description of their new technique, as the second paragraph in a letter on the beta decay of tritium. One week later, Bruno and two colleagues described a similar idea in *Physical Review*.[16] The Canadian group had submitted their paper in June, two months before the Glasgow team's letter appeared. The submission date of the letter is not known.

With implicit reference to Pontecorvo's visit the previous fall, the University of Glasgow record includes this comment: "One week after

Curran's letter to 'Nature' a rival paper from the Chalk River group (including Pontecorvo) appeared."[17] Given the similar expertise of the teams, it is possible that the ideas were in the air and would have developed independently in any event. As to who influenced whom, this we must add to the unresolved mysteries of this tale.

Bruno also visited Harwell "for a few weeks" during the fall, before flying to Italy on December 8. He gave a talk in Rome on December 17, which inspired the Italian state oil company, Agip, to use his neutron well-logging technique; Bruno supplied circuit diagrams of the electronics and other information.[18]

As part of Harwell's normal vetting process, the British security authorities were already building a profile of Bruno Pontecorvo. They noted that he travelled back to Canada via Paris, where he spent New Year's Eve in Montmartre with his former colleague Jules Guéron. The MI5 watchers reported that the trip seemed to be made for purposes of "jollification" and that "no other scientific contact was made."[19]

The British authorities, with traditional xenophobia, suspected the political allegiance of French scientists in general, and of Frédéric Joliot-Curie, a declared communist, in particular. Bruno spent just twenty-four hours in Paris, and then passed through England en route to Canada on January 6. The Harwell security officer reported to MI5 that Bruno was a "straightforward fellow with no political leanings."[20]

IN CANADA, AND SUBSEQUENTLY AT HARWELL, BRUNO PRESENTED himself as apolitical and unsophisticated in world affairs. In 1950, after Bruno's disappearance, his brother Gillo told MI5, "My parents as well as [our eldest brother] Guido told me that they had the impression that Bruno after the war was no longer interested in politics and was avoiding any connection with communism, both in the form of literature and in personal contacts. My parents attributed this to his position at Chalk River and Harwell." Gillo, who was especially close to his brother, saw through the charade. When Bruno visited Italy in June 1946, Gillo perceived that Bruno's professed lack of interest in politics "was phoney, because Bruno was so well informed of details of certain questions that were so much at the heart of communism." For example, Gillo pointed out that Bruno had a deep understanding of the motives of Yugoslavian dictator

Josip Tito. Gillo concluded, "He must have kept thoroughly up to date with communist literature."[21]

An example of Bruno's tact can be seen in his reaction to the news that the atomic bomb had been successfully tested in New Mexico on July 16, 1945. When the news came in, everyone on the Canadian team began to discuss global politics, in an attempt to understand the implications of the test. The big question was whether the bomb would be used on Japan. In the memory of one of those present, Bruno alone was "certain of the answer." He insisted that the Americans would have to use the bomb "for political reasons, before the Japanese surrender and before the Russians could play a role in their surrender."[22] As events transpired, this was completely accurate. Hiroshima was bombed on August 6 and Nagasaki on August 9. Late on the evening of the eighth, Stalin declared war on Japan. Soon after midnight Soviet troops invaded and annexed Manchuria, Inner Mongolia, and North Korea. Japan surrendered to the Americans on September 2, 1945.

NINE

MANOEUVRES
1945–1950

THE BOMBING OF HIROSHIMA AND NAGASAKI NOT ONLY BROUGHT A conclusion to World War II; it also ended the collaborative wartime work on nuclear physics, and heralded the start of national nuclear energy programmes—with both peaceful and military goals. In the UK, the development of nuclear power for peaceful purposes began at a new facility called the Atomic Energy Research Establishment, or Harwell.

Harwell's goal was to design and build the first nuclear reactors in the UK and, indeed, in Western Europe. It was operated by the United Kingdom Atomic Energy Authority, and on January 1, 1946, was established under the formal charge of the Ministry of Supply. John Cockcroft left Chalk River to become the first director of the British reactor project. Bruno had been invaluable in the design of the reactor at Chalk River, so in 1945 the British invited him to join their own new venture.

That same year, Bruno had been offered several highly prestigious and lucrative positions at American institutions, but in the end he chose Harwell. His decision raised eyebrows, including those of Emilio Segrè, who later told the FBI that he believed Bruno's choice to have been influenced by communist relatives in Europe, with dark hints of Soviet malevolence. According to Oleg Gordievsky, a KGB double agent who came over to the United Kingdom in 1985, Bruno Pontecorvo had been a valuable source for the KGB since 1943. If this is true, then Segrè's suspicions may have been correct.

Bruno had visited the University of Michigan, which wanted to hire him, and had received very attractive offers from the University of Rochester and the Radiation Laboratory at the University of California, Berkeley, home of Emilio Segrè. He also exchanged letters with luminaries such as Hans Bethe at Cornell, Robert Marshak at Rochester, Arnold Siegert at Syracuse, and the chairman of the physics department at MIT. Bruno dithered about his decision. In part this was the natural reaction of a talented physicist, much in demand, who wanted to weigh his options to achieve the best outcome for his career and his family. However, the correspondence also hints that Bruno was unsure about when he would be able to decide, as if he was not completely in charge of his own destiny.[1]

Several letters came from the University of Rochester. They offered Bruno a position as an associate professor at $6,000 a year, to begin "any time at your convenience or availability between now and Sept 1946." The facilities were superb: there was "already a small cyclotron operating" and the university "[hoped] to build a uranium neutron reactor." In addition to these obvious advantages, the authors of the letters pointed out that Rochester was only ninety miles from Cornell, home of leading nuclear theorist Hans Bethe. In the meantime, Bruno told Michigan that he would decide in January 1946, after seeing some people "at the British project." This referred to Bruno's discussions with James Chadwick about the restrictions Harwell would impose on him.

This was a difficult time for Bruno. For several months, he and other non-British members of the Canadian team had been severely restricted in their freedom. Security had become very tight as the test of the first atomic bomb approached. General Groves was suspicious of international collaborations, and the FBI was concerned about the number of foreign scientists involved, especially in Canada. This tension had been present for some time, but now it was growing worse. After the Normandy invasion in June 1944 and the liberation of Paris in August, several of the French scientists on the Anglo-Canadian project wished to return home and celebrate, but permission for these trips was denied.

In September 1944, Churchill wanted all visas for non-British members of Tube Alloys to be withdrawn. Around this same time, Halban visited France and talked to Joliot-Curie about developments in fission. He did this because he believed that he and Joliot-Curie held patent rights in the field. General Groves was furious about the visit, which created trouble for everyone. Bruno, like his French colleagues, wished to visit his relatives in Europe as soon as possible, but travel for all non-British members had been curtailed, and in the summer of 1945 Bruno's request to visit Italy was rejected.

At this stage, James Chadwick was directing British nuclear policy from Washington. He decreed that any offer of employment at Harwell should carry the condition that Pontecorvo was forbidden to travel to mainland Europe to visit his parents. Bruno, however, would not tolerate such a long-term lack of liberty, especially now that the war had ended, and just before Christmas 1945 he visited Chadwick in Washington, DC, to discuss this.[2] This initial meeting proved inconclusive.

In 1946, Bruno visited New York and Washington from January 23 to January 29 for the meeting of the American Physical Society, and it was then that his decision seems to have crystallized.[3] During the conference, he was courted by several potential employers. General Electric thought he was still in the market, and on February 7 they wrote to him with an offer. Bruno immediately declined, saying it would be a "pity to leave the Canadian pile just as research rather than engineering starts" and that he would stay in Canada "for at least one year." At the conference Arnold Siegert, a physicist at Syracuse University, also inquired about Bruno's availability, but on February 13 Bruno thanked him for asking but stated unequivocally, "I am staying here."

Meanwhile, on February 4, Otto Frisch had written from Los Alamos to extol the scientific excitement that the new laboratory in Harwell would offer, and encouraged Bruno to accept a position there. Frisch also mentioned "Uncle James" (Chadwick), and remarked that "the big show" in Canada—namely, the NRX reactor—was expected to start in mid-March.

Bruno finally informed Harwell that he would accept their offer only if he was allowed to travel. He stressed that one of Harwell's attractions for him was that it would allow him to be closer to his Italian roots and family, after ten years of absence. Harwell must have eventually agreed to his terms, as on February 21 Otto Frisch wrote of his pleasure that Bruno had accepted the job in England.[4]

Hiring Bruno Pontecorvo was a coup for British science, despite the government's reluctance to employ non-British nationals. This xenophobic attitude, which had surfaced when Bruno had been recommended to Chalk River in 1943, now came up again. Harwell needed him desperately, but his Italian origins threatened to kill the deal.

The question of his reliability was raised with General Groves. Chadwick confirmed that Pontecorvo would be employed by the British government at the Harwell nuclear establishment, and, to break the impasse, he suggested that Pontecorvo would "probably take steps to acquire British citizenship." In fact, Bruno became a naturalized British citizen later, on February 7, 1948, on the basis that he had been involved in a British project for five years. A report to Prime Minister Clement Attlee, specifically about Bruno, commented, "It is believed Pontecorvo had already taken out first papers in the US but had expressed a preference for British naturalization if he could get it." In hindsight, this may reflect Bruno's concerns regarding American attitudes toward communism.[5]

The lengths to which the British went to obtain Bruno Pontecorvo—even sending a report to the prime minister—shows how important the Italian physicist was for them. The number of quality institutions that had bid for him in North America also testifies to his international stature. Bruno held all the cards. In 1946 he could make a career wherever he wanted.

Despite claims that Bruno's move to Harwell was related to espionage, many of his colleagues have argued that there were innocent reasons for his choice. Given that John Cockcroft left Canada to become the director at Harwell in 1946, and that several other Canadian colleagues made the move too, continuity in research was one obvious advantage. One Harwell scientist, Godfrey Stafford, who later became the director of Harwell's neighbour, the Rutherford Laboratory, judged that the atmosphere at Harwell was key: "At Harwell at that time there was a blank canvas," he explained. He recalled that an attraction of Harwell was the implicit invitation, "Come here and do what you like."[6] So it is plausible that Harwell, with its close proximity to the leading physics departments at the Universities of Oxford, Bristol, and London, had a competitive edge, notwithstanding Bruno's North American offers. Being nearer to Italy after his decade-long absence was also attractive, as he himself claimed.

Segrè's allegations that Bruno's move had communist motives, which he made in 1949 to the security authorities, may have been sour grapes because Bruno had chosen Harwell over Berkeley; nonetheless they would later have consequences. Segrè's report coincided with the growing anti-communist hysteria in the United States, fueled by Senator Joseph McCarthy and the House Un-American Activities Committee, who pursued witch hunts against communists in government, science, and Hollywood. Eventually, their spotlight would shine on Bruno Pontecorvo.

MEANWHILE, ATTRACTIVE OFFERS OF EMPLOYMENT IN THE UNITED States continued to come Bruno's way. In 1947 Bruno visited Cornell, and Hans Bethe offered him a position as a tenured associate professor at a salary of $7,000 a year, requiring him to teach "no more than 6 hours a week."

The tone of Bethe's letter, written as if Bruno was still open to offers, suggests that even then Bruno's mind was not yet totally made up. This may indeed have been the case, as Bruno stayed in Canada until the end of 1948, while acting as a consultant to Harwell. The Production Pile Discussion Group was formed to advise the British reactor design team. James Kendall, the engineer responsible for pile design at Harwell, reported back from one meeting in Canada that the help from Pontecorvo "was worth that of all the others put together."

PRIVATE LIFE IN CANADA

Bruno's time in Canada was when fault lines in his public image first began to appear. For example, he presented himself as apolitical to friends and colleagues, but he was in reality a member of the Communist Party. His reasons for keeping this a secret are obvious, of course. The fact that he succeeded so comprehensively, however, shows that he was not naive in these matters. What's more, this ability to keep inconvenient truths hidden shows that he was more sophisticated than he often appeared to friends and colleagues. As a family man, his persona was that of a naive extrovert, the life and soul of the party, always pleased to be in the limelight; his home life, on the other hand, was more complex.

Laura Fermi recalled two incidents that reveal the nature of the man and his private life.

In January 1944, during a visit to Chicago, Bruno called on the Fermis at their home. He had broken his leg skiing, and was hopping about on crutches. This made him the centre of attention, which stimulated him to prance around even more dramatically.[7] Having gained people's attention, he milked it. Such was the public image of Bruno Pontecorvo, the showman. However, even though everyone perceived him as happy-go-lucky, he described himself as having an inferiority complex.[8] This he traced back to his childhood, when, as the fourth of eight talented children, he had to compete for attention. His handsome exterior and bubbly personality were thus cloaks that helped mask his insecurity.

Laura Fermi also hints that Marianne was less sociable than her husband. We have seen how the Fermis were upset not to have met Marianne and Gil in 1940, soon after the Pontecorvos arrived in America. Laura recorded that she was disappointed again in 1943, when Bruno visited the Fermis en route to Canada, but, as on the previous occasion, "his wife and son were not with him." The first and only time that the Fermis met Marianne was at the end of November 1948, shortly before Bruno moved to Harwell.[9] Bruno and Enrico had been attending a meeting of the American Physical Society in Chicago, and Enrico invited the family to dinner. The occasion was memorable for the impression that Marianne made.

Earlier, she had been in the city shopping, and had become lost on her way back to the hotel. This made the Pontecorvos late for dinner, which upset Bruno "out of proportion to the trouble it might have caused."[10] Throughout the evening, Bruno talked "as easily and volubly as ever," but was clearly annoyed. Marianne, by contrast, "kept silent," and in Laura's opinion was "painfully shy."

A colleague from Canada recalled Marianne at that time as a beauty. Of Bruno, he said, "Handsome, flirtatious, very Italian, he was the heart-throb of all the single women." These memories mirror those of David Jackson, who in 1947 was a graduate student, doing routine computations for the experimenters at Chalk River. Although he had been only a junior member of the laboratory, Jackson recalled, "we all were aware of Ponte, the glamorous Italian who played a lot of tennis and cavorted with the single young women at [the nearby town of] Deep River. I am sure Marianne was not happy, although she was also a very attractive woman then."[11]

There is a story from that era in which Bruno and his colleague Ted Hincks, while travelling to a conference in Montreal in June 1947, gave a ride to two attractive girls—in some accounts, secretaries from the laboratory chemistry division.[12] The two women wanted to visit their family in the Montreal area for a few days, and the two physicists obligingly provided transportation. The plan was to bring the women back to Deep River after the conference, but during the convention a science issue arose, which Bruno felt could be answered if he and Hincks consulted another physicist at MIT, outside of Boston.[13] The name of the physicist is long forgotten, but what happened next would become part of Chalk River folklore: Bruno and Hincks invited the two secretaries to join them on the trip to Boston, after which they would take the women back to Deep River. Meanwhile, other participants from the Montreal meeting returned to Chalk River, and word got around of Bruno's latest exploit. Marianne took offense, withdrew all of Bruno's money from the bank (a sum of $1,800), and went to the Canadian resort town of Banff with the three children—Gil and Tito now had a younger brother, Antonio, born in July 1945. Marianne and the three boys stayed in Banff for some time. When Bruno returned, three days late, he asked friends if they knew where Marianne had gone; finally, one of these friends managed to find Marianne and convince her to return home.

As a result of this adventure, for the rest of his time in Canada, Bruno was saddled with the nickname of "Ramon Novarro," a film actor who had become the latest sex symbol following the death of Rudolph Valentino. Novarro and Bruno bore a vague resemblance: the same dark, slicked-back hair, bedroom eyes, and Latin elegance. Marianne, with her fair skin, blond hair, and Nordic features, contrasted with Bruno physically no less than she did temperamentally. A colleague who knew the couple in both Canada and Harwell recalled Marianne as very quiet and beautiful, but also as a "mixed up Scandinavian, [who] seemed a funny choice for a randy Italian."[14]

Toward the end of his life, Bruno was interviewed by the Italian journalist Miriam Mafai. She concluded that Marianne's reticence and long silences were exacerbated by the fact that "Bruno was very much courted and wooed, and this could not please his wife." This, however, was only part of a more serious problem: during their time in Canada, Marianne began to show the first signs of a nervous condition, which peaked years later.

Bruno told Mafai he had noticed that Marianne had a "few oddities," which first manifested themselves when they were invited to social functions. For instance, when they were about to leave home to go to a friend's house for dinner (such as the Fermis' perhaps), Marianne would suddenly announce that she was not coming. Bruno, naturally cross, would say, "You're crazy," in what he described as "a normal voice," unaware that her intense shyness was a symptom of a deep malaise.[15]

These descriptions were given by Bruno later in life, after Marianne had suffered a catastrophic mental collapse in the USSR. However, signs of trouble had been there all along. As we've seen, in 1938, during her year-long separation from baby Gil, Bruno's letters refer to a mysterious illness, which she seems unwilling to discuss.[16] For a girl predisposed to depression, abandoning Gil to a French nursery when he was only weeks old could only have added to her misery. Let's now consider the ensuing years from Marianne's point of view: Reunited with Gil and Bruno, she flees to North America immediately after having a miscarriage. Her parents and siblings now thousands of miles away, her life is built around the demands of her husband's career. By 1943 she is living in the United States with a closet communist, who is about to work on a secret project in Canada. When such experiences are added to her inherent shyness, one can understand her reluctance to meet her husband's teacher, a world-famous Nobel Laureate. A woman in her situation merited sensitivity and support.

Marianne was as pretty as Bruno was handsome. Bruno, the peacock, enjoyed presenting himself in the company of his attractive wife. He was the charmer, everyone's best friend. Whether his scientific intelligence was matched by emotional maturity is less clear. On the other hand, it's possible that Bruno's decision to move to Harwell was influenced by a consideration of Marianne, who, like Bruno, would be within easy reach of her family, and might therefore feel more at ease. Although there is no proof that this was a factor in his decision, it could have been a powerful attraction.

Whatever the reasons for the move, by the end of 1948 the Pontecorvos' time in North America was coming to an end. From November 2 to December 3, 1948, Bruno used his accumulated holiday time to visit the western United States. This trip included visits to various West Coast universities and a side trip to Mexico. From that point on, Bruno's career moved ever eastward. The Pontecorvos left Chalk River for the last time on January 24, 1949, and flew to Britain, via New York. Bruno started work at Harwell on the first of February.[17]

INTERLUDE

WEST TO EAST

BACK IN 1942, WHEN THE AWESOME IMPLICATIONS OF FISSION WERE first realized, scientists in the USSR had doubted that an atomic bomb could be made in time to influence the war. Furthermore, Klaus Fuchs had sent information that the British and American nuclear teams were working on an industrial scale. Thus, by 1943, Stalin had received enough intelligence to know that, for the immediate war, a Soviet bomb was irrelevant. The relatively small atomic bomb project that he authorized was a "hedge against future uncertainties"[1]

When Igor Kurchatov took on the design of the Soviet atomic bomb, one of the first things he did was to find out what was known elsewhere. He spent several days in the Kremlin, where he studied material relating to the British atomic project. The information from Fuchs enabled Kurchatov "to bypass many labour-intensive phases of working out the problem."[2]

The news, gathered from spies, that a chain reaction could take place in a mixture of uranium and heavy water was invaluable for Kurchatov. Soviet scientists had believed this to be impossible, because they thought the chance of a neutron being absorbed by material before meeting a nucleus of U-235 was too high. Soviet physicists didn't have enough heavy water to perform the test for themselves, so "borrowing" the data from others was critical.

Other borrowed intelligence concerned plutonium. Kurchatov checked the last published papers before secrecy had taken over, and deduced that plutonium could be a novel route to the bomb, which would eliminate the need to separate U-235. He arrived at this realization due to the interest in plutonium revealed in these final open papers, as well as his general knowledge of nuclear physics and some technical calculations of his own. However, he lacked some critical data on the subject. He needed to know the answers to two questions: Is plutonium fission caused by slow neutrons or fast ones? And does it suffer from spontaneous fission?

The phenomenon of spontaneous fission limited the amount of uranium that could be kept in one place. Neutrons released by fission may induce further fissions if they hit appropriate nuclei, or alternatively may escape from the sample entirely without hitting anything. The latter is more likely for small samples than for large ones. The reason is that the ratio of volume (where fission can occur) to surface area (which enables neutrons to escape) grows with the radius of the sample. A tiny baby needs to be kept wrapped up even when a mature adult is lightly clad because, relative to its size, the baby has a greater surface area through which it can lose body heat. Analogously, small samples of uranium lose neutrons more easily than large lumps of the element. There is a critical mass of uranium or plutonium below which fission is not self-sustaining.

Fission might occur accidentally, for example due to a cosmic ray colliding with a uranium atom, or due to spontaneous radioactivity. This carries the danger of causing a spontaneous chain reaction and an uncontrolled explosion. To prevent this, the size of enriched uranium samples is kept small enough that that neutrons are more likely to escape through their surface than feed a chain reaction within their heart. To make an atomic bomb from U-235, two such "subcritical" samples need to be prepared; later, the samples must be brought together very rapidly to trigger the explosion. Kurchatov understood this principle in the USSR, and discovered that the West had understood it too.

Kurchatov prepared a research-and-development plan, which outlined the problems to be solved, and in March 1943 he drew up a list of laboratories in North America where solutions might have already been found. The Soviet intelligence agency, the NKVD (the forerunner to the KGB), sent the questions on to its agents abroad.[3]

The USSR was already setting up a network of informants scattered throughout the West's atomic project. At the end of 1942 or early in 1943, Peter Ivanov, an employee of the Soviet consulate in San Francisco, asked British engineer George Eltenton, who had worked in Leningrad but was now at the Radiation Laboratory in Berkeley, to obtain secret information about its research. Ivanov spoke to Haakon Chevalier, a communist friend of J. Robert Oppenheimer, who in turn approached Oppenheimer, but the latter would have nothing to do with the scheme. Ivanov then approached others at the Radiation Laboratory in search of information. He seems to have been successful.[4]

The Soviets spread their intelligence net very wide among the US laboratories. The Gouzenko defection and the arrest of Nunn May show that the net encompassed Canada too. Ever since Bruno Pontecorvo fled to the USSR in 1950, there has been debate about whether he too passed information.[5] The members of the congressional Joint Committee on Atomic Energy seemed to have no doubt. They dubbed him the "second deadliest betrayer," but no evidence other than his disappearance was produced. Their report, issued in 1951, included this cogent assessment: "Whether or not Pontecorvo in fact betrayed secrets before disappearing behind the Iron Curtain, his recollection of those secrets is now available to Russia. His unusual scientific mind is also available for Soviet reactor development."[6]

If Pontecorvo was indeed as important to the Soviets during the 1940s as some have claimed, then he must have left a footprint. By studying the known history of the Soviet atomic project, including the chronology of what the Soviets knew about Western progress and when, one might hope to identify his contributions, or lack thereof. For example, it seems plausible to dismiss one claim immediately—namely, that Pontecorvo gave the Soviets information about Fermi's successful nuclear-pile experiment in December 1942.[7]

In Kurchatov's memo, written on March 22, 1943, which specified the questions that agents should answer, he recorded that it was "still unclear if a natural uranium and graphite system is possible." As this was the very method that Fermi had already demonstrated with his Chicago reactor three months earlier, news of the Americans' success had clearly not yet reached the USSR. As we have seen, Pontecorvo visited Fermi in April 1942, and although they discussed aspects of the experiment, this was long before the outcome of Fermi's experiment was known. Pontecorvo didn't join the Anglo-Canadian project until 1943, and had no direct contact with Fermi's operation until 1944. Thus he initially had limited knowledge of the Chicago pile. Given that Kurchatov's own memorandum suggests his ignorance of Fermi's success, it seems unlikely that Pontecorvo had leaked the information.

During 1943 and 1944, several employees at the Metallurgical Laboratory in Chicago were suspected of having passed information to the Soviets and were summarily dismissed. It is more natural to suspect that one of these employees eventually informed the USSR about Fermi's operation, rather than the remote and disconnected Pontecorvo.

The US intelligence network suspected that information about the gaseous diffusion plant in Oak Ridge, Tennessee, was also being sent to the USSR. Pontecorvo had no involvement with Oak Ridge at any stage. If he passed any information to the Soviets at all, it would be restricted to knowledge gained during his time in Canada, from 1943 to 1948, or at Harwell, from 1949 to 1950.

OLEG GORDIEVSKY, THE FORMER KGB DOUBLE AGENT, ASSERTS THAT Pontecorvo passed significant amounts of information to the Soviets. The claims are fairly specific, but unsubstantiated. For example, in a book coauthored with the historian Christopher Andrew, who later wrote the authorized history of MI5, he stated that "at some point" after joining the Canadian project, Bruno had made contact with the Soviet embassy, "probably" in Ottawa. The tale becomes more detailed in claiming that Bruno's report was sent not to the GRU (the USSR's military intelligence agency) but to its "neighbour," the KGB. The KGB "resident" at the embassy initially ignored Pontecorvo, apparently suspicious that he was a provocateur working on behalf of the Canadians. Eventually convinced of Pontecorvo's reliability, the KGB adopted him. Years later, contacts in the KGB told Gordievsky that they "rated Pontecorvo's work as an atom spy almost as highly as that of Fuchs."[8]

When Gordievsky subsequently repeated this claim to me, he added, "Bruno Pontecorvo was an agent of the KGB for a long time. Probably he was recruited during the Spanish war. His information was very valuable."[9] When pressed, Gordievsky did not provide any source or even specific facts, such as what secrets Pontecorvo had supposedly passed. Nor did Gordievsky substantiate the allusion to Pontecorvo's "probable" recruitment during the Spanish Civil War.

This latter claim in particular is hard to reconcile with Gordievsky's previous account, in which the Soviet agents in Canada were initially suspicious of Pontecorvo's aims. Furthermore, the chronology does not fit well. The Spanish Civil War overlapped with Pontecorvo's time in Paris. During this period he had nothing to offer as a spy: there was no reason to suspect that he would become involved in a programme whose existence had yet to be imagined, and whose core focus (nuclear fission) had yet to be discovered; what's more, once fission had been discovered, research on the subject occurred in the open for some time. It is possible, even likely, that Pontecorvo attracted attention as a prominent communist sympathizer, with a glittering scientific career ahead of him. As such, he might well have been singled out by the Soviets, especially as he joined the Communist Party in 1939. However, this still does not fit easily with Gordievsky's original account of Pontecorvo's approach to the Soviets in Canada.[10]

ON APRIL 12, 1943, "LABORATORY NUMBER TWO" (THE CODE NAME for the Soviet Union's reactor project) was set up on Kurchatov's behalf. He estimated that a reactor would require two tons of uranium and fifteen tons of heavy water (for a heavy-water pile), or fifty to one hundred tons of uranium and one thousand tons of graphite (for a graphite pile). Graphite was easier to obtain than heavy water, so he chose the latter route. He didn't have the uranium but asked theoreticians to figure out how to optimize the construction of the reactor. One of these theoreticians was Isaak Pomeranchuk, a superb theoretical physicist who would later meet with Pontecorvo.

As it would take years to construct a practical reactor, Kurchatov built a cyclotron to make small quantities of plutonium for research as fast as possible. The cyclotron, which worked by accelerating a beam of deuterons (the nuclei of heavy hydrogen), began operation on September 25, 1944. Even with this advanced machine, however, the Soviets didn't succeed in separating plutonium from irradiated uranium until 1946.[11]

Kurchatov's biggest problem was obtaining uranium and ultrapure graphite for the pile. At the very start, in January 1943, the Soviet government asked the US for ten kilograms of uranium metal and one hundred kilograms of uranium compounds under the Lend-Lease arrangement. This was a clever move. The USSR was an ally, and, within the context of the war effort, the request might appear to be entirely innocent. For the US, of course, uranium was highly valued, its importance a jealously guarded secret. General Groves, the head of the Manhattan Project, agreed to the request in order to avoid raising suspicion, unaware that the Soviets already knew about about the West's "secret" atomic project. The search for local uranium in the USSR ramped up starting in 1943, but, given that Stalin was only hedging for the long-term future, it received a low priority. Not until 1945 did large-scale exploration begin.

In July 1943 Kurchatov wrote another report for the Kremlin. Like the March memo, it reveals that his knowledge about nuclear projects in the West was of a general nature and short on detail. Indeed, his knowledge at this time was less complete than it had been in 1941 and 1942, when he had received reports of British progress via Klaus Fuchs.[12] He stressed that he needed information about elements 93 and 94 (neptunium and plutonium), which were being researched by Emilio Segrè and Glenn Seaborg at Berkeley.

In December 1943, Fuchs travelled from England to New York; for most of the following year he worked on the gaseous-diffusion method for separating the rare U-235 from the dominant, and heavier, U-238. During this period he informed the Soviet Union that the United States had chosen gaseous diffusion as the means to separate the isotopes, and that this was occurring on an industrial scale. However, Fuchs knew nothing of the significance of plutonium, nor of the nuclear reactor pile in Chicago.[13] Nonetheless, by the start of 1945 the Soviets had a broad idea of the West's programme and were convinced that an atomic bomb was possible.

The Anglo-Canadian reactor project also was "penetrated by Soviet agents."[14] We have already seen one of these agents at work: during his time in Canada, Alan Nunn May handed over microscopic samples of U-235 and U-233 to the USSR. Soviet records show that Kurchatov was very excited to receive these samples, and issued a high-priority demand for more, ideally several grams' worth.

At some point, blueprints of the heavy-water reactor also made their way to the USSR, but their source is not known.[15] If Pontecorvo was responsible, this would substantiate Gordievsky's claim that knowledgeable KGB officers "rated Pontecorvo's work as an atom spy almost as highly as that of Fuchs."[16] If this description is justified, it raises questions such as: What was transmitted, to whom, and how? Pontecorvo's colleague Herbert Skinner shared his thoughts on the third question with MI5 in 1950,

IMAGES I.1 AND I.2. Two pictures of the senior scientists at Chalk River, with Bruno avoiding the camera. The scientists, from right to left, are Henry Seligman, Bruno Pontecorvo, Bertrand Goldschmidt, Jules Guéron, Hans von Halban, and Pierre Auger. (IMAGE I.1, WITH BRUNO SECOND FROM THE RIGHT, LOOKING AT HIS FEET, COURTESY PAUL BRODA AND ALAN NUNN MAY COLLECTION. IMAGE I.2, WITH BRUNO SECOND FROM THE RIGHT, LOOKING TO THE SIDE, COURTESY AMERICAN INSTITUTE OF PHYSICS.)

remarking on Bruno's "frequent" visits to the US-Canadian border, "where he could have met someone."[17] Furthermore, it is intriguing that when Pontecorvo was photographed with colleagues during this period, he invariably looked away from the camera. (See Images I.1 and I.2 on the preceding page.) Although it doesn't constitute evidence, this behaviour appears to be quite deliberate and evasive, and unique to this period in his life.

After Nunn May was arrested, scientists at Chalk River, including Pontecorvo, were vetted by Western intelligence agencies once more. The fact that Bruno passed this test is not really significant, as the vetting process seems to have been inefficient or even flawed. For example, Nunn May's colleagues knew he was a communist, yet this was missed by MI5.[18] Fuchs also slipped through the net, yet there were apparently many incriminating clues. The authorities' desire to have the best experts on call occasionally caused them to turn a blind eye to politics. Fuchs's case shows that this failing persisted, perhaps unconsciously, even after Nunn May had been exposed. Similarly, the vetting of Pontecorvo also failed to identify his communist affiliation.

Nonetheless, apart from Gordievsky's claims, there is no evidence that Pontecorvo was involved in transmitting blueprints, samples of uranium, or indeed any information from the Anglo-Canadian project. However, it is harder to dismiss the hypothesis that at some stage he was approached by the Soviet Union.

It seems unlikely that the KGB, under active instruction from Kurchatov to recruit agents at North American laboratories, would have been unaware that a member of the Communist Party was working at Chalk River—a leading expert on nuclear reactors and uranium, whose early research in nuclear physics had meshed so well with Kurchatov's own. Indeed, while Pontecorvo was working in Joliot-Curie's laboratory, he and Kurchatov had cited each other's papers. It is improbable that a "penetration by Soviet agents," resulting from Kurchatov's initiative, would not have included some approach to Pontecorvo. What reaction he had is unknown.

THE OTTAWA SPY RING

In June 1943, in response to Kurchatov's intervention with Stalin, the Soviet military intelligence agency, the GRU, sent a new team to Ottawa.

Its head was Colonel Nikolai Zabotin, who ended up in a labour camp following the defection of the team's cipher clerk, Lieutenant Igor Gouzenko. Other members included Lieutenant Pavel Angelov, who made contact with Nunn May, and Colonel Pavel Motinov, who later transported Nunn May's uranium samples to Moscow.

At the time, the head of the KGB was Lavrenti Beria, who also became overlord of the Soviet atomic bomb project. Under Beria's watch, atomic espionage was deemed so important that the KGB recruited its own agents using the Comintern.[19]

The KGB was also responsible for counterintelligence at Soviet embassies. Their operations in Ottawa, overseen by resident agent Vitaly Pavlov, were on a smaller scale than the GRU's, but were important nonetheless. The KGB kept a watch on the GRU. The GRU, for its part, had no access to KGB communications.

Stalin usually required agents of both the GRU and the KGB to supply information independently before he would trust it. For example, information about the bomb work at Los Alamos, sent by Ted Hall, confirmed the information sent by Klaus Fuchs, and vice versa. Stalin "distrusted intelligence" until it was received from "at least two independent sources."[20]

This is significant if one accepts Gordievsky's claims about Pontecorvo. Bruno's cousin Emilio Sereni was well connected in the Comintern, which the Soviets used as a vehicle to recruit contacts. If Pontecorvo was passing information on behalf of the KGB, Gouzenko—a GRU operative—would not have known. And, as we shall see later, the KGB used a uniquely successful courier in Canada whose history had some intriguing overlap with Pontecorvo's own.

The GRU's Colonel Zabotin had not been in Ottawa long before he had earned a reputation as a lover of the high life, an eager party host, ever ready to charm the wives of diplomats. Boisterous and charismatic, he became a great favourite on the diplomatic circuit. As a means of getting to know the inner secrets of a community, this was ideal. This frenetic pace slowed, however, when the wives of the team members arrived from Russia. The presence of the KGB also sobered the GRU team, at least metaphorically.

The Soviets became aware of the Chalk River laboratory but were unable to penetrate its security, and so initially had no real idea of its purpose. Nunn May's evidently communist sympathies were known in Moscow, since the GRU chief in Moscow, Fyodor Kuznetsov, instructed Zabotin to make contact with the scientist. Zabotin assigned Angelov to be Nunn May's contact, and Nunn May was given the code name *Alek*, which was a somewhat transparent cover for his real name, *Alan*.

Following Nunn May's exposure, the Canadian Royal Commission that investigated the matter noted, "In view of his background and the position he occupied, he was a logical person from whom the Russians could expect to obtain the available knowledge on atomic energy."[21] Nunn May himself told his stepson, Paul Broda, that he was approached by the Soviets early in 1945. At that stage, Angelov was completely unaware of the Chalk River laboratory's role in nuclear physics, and erroneously thought it to be a factory for making conventional explosives.[22] Even allowing for the firewalls between the GRU and the KGB, this makes it unlikely that anyone, let alone Pontecorvo, had been passing significant information since 1943.

Over several weekends, Nunn May borrowed reports from the Chalk River library on Friday evenings. He would hand them to Angelov, who would hand them back on Sunday evening, enabling Nunn May to return them to the library on Monday morning. In this way, they worked through much of the "basic material" on the chemistry of uranium and plutonium, the design of the Chicago graphite reactor, and basic nuclear data.[23] Nunn May did not pass any specific details or blueprints of the heavy-water reactor, however. As we saw earlier, Kurchatov had decided to pursue graphite rather than the heavy-water approach. At the time, ZEEP was near completion, while the NRX was still three years away.

Nunn May delivered the fateful samples of U-233 and U-235 to Angelov in the first week of July 1945.[24] The Soviet embassy informed Moscow by telegram: "ALEK [Nunn May] handed over to us . . . 162 micrograms of uranium 233 in the form of oxide in a thin lamina." The news must have caused a sensation in Moscow, because Colonel Motinov—the assistant military attaché in Ottawa—was ordered to bring them from Canada to Moscow personally.

On July 28, 1945, "The Director" in Moscow sent a telegram to Colonel Zabotin in Ottawa: "Try to get from him [ALEK] before [his return to England] detailed information on the progress of the work on uranium. Discuss with him: does he think it expedient for our undertaking to stay

on the spot; will he be able to do that or is it more useful for him and necessary to depart for London?"

In August, Zabotin visited a friend who lived near Chalk River, and
took a cruise along the river so as to take a look at the plant himself. He
reported back to Moscow. His superiors replied on August 14, asking
him if the friend had contacts in the plant. They followed up on August
22 with the instruction, "Take measures to organize acquisition of documentary materials on the atomic bomb! The technical process, drawings,
calculations."

So far, Zabotin had not received any assessment from Moscow about
the value of the information that he had already sent, so he telegraphed his
superiors as follows: "I beg you to inform me to what extent have ALEK's
materials on the question of uranium satisfied you and our scientists (his
reports on production etc.). This is necessary for us to know *in order that
we may be able to set forth a number of tasks on this question to other clients.*"[25] (Italics added.) The GRU clearly had identified Chalk River as a
priority. It is inconceivable that the KGB had not done so also.

Igor Gouzenko's defection from the Ottawa embassy in September
1945 had a huge impact in Moscow. Stalin ordered Lavrenti Beria to initiate damage limitation. Both the GRU and KGB immediately severed contact with their agents, without explanation.[26] Better to preserve agents for
the long term, even if it meant experiencing a slump in intelligence for a
while.

According to Gordievsky, Gouzenko's defection did not compromise
all the Soviet agents in Canada. Nunn May had been working with the
GRU network. This was shut down by the Soviets following his exposure.
A network of KGB agents remained, however, relatively unscathed.[27]
Gordievsky's contacts in the KGB insist that one of these was Bruno
Pontecorvo.[28]

THE UBIQUITOUS KIM PHILBY

Gouzenko's defection, which occurred just three days after the Japanese
signed the documents of surrender, was the moment when the Western
powers knew for sure that their former ally had become an adversary.[29]
The most surprising and important name that Gouzenko revealed was

that of Bruno's colleague, Alan Nunn May. The news emerged around this time that Bruno was weighing offers from Harwell and various American institutions. Whether or not his decision was influenced by these events, there is one common thread between them: the role of the infamous double agent Kim Philby.

At the time, Philby was the head of Soviet counterespionage in London—or at least this was his official job. In reality, Philby was a double agent who worked for the Soviet Union but was paid by the British and would not be exposed until the 1950s.[30] When he learned of Gouzenko's defection and its aftermath, Philby alerted Moscow that Nunn May was under suspicion.[31] We know this because later decrypts[32] of Soviet diplomatic cables show that, on September 17, Pavel Fitin, a high-ranking NKVD official in Moscow, sent a message to the NKVD resident in London, asking for verification of Philby's news about the "GRU in Ottawa."[33]

Nunn May was due to return to the United Kingdom in the fall of 1945, and take up a position at King's College London. In addition to his duties at the university, one of his tasks would be to advise the government on the incipient British nuclear research programme, including the new laboratory at Harwell. This would have placed him in a powerful position, with significant interest for the Soviets. At that stage the British did not have solid enough evidence to prosecute Nunn May, so, following his return to England on September 16, security officers from MI5 followed him, hoping to catch him in the act of passing information to a contact.[34] Philby was aware of this. He succeeded in alerting the Soviets, who cancelled a meeting between Nunn May and his Soviet contact, planned for October 7 in London.[35] This was the first of three crucial interventions that Philby would make with regard to atomic scientists, as we shall see. It wasn't until March 1946 that the British felt confident enough to arrest and prosecute Nunn May.

Following Gouzenko's explosive revelations, Western authorities were worried that the Soviets might have targeted other scientists. Pontecorvo was vetted again, along with others at Chalk River. However, these inquiries produced no evidence that anyone else had collaborated with the Soviets, or shared any secret information with unauthorized people—at least, as far as the Canadian team was concerned. As we have seen, Pontecorvo had been weighing various attractive job offers from American institutions during the latter half of 1945, only to turn them down in favour of Harwell, a fact that Segrè later regarded with suspicion. If Segrè's worries have any basis in fact, then the timing of this choice, which came just as Philby had made the Soviets aware of Nunn May's exposure, is intriguing. And these weren't the only pieces moving on the chessboard of Pontecorvo's destiny at this time.

Klaus Fuchs, for example, was still at Los Alamos, free from suspicion. In September 1945, Fuchs told his Soviet controller that he would return to England after the war. In November he was interviewed for a position at Harwell, and his appointment there was arranged. Fuchs then prepared to leave the United States, which he did in June 1946, eventually starting work at Harwell in August.

In the meantime, with the Nunn May affair known to the authorities, but not yet to the public, Chadwick agreed to hire Pontecorvo at Harwell, with no constraints on his travel. In response, Bruno suddenly announced that he wanted to stay in Canada, at least for a while.[36]

As the facility at Harwell was still under construction, and many decisions about its personnel were being made, it is possible that these intermingled events were nothing more than chronological coincidences. However, if one gives credence to the claims that Pontecorvo had been working for the Soviets since 1943, the timing becomes more tantalizing. In any event, the role of Philby—who first alerted the Soviets about the Western intelligence community's interest in Nunn May, then later tipped them off about Fuchs, and later still, as we shall see, about Pontecorvo—will become central to the whole affair.

LONA COHEN A.K.A. HELEN KROGER

Nunn May had worked exclusively with the GRU. Meanwhile, the KGB had its own ring of spies. At Los Alamos these included the brilliant young prodigy Ted Hall. Hall had entered Harvard at age sixteen, and two years later joined the team at Los Alamos, where he was one of the youngest scientists to work on the bomb. Unknown to his colleagues, discussions with his roommates at Harvard had crystallized Hall's interest in Marxist ideology. He felt that it was essential for the USSR to build a bomb, to prevent a US monopoly.

Hall probably had a deeper knowledge of the bomb's dynamics than Klaus Fuchs.[37] Fuchs is well known because in 1950 he was exposed and then convicted with much publicity. Hall's name is not as well known. He was only identified as a result of VENONA decrypts that mentioned an agent code-named MLAD, Russian for "youngster." The decoded messages also revealed the details of MLAD's travels, which enabled his true identity to be pinpointed in the spring of 1950.[38] Nonetheless his name only became public knowledge in the 1990s.

The reasons Hall was never prosecuted are complex, but a crucial factor was that he refused to confess, unlike Fuchs and Nunn May. Another key to his success was his KGB contact: the American Lona Cohen. Lona Cohen and her husband, Morris, were arguably the most successful of all the couriers and organizers that the Soviets had in North America, and were later celebrated as "heroes of Soviet intelligence."[39]

Born in Connecticut just four months before Bruno in 1913, Lona was born to Jewish immigrants, the fifth of ten children, just as Bruno was number four out of eight. Lona's parents worked in textile mills, whereas Bruno's father owned one. Bruno led a life of relative privilege, leaving at age twenty-three for Paris, where his intellectual curiosity led him to communism. Lona grew up in the hard reality of the Great Depression, and watched her parents scrimp and save to raise the children. Lona left home at age thirteen to find work in New York, where she witnessed social injustice firsthand, became a committed leftist, and in 1935 joined the US Communist Party.

Bruno met Marianne in 1936, and the course of their lives was ultimately determined by his commitment to communism. Lona met Morris in July of 1937, and was later recruited by her husband as an agent for the Soviet Union.[40] She was given the code name *Helen*, which would later have ironic resonance.

When the United States entered the war, Morris was drafted into the US Army, so "Helen" took over his network of seven agents. She was one of the mules of Soviet espionage. Her job was to pick up clandestine copies of documents that had been smuggled out of research centres, carry them secretly across the country, and deliver them to the central controller—in her case, the Soviet resident in New York. This separated the source of information from the central controller, who would know the source's nom de guerre but not always his or her true identity. Lona was so successful that the resident soon assigned her more sources; in the spring of 1945, she was put in contact with the young Los Alamos physicist Ted Hall.

Some of the most important atomic papers received by the Soviets originated with Hall, and were couriered by the woman he knew only as Helen.[41] The fact that Hall remained undetected for so long (and ultimately refused to confess) also protected the identity of the Cohens throughout the war and for nearly two decades thereafter.

Before "Helen" began her work as a courier to and from Los Alamos, she travelled to the Canadian border to meet a contact who handed over papers that had originated in Canada. This occurred in early 1945. The name of her contact has never been revealed. Although it is tempting to speculate that her appointment might have been with Bruno Pontecorvo, especially given Skinner's observation that Bruno travelled to the US-Canadian border regularly, the dates when Pontecorvo is known to have made such trips do not mesh easily with those of Lona Cohen. Pontecorvo visited the United States briefly during the 1944 New Year, and returned to Canada on January 2; however the KGB didn't activate Lona Cohen in Canada until January 11.[42] When interviewed years later, she remembered her first courier visit to the Canadian border as being "sometime in the first chilly months of 1945." Pontecorvo's next recorded trip into the United States occurred in May, which does not fit with Cohen's description, although it's possible her memory of the time frame is unreliable.

By the end of the war, Lona Cohen was deeply involved in smuggling information both from Los Alamos and from Canada. In November 1945, Morris was discharged from the army and immediately signed on again with the KGB. Although Gouzenko's defection in September had directly compromised only the GRU networks, the KGB took care to protect itself by putting its agents into effective hibernation for two years. Thus, the Cohens went dormant, and carried on with their lives like model US citizens.[43] Years later, when a historian of the KGB interviewed Lona and Morris in Moscow, she turned to her husband and said, "Remember the Canadian case? We are connected."[44]

By the end of 1947, Lavrenti Beria was desperate to know about recent Western progress with nuclear weapons. It was around this time that he reactivated the wartime networks. In England, Klaus Fuchs, who had been quiet since 1946, linked up again with a Soviet contact and informed him that Enrico Fermi and Edward Teller, now at the University of Chicago, were interested in creating a thermonuclear "hydrogen" bomb, which used tritium and deuterium. In North America, the Cohens also resurfaced. As it happened, Ted Hall was working at the University of Chicago, and the Cohens, sensing a pathway to news about Fermi and Teller, renewed contact with him.

Ted Hall's wife, Joan, recalls a meeting that she and Ted had with the Cohens in 1949. Morris Cohen had met Ted Hall the previous year, and put great pressure on him to become active again as a spy.[45] Cohen was successful, but within a year Hall wanted to disconnect once more, which led to the meeting of Joan and Ted Hall with Lona and Morris Cohen in a New York park.[46] In Joan Hall's opinion, the Cohens were trying hard to renew their old contacts, in part to prove to Moscow that they, Lona and Morris, were still a powerful force.

The Cohens were able to renew their link with Hall, at least in 1948 and 1949. They also succeeded with an agent based in Canada.

Lona Cohen's official KGB biography reveals that, around this time, she "obtained a sample of uranium from Canada," which she transported to her Soviet contact in New York.[47] V. N. Krasnikov, the KGB's resident deputy in New York, recalled that in the later part of the 1940s he was not only interested in the H-bomb, but also in uranium ores: "We contacted people on this issue. In my safebox in New York there was an envelope with powdered uranium." He added, "The only uranium known to have come to New York is credited by Russian intelligence to Lona Cohen."[48]

Lona Cohen's uranium sample was not the same as Nunn May's, which had been handed over in the summer of 1945 and taken under close guard from Ottawa to Moscow. After Nunn May's uranium arrived in Moscow, Igor Kurchatov asked for more samples. Cohen's uranium was then obtained "on orders of the Centre."[49] Combined with Krasnikov's account given above, this suggests that Cohen's uranium was obtained later than 1945. Furthermore, uranium from the Canadian reactor was only available after 1947, and the Cohens had gone underground from late 1945 until the start of 1948.

By 1948, the NRX reactor at Chalk River was working. It produced both plutonium and U-233. Joan Hall's impression that the Cohens were under pressure to reactivate their networks at this time would suggest that, in addition to approaching Ted Hall, they also contacted their former source or sources in Canada. This period of the late 1940s is also when the Soviets began to show interest in developing a hydrogen bomb; this requires tritium, which can be produced in a heavy-water reactor. Hence the Soviets would have had a real interest in obtaining the blueprints of the Canadian reactor, as well as uranium samples, at this stage.

Lona's source in Canada has never been identified. If Gordievsky is wrong about Pontecorvo being the source, some other individual has managed to keep their secret ever since.

HALF TIME

TEN

CHAIN REACTION
1949–1950

IN 1949, BRUNO'S ODYSSEY BROUGHT HIM TO HARWELL, IN THE HEART of England, to work on the British nuclear reactor programme. He also devoted some of his time to fundamental work studying cosmic rays. However, his other love, the neutrino, would remain tantalizingly out of reach—the reactors being designed at Harwell would be unable to generate enough neutrinos for his needs.

Harwell village, with its half-timbered, thatched cottages, could serve as a model for a classic English picture postcard. The Downs cross the landscape about ten miles to the south. For most of the 1930s this area was a peaceful idyll, which was shattered in 1939 when an airfield was built to house Wellington bombers. After the end of the war, in 1946, this site was taken over by "the Atomic," as Harwell Laboratory was affectionately known.

The Atomic had an air of mystery, even of menace. Tall chimneys, tower blocks, and offices of red brick, prettified with sash windows, gave the place the appearance of an industrial site, which in effect it was. The runways of the former airfield became roads, and the vast areas of concrete where the planes had once taxied became bus terminals: large numbers of coaches were needed to ferry workers between the laboratory and the surrounding villages, which were several miles away. Behind the security fence, which was manned by armed police officers, the abandoned aircraft hangars became laboratories. Those allowed inside would discover the postwar state of the art in big science. The walls within the hangars housed three stories of offices and labs, which were reached by metal staircases. From walkways, high above the hangar floor, you could look down on piles of graphite and concrete blocks, which eventually would house a nuclear reactor.

The scientists wore suits and ties. Many completed the ensemble with a waistcoat, within which a pencil or fountain pen would nestle, ready to record data. They smoked pipes as they watched lights flicker on monitors, recorded the readings from dials, or adjusted Bakelite knobs on the electronics. That was how science was done in the Britain of the 1940s.

Seen from sixty years later, there is inevitability to Bruno Pontecorvo's disappearance, a domino effect as world events cascaded through the first half of 1950 toward a terrible personal climax.

Bruno had, after all, always felt pursued: by the fascists in Italy, by the Nazis in France, by the FBI in America; he later sensed a general atmosphere of suspicion in Canada. He had only been at Harwell a year when, in February 1950, his colleague, Klaus Fuchs, was arrested. Pontecorvo and Fuchs were independent colleagues, but shared common ideals, and the effect on Bruno can't have been calming.[1]

When tectonic plates shift on the ocean floor, waves spread imperceptibly until, perhaps thousands of miles away, a tsunami reaches land and wreaks havoc. For Pontecorvo, the disaster struck in September 1950. Unseen waves had been building for months, but their source, the metaphorical quake that set the whole saga in motion, took place back in 1949, in the United States. It was Bruno's old friend, Emilio Segrè, who set the fateful events in motion.

In the US, Senator Joseph McCarthy's witch hunts and the persecution of "Reds" initially began as a Kafkaesque loyalty test: to prove your loyalty, give us the names of people known to you to be communists, or to have professed views that are communist, or left of centre, or. . . . And names would be duly provided. Anyone who refused to take the test was immediately damned. It is a marked irony that the US was adopting practices that were de rigeur in the USSR, the very regime that was being painted as the new Satan.

Which is how Emilio Segrè was cast as Judas Iscariot.

Segrè, like Pontecorvo, was one of the Via Panisperna Boys. During the war, as we saw, their research was appropriated by the military. Now, postwar, the US government's newly formed Atomic Energy Commission (AEC) was taking over this area. The slow-neutron method, developed and patented by the Via Panisperna Boys, was pivotal to their vision. Segrè was financially sophisticated and eager for the monetary rewards that this patent promised. Negotiations between the AEC and the patent holders were dragging on, however, and the US government was making strategic investigations into the political background of the "Boys," looking for dirt to bolster its case against them.

Segrè felt compromised. He was now working at the University of California, which became infamous for its enthusiastic application of the loyalty test to its employees. Segrè knew Pontecorvo's family well. He was aware that Bruno's siblings included professed communists, and that his cousin Emilio Sereni was a communist member of the Italian government. Segrè feared that his former association with Pontecorvo could call his own loyalty into question. So on November 9, 1949, Segrè met with Robert Thornton, an old friend from Berkeley who was now an official at the AEC, and told him about Bruno's communist associations. Thornton passed this information on to the FBI.[2]

The FBI checked their files and, realizing that Pontecorvo was working at Harwell in the UK, duly informed the British security services. The FBI also gave MI5 a second piece of information that corroborated what Segrè had said. Sent on December 15, the note "regarding possible communist or pro-communist tendencies of two nuclear physicists and a biologist,"[3] keeps the names of its sources secret, but the content shows that the informant is not Segrè.[4] The note reads, in part:

Informant A of proven reliability on communist matters vouched for the reliability of informant B who said he was acquainted with three individuals in Paris under Prof Langevin who were exposed to the virus of communism. These were Bruno Pontecorvo, Sergio de Benedetto and Salvatore Luria. [B] said that he later met Bruno Pontecorvo on the ship Quanza and through him met Benedetto and Luria in New York City socially. [B] reported that he gained the impression in Paris that these three were either pro-communist or outright communists. He said that conversations that he had with them in the USA tended to strengthen his belief. The last time he saw them was in 1944. [B] reported that all three were friendly with Mr Sereni, a communist.

MI5 took note. Someone highlighted the above paragraph in Pontecorvo's file.[5]

The testimony continued: "[B] reported Ambroglio and Berti had returned to Italy and Sereni had not been in the USA. [B] reported that Pontecorvo at one time was definitely aligned with the communists. In addition he reported that his brother-in-law (Ducio Tabet) was also pro-communist and had returned to Italy." The FBI followed up on Bruno's family and discovered that Bruno's sister Giuliana, who had also come to their attention as a possible communist, lived a couple of doors away from an alleged member of Comintern.[6]

The emerging picture of Bruno Pontecorvo at the FBI was of a nuclear scientist who was a communist, in an extended family of communists whose members were in all likelihood well connected with Comintern. If Segrè had hoped to smooth his own vetting process, and his claim on the patent, the attempt had backfired. Segrè and "Informant B" merely fed the anticommunist hysteria already rampant within the Atomic Energy Commission and the FBI. Segrè's attempt to provide security for himself, by naming Pontecorvo, instead encouraged the AEC to dig deeper into Bruno's activities. During 1949 and early 1950, the FBI doggedly investigated Pontecorvo's politics.

In the United Kingdom, by contrast, MI5 did nothing. Their lack of action at this time is possibly due to the fact that when the news of Pontecorvo's communist associations arrived in 1949, the Harwell security team and MI5 were heavily occupied with the case of Klaus Fuchs. Fuchs was interrogated on December 21, again on the thirtieth, and for a third time on January 13.[7] By the end of January, the investigation of Fuchs was complete, and he was arrested on Friday, February 3, 1950.

The following week, starting on the tenth, Harwell hosted a four-day conference to decide which items in the field of nuclear physics should remain secret and which could be declassified now that the war was over. It was during this gathering that news of America's growing concerns about Bruno reached Harwell and sealed his fate.

On the Saturday evening there was a reception at Ridgeway House—an austere, government-standard brick building, which was used as a hostel for visitors, and doubled as an event centre thanks to its large hall. Bruno had been invited to the reception by Sir John Cockcroft, "to meet the US and Canadian Delegates."[8] The American delegation from the AEC included Robert Thornton. Although on the surface the occasion was gregarious, with colleagues recalling their times together during the war, the undercurrent must have been cool. Thornton had already transmitted Segrè's information about Bruno to the FBI. Now, at some point during the conference, he told Cockcroft that Pontecorvo and his family were communists.[9]

This news worried Cockcroft. He had known Bruno as both a colleague and a friend during their time in Canada, and had expressed concerns about his background in 1946 when his transfer to Harwell was first mooted. However, there had been no direct linkage of Pontecorvo to communism at that stage—the FBI's 1942 message about the communist literature in Pontecorvo's house having gone astray. Now Cockcroft passed Thornton's news to his security chief, Wing Commander Henry Arnold, and instructed him to get to the bottom of it.[10]

Arnold, slightly built and suave, was a model "English gentleman."[11] Lorna Arnold, the distinguished historian of the British hydrogen bomb (and no relation to Henry), described him as the "perfect intelligence officer." At the end of the 1950s, Henry and Lorna worked in the same branch of the UK Atomic Energy Authority, and the internal mail service kept getting their correspondence mixed up. As a result, Lorna recalled, "Henry was always coming to my office." She remembered him as "very kindly, pleasant and courteous; average in height and build." He was the perfect intelligence officer, she explained, because "his unobtrusive build and his personality helped him merge into the background. He was very observant and very tuned into people." He was also very good at understanding the psychology of suspects, winning their confidence and then destabilizing them.[12]

Arnold's style was "softly, softly." He maintained discretion by meeting with Pontecorvo in connection with his regular work. At the time, Pontecorvo's research was focused on cosmic rays, and he was hoping to go to the Swiss Alps to work on an experiment being conducted on the Jungfraujoch. Trips abroad required approval from the security division, so Arnold used this formality to make seemingly innocent conversation about Pontecorvo's trip. Arnold asked whether he would be visiting his parents, who lived in Milan, or his siblings elsewhere in Italy. Bruno confirmed that he had last seen his brother Gillo at an international conference in Lake Como in 1949, and he expected to meet him during the trip. During this discussion, Arnold wheedled out significant information, as he reported on March 1 in a phone call to MI5: "Pontecorvo disclosed that his brother, who is also a scientist with an international background, was an active communist."[13]

This set alarm bells ringing, if only quietly at first. Arnold went on to report that Bruno had recently been offered a job at the University of Liverpool. An easy way to eliminate any possible security risk would be to transfer Bruno Pontecorvo there, away from the classified work at Harwell. Arnold promised to send MI5 the latest facts about the Pontecorvo family, "so that [MI5] could offer advice in due course."

More information about Pontecorvo's family arrived at MI5 the next day. This news came not from Arnold but from an unexpected quarter. The Special Branch of the British police, which focused on security issues, reported that a source in Sweden had unequivocally asserted that Pontecorvo's wife, Marianne, was a communist: "In Enskede, a suburb of Stockholm, there lives a Fru PONTE CORVO a Swedish national married to Italian born subject whose present nationality is unknown. Mr PONTE CORVO lives in England where he works in one of the British Atom Installations. Fru PONTE CORVO was in England in summer 49 with her two [sic] sons. Both she and her husband are described as avowed communists."

This last remark has been highlighted with two solid lines in the margin. The letter commented, "This report would appear to refer to Bruno Pontecorvo employed in [Harwell]."[14]

Although the broad details are right, the letter is a classic case of the game of "Chinese whispers." Somehow the Swedish authorities thought Bruno had six sisters and one brother, rather than three and four, and that he had two rather than three children. More significant is that in reality "Fru PONTE CORVO" lived in England and had only been visiting her parents in Sweden. When such errors are discovered, the recipient has to decide whether they are trivial flaws in transmission, or instead constitute evidence that the subject of the report has been misidentified. A few weeks later these errors were cleared up, with an apology from Special Branch. In addition, perhaps to save face, the writer insisted that their source "re-emphasises the authority of the informant."[15]

In March, Arnold must have been busy with other matters as he sent MI5 no further information about Pontecorvo. In the meantime the directors and their deputies in the MI5 divisions of Counter-espionage (section B), Examination of Credentials (section C), and Security (section D) were exchanging minutes and papers about Pontecorvo. They had to assess the level of threat, and then decide what to do about it. On March 20 Colonel John Collard of section C judged that the credibility of the Swedish source needed to be determined. Collard's opposite number in Counter-espionage, J. Robertson, decided that further investigation should be undertaken by his own branch. Robertson's colleague W. S. Mars, concerned about the lack of progress, also added a note to the file on March 20: "Is any news expected from Arnold soon?" On March 26 Collard spoke by phone with Arnold, who "undertook to see Pontecorvo again at a suitable opportunity."

The language of these exchanges gives a mixed impression of the government's sense of concern and urgency. On one hand, the information that trickled in from sources such as Segrè, the FBI, and the Swedish informants—sources whose identity and reliability were unknown except to a handful—caused a great deal of activity in the London headquarters of MI5. On the other hand, at Harwell, where Pontecorvo was based, Henry Arnold continued his soft touch with little sign of rapid action.

During the March 26 phone call, Collard told Arnold that MI5 was concerned about Pontecorvo's reliability, and urged Arnold to have a more formal meeting with him. The "suitable opportunity," which Collard suggested Arnold look for, arose about a week later. Whether or not Bruno suspected anything previously, this time there was no doubt that the authorities were seriously interested in him. On Wednesday, April 5, Arnold had a "long talk with Pontecorvo." His ability to wheedle information out of people is evident in his report, which states that Pontecorvo once again yielded up "quite freely" the information that his "brother Gilberto [Gillo] is an openly confessed Communist." Furthermore, Arnold stated, "I understand that Gilberto's wife who is French is also of the Brethren."

Arnold must have asked Pontecorvo directly about his personal beliefs, as his report states:

Pontecorvo emphasized that he himself is not a communist and that his politics, if any, are Labour. His views are definitely "Left" but I found it difficult to assess exactly how "Left" they were. His wife Helene [Marianne] . . . holds very much the same political opinions as her husband, but here again Pontecorvo insisted that neither he nor his wife were communist. [Pontecorvo] expressed the opinion that with such a "Red" family background the Security authorities must naturally entertain certain doubts about himself, but he pointed out that on the continent the proportion of communists is far greater than in this country in that it is quite difficult to find an educated family that has not got some communist connections.[16]

Later, after the Pontecorvo family disappeared, MI5 homed in on the fact that Bruno had volunteered information about his communist relatives while at the same time denying that he himself held such views. MI5 viewed Bruno's move as a proactive attempt to defuse any suspicion that might have arisen if the authorities had discovered the existence of these communist relatives for themselves. This interpretation seems likely, as today we know that Bruno's statement was false. As we have seen, he had been an enthusiastic communist ever since his early days in Paris, and at the time of the Soviet-German nonaggression pact in 1939 had become an active member of the French Communist Party. Indeed, it was none other than Bruno who had introduced his siblings Giuliana, Laura, and Gillo to communism.

Arnold's perceptive insight is on display in his assessment of Pontecorvo's likely actions, written for MI5 in April 1950: "My personal view of Pontecorvo is that whereas he has obtained British nationality he would quite readily change it again should it be to his scientific advantage to do so." Then, with wonderful irony, he adds, "Naturally I do not include countries which are under Russian domination [here, someone has penned an exclamation mark] but he has already toyed with the idea of an appointment in Rome University and is at present turning over in his mind an offer which has come to him from America."[17]

Arnold's assessment: "From a security point of view it is difficult to regard a person with Pontecorvo's international outlook and family history as reliable and I feel it would be a good thing if he were able to obtain a post at one of the British universities where we might continue to avail ourselves of his undoubted ability as a Consultant in limited fields."[18]

Colonel John Collard, the case officer at MI5, assessed Arnold's report and reviewed Bruno Pontecorvo's file. Collard identified three new developments since Pontecorvo's appointment. First: Bruno's admission that he had communist relatives abroad, and the "unsubstantiated" Swedish report that Bruno and Marianne were themselves communists. When Bruno's suitability for employment at Harwell had first been assessed in 1946, Arnold had given MI5 the opinion that Pontecorvo was a "straightforward fellow with no political leanings." Collard wryly noted the contrast with Arnold's 1950 report.

Second: Pontecorvo's reference to Rome and the United States showed him to be willing to consider work outside Harwell, even outside the UK "despite [his] recent naturalization." Between the lines, Collard gives a hint of being let down in his memo, disappointed that Pontecorvo is not really a team player, or at least is not committed to the British club.

Third: Harwell no longer regarded Pontecorvo as "indispensable."

Based on this trio of new insights, Collard concluded that the security services should now reassess Pontecorvo's case, after which they should discuss possible actions with Harwell.

By the end of April, officials at the highest levels of MI5 were increasingly worried. On the twenty-seventh they received a clarification about the earlier Swedish report. Mrs. Pontecorvo, it was now agreed, lived in England and had been visiting her mother in Sweden. At least these facts were now correct. This latest missive stressed that "Mrs Pontecorvo is known personally to our source" who reiterated that "among friends she openly expresses communist sympathies." Roger Hollis, the future director general of MI5, saw this note and wrote on the comment page, "Please discuss this case. Surely we should see if he can be moved." Four days later, on May 10, Hollis recorded that he had "discussed with [Director of Counter-espionage] and we agree there is a security risk here." Hollis concluded that Pontecorvo's case was "not up to purge standards, but nevertheless we cannot feel happy that he should remain where he is."[19]

The next day, Martin Furnival Jones, another rising star of the security services who became the director general of MI5 after Hollis, discussed this with the senior echelons of the security services' legal branch. They

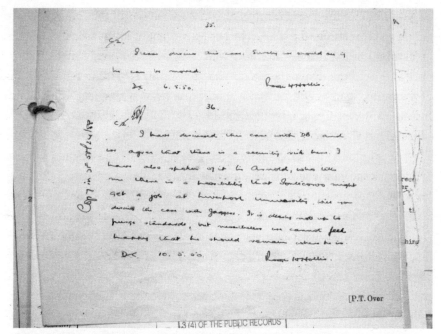

IMAGE 10.1. Written comments by Roger Hollis in Bruno Pontecorvo's security file, May 1950. Hollis agrees that there is a security issue but that it is "not up to purge standards." He adds that "nevertheless we cannot feel happy that he should remain where he is" and welcomes the possibility of a move to the University of Liverpool. (AUTHOR, THE NATIONAL ARCHIVES.)

were concerned that the case, which was already a headache, could create even more problems if Pontecorvo—a probable communist sympathizer who had somehow slipped through the net—accepted a job in the United States. They proposed that a note be sent to the home secretary, and that Cockcroft transfer Pontecorvo within the UK. Arrangements were thus made to transfer Bruno from Harwell to the University of Liverpool, away from secret work. Although Bruno was not aware of these machinations, Arnold's interviews, supplemented by more general inquiries from laboratory managers in the wake of the Fuchs affair, certainly made him feel persecuted.[20]

AT THE END OF APRIL, BRUNO VISITED PARIS TO JOIN THE CELEBRATIONS for Frédéric Joliot-Curie's fiftieth birthday. What transpired would add to his sense of persecution.

By 1950, Joliot-Curie was the high commissioner of the French atomic energy commission, the CEA. The commission's goal was to build nuclear reactors. Joliot-Curie wanted to produce abundant power for the nation but abhorred the unavoidable consequence—the production of plutonium for bombs. As a result he was a vociferous campaigner against nuclear weapons. Joliot-Curie was in Stockholm on his birthday, March 19, for a meeting of the World Committee of Partisans for Peace, where he signed the Stockholm Appeal, calling for an international ban on nuclear weapons. Millions around the world subsequently signed this document. The American media duly attacked him, and lumped Joliot-Curie's name together with that of the recently arrested Klaus Fuchs. The CIA listed Joliot-Curie and half of the CEA's scientific committee as "communist sympathizers."

As Joliot-Curie was away from Paris on his birthday, the celebration was postponed. His closest colleagues at the CEA organized an intimate and sumptuous banquet at the Popote des Ailes restaurant in Viroflay, near Versailles, for the evening of April 26. Bruno was one of a small number of associates invited to the celebration. In the meantime, many critical events would transpire.

On April 5, Joliot-Curie appeared at the National Congress of the French Communist Party, where, to loud cheers, he declared that "progressive and communist scientists shall not give a jot of their science to make war against the USSR."[21] He added that if the French government asked him to make nuclear weapons, he would refuse. On April 26, on the afternoon before the banquet, he was summoned to see the French prime minister, Georges Bidault. Bidault, who had been a comrade of Joliot-Curie's in the Resistance, was distraught as he fired Joliot-Curie from his position as head of the CEA.

The news was not announced at the start of the banquet. The food and wine were superb, the atmosphere lighthearted—until Joliot-Curie rose to speak. Then everything came out. He rambled on "for hours, it seemed he could not stop himself."[22] Joliot-Curie was a broken man. In witnessing his former colleague's downfall, Bruno Pontecorvo, already under intense pressure from Arnold and MI5, began to fear that the tentacles of McCarthyism had now reached Europe.

BACK IN ENGLAND, MI5 WOULD SOON RECEIVE MORE GOSSIP ABOUT Pontecorvo. In the first week of June, Superintendent Evan Jones of

Special Branch received a letter from someone whose name is redacted, but who is clearly a current or former member of Special Branch as he mentions having enjoyed the Branch Dinner. There is a certain quaint charm in the letter's vague allegations, which, though wrong in several details, are nonetheless fundamentally accurate:

> Dear Evan,
> An old informant, reliable, told me a few days ago that an Italian scientist is employed in, or has some close connection with, a hush-hush factory or laboratory near Bristol. Although he does not profess it, he is a communist, acquainted with Fuchs and when in Town frequents the same restaurant as Fuchs did. My friend and others suspect him. Not much to go on and it might be a wild goose chase but I am sure my friend would not have mentioned it to me unless there was something in it. . . . I have no knowledge of any factory etc near Bristol and so cannot assess the truth or value of this. It is likely the Italian scientist is British born or is naturalized as I cannot conceive an alien being given access to anything in the nature of a research station, especially after the Fuchs affair.

Harwell is in the country, midway between London and Bristol, which suggests that the informant was not strong on geography. The phrase "in Town" was the colloquial way of referring to London in the 1950s, but "frequenting the same restaurant as Fuchs" seems fanciful and irrelevant, as Fuchs was never known to have used a restaurant as a meeting place.

As in the case of the Swedish informant, there were hints that the information in this letter might be dubious, but in any event it was duly noted by MI5 on June 9, and added to the growing file of anecdotal evidence about Pontecorvo. Michael Perrin, the former ICI administrator who by 1950 had become the director of atomic energy for the Ministry of Supply, wanted to assess the strength of these various claims, which clearly shared a common theme. He phoned MI5 to inquire about the reliability of the source from Sweden, whose evidence was the most explicit. Internal inquiries were then made within MI5. The record of these inquiries, even today, is heavily redacted, but reveals that an "R Badham" of section C "spoke to [X] on the phone today, and asked if he could let me know the reliability of their 'source'. He phoned me back later to say that the source was 'absolutely reliable.'"[23]

Badham then phoned Perrin and told him that he could confirm the source to be "absolutely reliable." Perrin, wanting to know more before passing judgment, then started asking questions: Did the source know Pontecorvo personally? Would MI5 guarantee the source as unimpeachable? Badham replied that he had told Perrin as much as he knew. Perrin, frustrated, decided that in order to form a solid assessment he had to get past these internal firewalls and speak directly with Roger Hollis. However, upon further reflection, he backed off and decided that this was not necessary "at this stage."[24]

And it would appear either that Perrin never took this any further with Hollis, or that Hollis himself took no action.[25] The next entry in the MI5 files is a letter from the British embassy in Washington, DC, received on July 19; after that, it would seem, everyone took off for the summer. Nothing more was added to the Pontecorvo files until October, when news of his disappearance broke.[26] From that point forward, as we shall see, the files are full of attempts to shut the barn door after the horse has bolted, along with discussions of how to minimize the political damage. It is clear that MI5 and practically the entire British security apparatus were taken unawares. It is my judgment that, in the summer of 1950, not even Bruno Pontecorvo anticipated that he was about to leave England forever.

ELEVEN

FROM ABINGDON— TO WHERE?

1950

HARWELL IS A COUNTRY VILLAGE, FIFTEEN MILES SOUTH OF OXFORD, surrounded by farmland, orchards, and racehorse stables. In the 1940s it was a rural backwater.[1] The location suited the government's purpose: it was accessible to London and Oxford University, yet sufficiently isolated that secrecy could be maintained. Except for the small village of Harwell, from which the laboratory took its name, the nearest towns formed a ring about five miles distant. Abingdon, a small market town to the north, midway between the laboratory and Oxford, was a favourite place to live among employees.

Abingdon is located on the River Thames, about fifty miles inland from London; in Tudor times King Henry VIII had briefly lived there to escape the plague. Among the town's several historic buildings was a boys' high school, Roysse's, that had been founded in the twelfth century and whose students ranged in age from eleven to eighteen. Roysse's rose to prominence following the arrival of James Cobban as headmaster in 1947. Although he had missed out on D-Day due to appendicitis, Cobban had a distinguished war record and had risen to the rank of lieutenant colonel due to his work in military intelligence.

When Cobban arrived, the school was on the periphery of the town, which was about to be revolutionized by the work at Harwell. The arrival of hundreds of scientists and their families led to a sudden expansion. Abingdon was being transformed into a company town for those associated with "the Atomic."

To accommodate the influx, an estate of red brick houses was built across the road from the school. The Pontecorvos moved into one of these houses: number 5 Letcombe Avenue, just a five-minute walk from the local tennis club and immediately adjacent to Roysse's. In 1949, many sons of Harwell employees enrolled at the school, Gil Pontecorvo among them.

For Gil, now eleven years old, the first year at Roysse's was difficult. Having made friends in the school at Chalk River, he had been moved away to a different continent and had to start afresh. In England some students thought he was American, and he remembers them teasing him: "You Americans! You want to drop bombs on us." He also remembers a talk about Moscow he heard at one of the school clubs, which was like "a story from the Arabian nights, with merchants sitting cross-legged in the street selling their goods, the difference being that instead of streets paved with gold, Moscow was dirty and cold."[2] The image of Moscow that this left in his mind was of a Wild West frontier town, lawless, in the badlands. Although this image is extreme relative to my own memories from that era, it was indeed the case in the West that Moscow, and all the lands beyond the Iron Curtain, seemed like some black night, filled with unknown spectres: terra (and terror) incognita.

On September 19, 1949, Gil joined the youngest class of students, along with a boy named Anthony Gardner.[3] Anthony and his younger brother Paul, who lived two doors down from the Pontecorvos, at number 9, became good friends with Gil and his brothers. Their father, John, was an administrator at Harwell, and a very good tennis player, who had formed an immediate kinship with Bruno. Around the corner lived Egon Bretscher, who had come to Harwell from Los Alamos. His son, Mark, who was younger than Gil, was due to start at Roysse's at the beginning of the following school year, in September 1950. Gil and Mark were not close friends but the families knew each other. In the summer of 1950, Gil, who had already had a year's experience at Roysse's by then, agreed to accompany Mark on his first day of school, to introduce him to the strange surroundings.

Today Mark is a distinguished microbiologist, Anthony Gardner is a successful actor, and his brother Paul, now retired, is a long-term Abingdon resident. In interviews with me, all three recalled their memories of Abingdon, in that tense period just after the war when provisions were rationed; families kept chickens, ducks or even goats for eggs and milk; and young boys could cycle in the surrounding countryside on roads almost free of cars.

IN JANUARY 1950, WHEN FUCHS WAS ALREADY UNDER SUSPICION BUT the security services had yet to establish if others at Harwell were involved, an MI5 agent arrived at Roysse's in the guise of a music teacher named Royd "Doggie" Barker.[4] At least, that's what several former pupils believe. This fantasy—if fantasy it is—would hardly have been dampened by the events of that year, starting in February with the arrest of Klaus Fuchs, who for a few months had lived on the school premises.

Immediately adjacent to Roysse's campus was an old house called Lacies Court, which had been donated to the school, who used it as a residence for unmarried teachers.[5] During part of his time at Harwell, Fuchs too had lodged there, though he had no formal association with the school itself. His time at Lacies Court was a result of the school's links with Harwell, which helped a lone scientist find temporary lodgings. However, his arrest in February 1950 created excitement among the boys, and led to the whispers about the music teacher.

Such fantasies about the cloak-and-dagger world of spies were standard stuff for teenage boys in the 1950s, with the Cold War raging. For pupils at Roysse's, who within the space of a few months would twice find themselves associated with intrigue (first with Fuchs, then Pontecorvo), such rumours could easily grow from very little. However, none of the former students I interviewed could recall why Barker had been singled out in this way.

In any event the boys were remarkably prescient, as Barker subsequently had a stellar career in MI5, rising to become the director of section A, the branch "responsible for managing the Service's operational capabilities such as its technical and surveillance operations," also known as bugging.[6] When he retired in 1984, Barker was honoured by Margaret Thatcher for "conspicuous service to the Crown."[7]

Royd Barker had a peculiar loping stride, which made him rather noticeable. This is not helpful for an intelligence officer who wants to merge into the background, but has advantages for one who wishes to be welcomed and absorbed by a community. He was an extrovert, but also a disciplinarian. Some boys described him as a popular teacher, while others recalled that he always carried a stick, which he used to keep order.[8]

The arrival of large numbers of clever, talented, highly educated people from around the world injected large amounts of creative energy into the sleepy country town of Abingdon. Theatre and music societies thrived. Barker, the musician, took to these like a fish to water. A musician and schoolteacher in that singular community, where he could meet the sons of scientists during his day job and mingle with Harwell employees in choral societies during his leisure time, would be well placed to hear local gossip about "the Atomic."

As we saw earlier, Roysse's headmaster, James Cobban, had joined the school in 1947 after a distinguished war career in the intelligence services. It is not known if Barker was already a spook, and was placed on the school staff as part of the national security effort, or instead was recognized as a potential recruit for the security services by Cobban. In those days, recruitment to MI5 still happened through the "old boy network," so Cobban could have acted as a scout who recognized talent in one of his employees and recommended Barker to MI5.[9]

One friend who knew Barker well for many years told me, "I wondered how he was recruited and thought it was [while a student] at Oxford."[10] If this is the case, it would seem that Barker's teaching provided a cover for his MI5 role.[11] Indeed, we now know that this would have been consistent with MI5 policy. In 1946, following Nunn May's exposure, Guy Liddell had recommended that, for security purposes, it was "most important" that MI5 "should place informants" in the atomic energy laboratories, whose charge would be to know about "the general mode of living and political views of young scientists."[12]

Further adding to the intrigue, in 1952 Barker requested, and was granted, leave from the school for several weeks "in order to go off on a musical tour in Yugoslavia." There is no record of any musical contacts between Abingdon and communist Yugoslavia in 1952. And indeed, former pupils insist that Barker's link to MI5 was common knowledge at the time and not a later creation.[13]

If this is correct, Barker's presence at Roysse's in 1950 is intriguing. As events would turn out, however, he seems to have had no forewarning of the Pontecorvo scandal, which was about to hit the community.

THE GARDNER BROTHERS REMEMBER BRUNO AS A CHARMER, AN ADULT who had retained the gaiety of childhood and could relate to them. Letcombe Avenue sloped gently from one end to the other, and Anthony later recalled "Bruno riding his bicycle up the rise—backwards."[14] This was a favourite among Bruno's many party tricks, and gave credence to his boast: "I could have been in a circus."[15]

The Gardners often went to Gil's home for tea. In those days, houses tended to remain unlocked and the boys went in and out regularly. They remembered Marianne as "quiet, nice" and "elfin-like with a sweet face," always happy to provide teas or, on the occasions when the two families met for dinner, plates of spaghetti—an exotic dish in the middle England of the 1940s.[16]

Another friend, David Lees, recalled that he was often invited to the Pontecorvos' home on Sundays. He remembered that the house was untidy, with children's toys scattered around, and that Bruno would parade through the house in his pajamas, while speaking a strange language. David was informed that this was Italian, "the language we speak at home on Sundays."[17]

Bruno played tennis at the local club in Albert Park, where grass courts nestled among the trees and bushes. Large Edwardian houses fronted the park on three sides; the school completed the square. The entire area embodied the fantasy image of England as propagated in Agatha Christie novels. The grass courts didn't open for the season until late April or May, so it must have been in the weeks immediately prior to the Pontecorvos' departure that the following event took place.

In the early summer of 1950, Bruno and John Gardner were in the middle of a tennis game when Bruno suddenly stopped. He had noticed someone standing among the trees. "I have to go and speak to that man,"

Bruno exclaimed, and immediately left the court. A few minutes later, he returned and apologized, but gave no further explanation and carried on with the game. Anthony Gardner's impression is that his father later mentioned this "because it was so odd."[18]

Clearly the stranger was not the Harwell security officer, Henry Arnold. There would be no need for Arnold to make such a secretive approach, and in any case Gardner would have recognized him. It might have been an innocent encounter, elevated in Gardner's mind after Bruno's disappearance, though why Bruno declined to provide an explanation is harder to understand. It's also possible that the incident might have been embellished in the retelling.

BRUNO'S YOUNGEST BROTHER, GIOVANNI, LIVED IN AMERSHAM, ABOUT thirty miles from Harwell, and they met in Amersham or Abingdon about once every three weeks. After Bruno disappeared, MI5 would become suspicious about these visits to Amersham: "In the event of him being contacted by an agent of a foreign power it is unlikely the contact would be near Harwell. It is therefore quite probable that some contact was made with Pontecorvo in this district."[19]

On one occasion, Giovanni and his fiancée went to see a play in London with Bruno. Bruno, who was in one of his usual extroverted moods, had acquired a jar of bubble solution and he proceeded to blow bubbles while they waited in line at the theatre. Not everyone was amused, as Giovanni's fiancée later described Bruno to MI5 as a "queer fellow" who was "very childish."[20]

Giovanni ran a poultry farm, and money was tight. In June 1950 he asked Bruno for a loan of thirty pounds. Bruno didn't have the money on hand, but promised to stand as a guarantor for that amount. The brothers met for the last time at Bruno's house at 5 Letcombe Avenue on July 7. On that occasion they discussed Bruno's upcoming camping trip to France, Switzerland, and Italy, and Bruno promised Giovanni that on his return to England they would have a long talk about his poultry business. Bruno explained that he anticipated getting a significant amount of money from the neutron patents, which were currently being argued over by lawyers in the United States. "When my claim is settled we wont [sic] have to worry," Bruno assured his younger brother. "I will set you up in business."[21]

In June Bruno and Marianne visited Liverpool to see the university and decide where to live after his transfer, which was planned for January 1951.

The head of the physics department at Liverpool, Professor Herbert Skinner, knew the Pontecorvos well, from both Canada and Harwell. He was enthusiastic that Bruno might join his department. He recalled that Bruno made an "obvious effort to sell the job to Marianne," who disliked the city because it was "cold." Skinner felt sure that Bruno's acceptance of the post was genuine.[22]

The University of Liverpool's physics department was building the largest accelerator in Britain, or indeed Europe, at the time. Those of Bruno's colleagues who knew him best agreed that the scientific possibilities at Liverpool would have excited him. Bruno himself told his brother Giovanni that the laboratories were wonderful. Guido, the eldest of the Pontecorvo siblings, visited Liverpool years later and remarked that the house where Bruno and Marianne planned to live was "very grand."[23]

Everything seemed on course for the move to Liverpool. On July 24, the day before the Pontecorvos left for their summer holiday, Bruno wrote to the vice chancellor of the university. He accepted the professorship, and declared that he looked forward to joining them.

The end of Gil's first year at Roysse's came on the same day that Bruno sent his letter. With his transfer to Liverpool arranged for the following January, Bruno and his family prepared for their summer holiday. The plan was to visit the continent by car, along with his sister Anna, who lived in London.

One bonus of Bruno's enforced departure from Harwell was that he had accumulated several weeks of leave time. He didn't want to squander this bonanza so he planned the family's holiday to last six weeks. They would camp, visit his parents and siblings in Italy, and return to England by ferry from Dunkirk on September 4, just in time for a physics conference in Harwell on September 7.[24] Gil would start school again on the nineteenth.[25] During the fall, Bruno planned to spend time in Liverpool, in preparation for his permanent move there the following year.

Among the decisions that had to be made was whether Gil would stay at Abingdon as a boarder, or move to yet another new school in Liverpool.

Of course, even as a day pupil, Gil's education wasn't free. The bill for the summer term, which he had just completed, totalled fourteen pounds, four shillings, and seven pence in the old British currency of the time (about £500 today).[26] The invoice arrived at 5 Letcombe Avenue after the family had left on their trip.

The Pontecorvo family owned ducks. These birds were more than pets, serving primarily as a means to produce eggs to supplement their diet, as postwar rationing was still in force. The Gardners, who kept chickens, agreed to look after the ducks while the Pontecorvos were away.

On July 25 the Pontecorvo family—now including Bruno's sister Anna—crammed into Bruno's precious Standard Vanguard car. With camping equipment, three adults, and three boys all crammed into the car, luggage was limited to essentials, and carried in collapsible canvas sacks. The group carried with them two army-surplus satchels, a floppy zip-up bag, which contained their underclothes and a few outer garments, and a small zip-up briefcase, which Bruno always kept close to him.[27]

Paul Gardner later recalled, "They were driving up the Avenue in their pale-coloured Vanguard car and I waved to say cheerio." The car turned left onto Bath Street, and headed south and then east toward Dover and the overnight ferry to Dunkirk.[28] That was the last Paul ever saw of them.

THE PONTECORVOS CROSSED THE ENGLISH CHANNEL BY FERRY ON THE night of July 25. After arriving in France, they drove through Arras and Dijon, and arrived in Neuchâtel, Switzerland, on the twenty-eighth. On July 31 they reached the Italian town of Menaggio on Lake Como. There, they bumped into the wife of Professor Piero Caldirola—a nuclear physicist from Milan—who suggested they all meet her husband. Bruno invited Professor Caldirola to the physics conference in Harwell on September 7, gave him his address in Abingdon, and arranged to see him there in the fall. The family camped in Menaggio until August 4. Anna then left the group temporarily, taking a boat to the city of Como, while Bruno and his family went to the Dolomites for a couple of days.[29]

The holiday continued to be idyllic as Bruno took his family to visit his parents in Milan, on or around the twelfth.[30] Here, Anna rejoined the group as they prepared to complete the final leg of their journey, which would take them to Rome. Before Bruno left his parents' house, he told them that, on his way back to England at the end of the month, he was due to visit the cosmic ray experiment near Chamonix. As this would be an opportunity for the family to be together again, and for his parents to visit the Alps, they arranged to meet in Chamonix on August 24.[31]

Bruno's mother's letter to Guido, written after his disappearance, gives her view of the situation in Milan: "When Bruno and his family came here to see us . . . they were so happy, serene and normal. They could not have had such a great step in mind. They said goodbye, just as one says goodbye to someone one is going to see again very soon, which is what we had arranged."[32]

Bruno would never see his parents again.

MISADVENTURES

Around August 17, the group arrived at Ladispoli, a small seaside resort near Rome where Bruno's sister Giuliana had rented a summer house. Bruno and Marianne then drove on to Circeo, another seaside town about two hours to the south. There they would camp, swim in the Mediterranean, and catch fish. However, they left their son Antonio at Ladispoli with Anna and Giuliana, as he appeared to be suffering from sunstroke. Anna and Antonio stayed on at Ladispoli until the twenty-first, when they accompanied Giuliana to her home in the San Giovanni district of Rome. Bruno's brother Gillo and his wife, Henrietta, were there also.

August 22 was Bruno's thirty-seventh birthday. Gillo was rather annoyed, as Bruno had passed by Rome en route from Ladispoli to Circeo without making the small detour to see them. So Gillo, Henrietta, and Anna drove to Circeo for Bruno's birthday, which they spent scuba diving in the warm waters of the Mediterranean. They spent the night, and drove back to Rome on the twenty-third. Bruno and Marianne stayed in Circeo.

Up to this point, everything appeared normal. Then events become strange, and the exact timeline becomes hazy. On the evening of August 23, Bruno sent a telegram from Circeo to his parents, cancelling his visit to Chamonix. He gave no reason for the change but promised to expand in a letter.[33] In the letter he reported that he had got into an accident in his car, "which bent its mudguard and smashed a front light." However, none of his siblings in Rome seems to have noticed the damaged car. The location and date of the accident are also confused. It seems to have happened somewhere between Circeo and Rome, on August 23. As it happened, Bruno's temporary car insurance for mainland Europe expired that day.[34] Unless this is the sole clue that he never intended to return, it would seem that Bruno was planning to drive back to England uninsured.

This oddity pales into insignificance, however, given what was about to transpire. The circumstances of this accident, if there was one, seem connected with Bruno's decision to flee. They certainly reveal deception on Bruno's part. As we shall see, Bruno's parents—as well as MI5—would identify inconsistencies in his version of events.

BRUNO, MARIANNE, AND THE BOYS WERE DUE TO MEET HIS PARENTS in Chamonix on August 24.

Here we see the first hints of duplicity. Only a week earlier Bruno had made the arrangements. Some event must have happened on the twenty-third that forced an urgent change of plans. The journey to Chamonix in 1950 would have required a full day in the car.[35] Yet on August 23, Bruno was alone in his car, and had an accident. At least, that is what he told his parents later. On the day of the alleged accident, in the telegram sent at 6:40 p.m. from Circeo, he bluntly stated, "Very upset to have put off visit to Chamonix definitely. Please go yourselves, will do you lot of good. Letter follows."[36]

The full text of Bruno's letter, which he sent on August 25, is presented below. It implies that some time has elapsed since he sent the telegram, which he tries to excuse. First, he refers to his telegram of yesterday, whereas two days have actually passed. He also refers explicitly for the first time to an accident, and gives details about it. Furthermore, he claims that the children have become ill, which is another reason for the cancellation.

> This letter follows my telegram of yesterday from Circeo. I had not
> written before because I had hoped to be able to come to Chamonix,
> but with the children as they are, it was impossible. Yesterday [August
> 24] I returned from Rome to have the car mended and saw your cards
> saying you had left just when I had sent the telegram. Well finally
> we had trouble with the car having had a fairly serious car accident
> happily without consequences for us though with consequences for

the car. In trying to avoid a cyclist who ran across in front of the car I hit a tree. I was not injured (I was alone at the time) but the car was seriously damaged and is now having repairs. On the other hand the children and I benefited a lot from the sea and they were well until the 23rd but they are now in bed with tummy trouble. We think that one day we caught the sun too much or that they have eaten something. Now we have left Gil [*sic*] at Ladispoli where he is enjoying himself no end and we are in Circeo which is a delightful place and where we have taken a room. As soon as the children are well enough we shall return to England. It is not possible to come to Chamonix as we shall have no time and it would tire the children. I am very sorry not to have been able to meet you or to warn you in time but there was no means. I am slightly consoled that even without us at Chamonix would do a lot of good [for you]. I would hope you don't mind too much. Again a thousand excuses.[37]

Bruno's mother realized immediately that something was amiss. Bruno never saw his parents' reply, which they sent a few days later. This letter arrived at his home in Abingdon, where it remained unopened in the empty house until MI5 found it in October. His mother had noticed an inconsistency in Bruno's claim that a sudden indisposition of the children had played a role in his change of plans, and admonished him appropriately: "But is this <u>REALLY</u> what happened? Because there is something that is not clear to us. Antonio was already ill at Ladispoli [on August 17]. Pitiful lies Bruno is not a good policy."[38] This letter is an important clue in our attempt to disentangle the chronology of Bruno's sudden decision to flee.

Without a doubt, something significant happened to Bruno on August 23. His birthday on the the twenty-second was carefree. By the twenty-fourth, he was sending duplicitous messages to his parents rather than meeting them in Chamonix, and was also apparently spinning yarns to Anna. Now at Giuliana's in Rome, Anna did not see the damaged car herself. Bruno simply told her that "a front light was smashed and a mudguard was bent and while the car was in good enough condition for minor trips, it was not good enough for travelling about Europe."[39] The fact that events were moving out of Bruno's control is obvious from his terse response to Gillo, who remonstrated him for the way he was treating their parents. Bruno replied that "they must get used to it for once." Gillo's wife, Henrietta, repeated this quote to MI5 but in a sharper form: "They will have to get used to it."[40]

WHILE IT SEEMS THAT BRUNO MAY INDEED HAVE DRIVEN SEVENTY miles from Circeo to Rome on Thursday, August 24, or Friday, August 25, in all probability his goal was not to take the car to a mechanic. Car repairs of the relatively cosmetic form that he described hardly merit such a trip, and there is no evidence that he contacted the Standard Vanguard dealership in Rome.[41] Furthermore, he seems to have used his car in the following days before finally dumping it in Rome prior to his disappearance. In any case, Bruno's behaviour was growing increasingly erratic. Something singular appears to have happened on Wednesday, August 23, and caused a radical change in his life.

One event that affected Bruno certainly happened that day. In Giuliana's house, news arrived via *l'Unità*, the organ of the Italian Communist Party. Over the past month, the stories of witch hunts in America had gone from bad to worse. World War III threatened to erupt as the military confrontation in Korea escalated.[42] Then, on the twenty-third, the patent dispute between the Via Panisperna Boys and the US government made headlines. In the version of the story promoted by *l'Unità*, the government had defrauded the team of Italian scientists.

Suddenly Pontecorvo found himself in the spotlight, in conflict with the US government. While this was a big story in Italy due to the involvement of Fermi's team, it was hardly commented upon in the UK. However, Bruno might have feared that the McCarthyist witch hunters would use his communist associations to portray him as a traitor, and that a major public scandal was about to ensue. According to this thesis, Bruno foresaw himself being subjected to further security vetting, which would threaten his ability to continue working as a scientist. And thus he panicked.[43]

Although it is possible that the patent crisis set in motion the events that culminated in Bruno's flight to the USSR, this could not have been the sole cause of his disappearance.

The most influential member of Bruno's family was his cousin Emilio Sereni. Sereni was a leading communist and, more significantly, by 1950 had risen to become a minister in the Italian government as well as a member of Cominform (a successor organization to Comintern). He was also well connected with senior members of the Soviet administration.[44] Sereni had been in Prague from August 14 to August 18, and, it is believed, had contact with Bruno during the final days of the month.[45]

The political implications of the patent crisis may have been what led Bruno to consult Sereni in Rome on the twenty-third or twenty-fourth. However, far from being reassuring, the news Bruno received was worse. Much worse. While in Rome, possibly during a meeting with Sereni, Bruno learned that the arrests of Fuchs, Greenglass, and the Rosenbergs had not been the end of it: the FBI was interested in Bruno Pontecorvo.

We shall come later to the story of how this singular information reached the KGB. In any case, this news alone would probably have been sufficient to inspire panic in Bruno. Coming on top of everything else, it precipitated his sudden flight.

Today, we know that by this stage the Soviets were already making preparations for Bruno's future.[46] Thus it seems probable that his trips to and from Rome immediately after his birthday had more to do with Emilio Sereni and other communist contacts than with car repair. It was almost certainly Sereni who both encouraged and arranged the details of Pontecorvo's defection to Russia, as Ronnie Reed of MI5 subsequently concluded.[47]

THAT WEEKEND, FROM AUGUST 26 TO AUGUST 27, THE CLAN GATHERED in Ladispoli. According to Anna's interview with MI5, Gil and Tito came to Ladispoli from Circeo—Ronnie Reed later commented, "It's not clear how."[48] On Sunday afternoon, Giuliana said she wanted to go to see some friends a few miles away. Bruno offered to take her in his car, but "when Giuliana told him that her friends were communist or near-communist he refused to take her." Later, Reed regarded this as significant, but drew no conclusions from it.

On Sunday evening, Anna had a spare ticket for the opera. There was some debate about whether Bruno or Marianne should accompany her; Marianne said she would prefer to go to a dance, as she was "not the intellectual type."[49] Anna remembered this event clearly when interviewed by Ronnie Reed, and the chronology agreed with the record of Sunday the twenty-seventh in her diary.[50]

On Monday, August 28, which was to be Anna's last day in Italy, Marianne went shopping with her in Rome. Anna later told MI5 that Bruno was separated from the group, but joined them at Giuliana's in the afternoon for a family conference about their sister Laura.[51] There was an argument with Bruno about the shopping expedition as the family was extremely short of currency by this time; Bruno had been rationing the milk for the children very carefully, and they had no spare cash at all. In those days there were strict limits on how much currency British tourists could take abroad. The maximum allowance for the Pontecorvo family was £205 plus £10 for the car, and when they left England Bruno had declared only £150 in traveller's cheques and £10 in cash. His bank accounts in England showed that he had not spirited away large sums of cash.[52] The group's lack of money near the end of their holiday is significant, because within two days Bruno would purchase airline tickets for the whole family, on a flight from Rome to Stockholm, for which he paid 602 American dollars in cash.[53]

Meanwhile, Bruno visited the University of Rome, where he was seen by Giuseppe Fidecaro, who was then a twenty-four-year-old physics student. Bruno had called there in the hope of seeing Edoardo Amaldi, his old colleague. However, Amaldi was away in the United States, so Bruno met with Mario Ageno, another former colleague and tutor to Fidecaro.

Fidecaro told me that Bruno came to the institute once during the morning, and then returned for a second visit in the afternoon. He remarked, with a smile, "I couldn't see Bruno wasting hours of his time running behind [Marianne and Anna] shopping! Conversely I do not see two ladies going around shopping for a full day with a man on their shoulders!" The university was near Giuliana's house, which enabled Bruno to take part in the family conference that Anna recalled, and then return to the university later.

Fidecaro recalled that late that afternoon he accompanied Bruno and Mario on a stroll: "At the end of the visit we left the Institute walking together 'lento pede' in the warm light of a late summer afternoon in Rome. I think it couldn't be earlier than 6:30–7:00 p.m." Bruno and Mario were talking "*del più e del meno*"—engaging in casual and personal chatter. Fidecaro was not party to their conversation. "After we separated, in the vicinity of the *Stazione Termini*, everybody went on his own way. Mario and I went home independently." Giuliana's house in the San Giovanni district, the Roma Termini railway station, the university, the shopping district, and the garage in Piazza Verdi (to which Bruno would deliver his car the next day) are all in the same area.

This chronology is consistent with Bruno's memory, decades later, that he took his car in for repairs on August 29.[54] Fidecaro remarked that Mario could have recommended the garage in Piazza Verdi, where Bruno took the Vanguard—and where it remained, abandoned. Sixty years later, Anna would insist that "there was never any hint of anything unusual" in Bruno's behaviour. Up to the time that Anna left Rome, she saw no change in Bruno or Marianne's manner; Bruno in particular seemed "relaxed and natural."[55] Anna slept at Giuliana's and left at seven thirty on the morning of the twenty-ninth. She did not see Bruno or Marianne that day. When she'd said *arrivederci* to her brother, there was nothing to suggest that thirty years would pass before they would meet again.

ONCE HE'D DECIDED TO FLEE TO RUSSIA, BRUNO NEEDED TO "LOSE" THE car without raising suspicion. Giuliana, as an innocent party, would have raised the alarm at once if it had remained at her house; if she were involved in the deception, and failed to alert the authorities that Bruno had disappeared, the car's presence could have created serious problems for her.[56] The story of the "accident," which provided an alibi for Bruno's trip to Rome on the twenty-fourth (not to repair the car but to meet with Sereni) now could be put to further use.

On August 29 Bruno took his car into the garage at Piazza Verdi.[57] With the car taken care of, that same day he visited the office of Scandinavian Airlines (SAS) in Rome and inquired about tickets for Stockholm.

ANNA RADIMSKA, A DARK, SMALL, ATTRACTIVE POLE FROM WARSAW, sold him the tickets, as we know from the subsequent investigation by Ronnie Reed of MI5. She remembered Bruno and Marianne well, because they caused so much trouble. They came to the office on the twenty-ninth and made provisional bookings for themselves and their three children to travel to Stockholm.[58] Bruno booked his ticket under the name of Pontecorvo, and the tickets for Marianne and the children under the name of Nordblom-Pontecorvo. Marianne was "noticeably upset and tearful and on one occasion drew [Bruno] back from the counter by tugging on his coattails."[59] The MI5 report adds that "[Bruno] Pontecorvo seemed quite unconcerned."[60]

Marianne missed her mother terribly, as is clear from letters found in the Abingdon house by MI5.[61] Originally she had hoped to spend two weeks in Sweden that summer, and the rest with Bruno and the boys in Italy, but her mother had insisted that she stay for at least a month, so this plan had been abandoned.[62] Under the circumstances, the prospect of flying to Stockholm with no possibility of seeing her mother would be understandably upsetting. As events transpired, five years would elapse before Marianne had a chance to write home to Sweden, and she would never see her mother again.

The usual practice was for airline reservations to be confirmed the same day, but Bruno asked for them to be held until the thirtieth, when he promised to confirm and pay. This suggests that there was still some uncertainty about the venture. In response to police inquiries made a few weeks after the Pontecorvos' disappearance, Anna Radimska said that the Pontecorvos didn't look very impressive: "I was struck by their shabbiness. They seemed kind of crumpled."[63] Because she thought they were too poor to afford the fare, she bet a colleague that they would not return. When Bruno failed to appear by noon the next day—the deadline for confirming the reservations—the bet was won and the bookings were cancelled. At 4:00 p.m., however, Bruno showed up again, alone. Anna had to get busy and renew the bookings.

The price of the tickets was equivalent to £175 and Bruno produced it in lira. But there was a problem: foreigners with less than six months' residence in Italy had to pay in US dollars—$602 in this case.[64] Anna remembered that when she told Bruno of this requirement, the news made him very angry.[65]

Bruno, still agitated, told Anna to hold the reservations, and left. He returned shortly before the office closed, with a handful of hundred-dollar bills. Such high-denomination bills were rather rare in Rome at the time, except among US citizens, or the members of clandestine organizations.

This suggests that Bruno's exodus was organized by the Soviets, at a high level. There would ordinarily be no way for Pontecorvo—a British tourist, subject to strict currency regulations—to obtain so much cash after touring for five weeks abroad. His sister Anna was surprised when she learned of this later, recalling how short of money Bruno had been toward the end of the trip.

Bruno was clearly able to change the money from lira to dollars without much difficulty, in a short span of time. In 1950, MI5 saw this as the first solid clue that some third party had orchestrated his flight. No one at the time, of course, suspected that his cousin Emilio had a link to Soviet finances.

On August 31, Bruno sent a postcard to his colleagues at Harwell, which they received on September 4: "Had a lot of fun with submarine fishing but I had a lot of car trouble. I will have to postpone my arrival until first day of conference [September 7]. Can you tell Egon Bretscher. Hope everyone has prepared his talk and done good work at Chamonix. I am sorry I missed Chamonix but I could not make it. Goodbye everybody. Bruno."

Later, the timing and phrasing of this postcard was forensically debated. The message ended with the phrase "Goodbye everybody." This sounded terminal, whereas the comment that he would "postpone [his] arrival" implied a temporary delay. As Bruno knew that he was en route to the USSR at this stage, the postcard has been interpreted as further duplicity. However, this supposes that he knew he would be staying in Russia for good, which is not necessarily true.

That same day, Bruno sent a final telegram to his parents: "From your card it is apparent that you have not received my wire to Milan and poste restante [general delivery] in Chamonix. . . . excuses I could not meet you . . . indisposition children damage to car hope mountains did you good. We are now well."[66]

At eight o' clock, on the morning of September 1, 1950, Bruno, Marianne, and the three boys flew from Rome to Munich. In Munich they transferred to a direct flight to Stockholm. Their luggage, which weighed a total of sixty kilograms, was contained in ten bags, each of which would be considered carry-on size today. This is because they were carrying nothing more than the possessions they had taken on their camping holiday, all of which were stashed in small canvas bags.

The airline manifest reveals more clues about the family's arrangements. The tickets for Marianne and the boys were in one number sequence, but Bruno's ticket came from a different block, and was issued from a different office.[67] There were two male passengers on the flight,

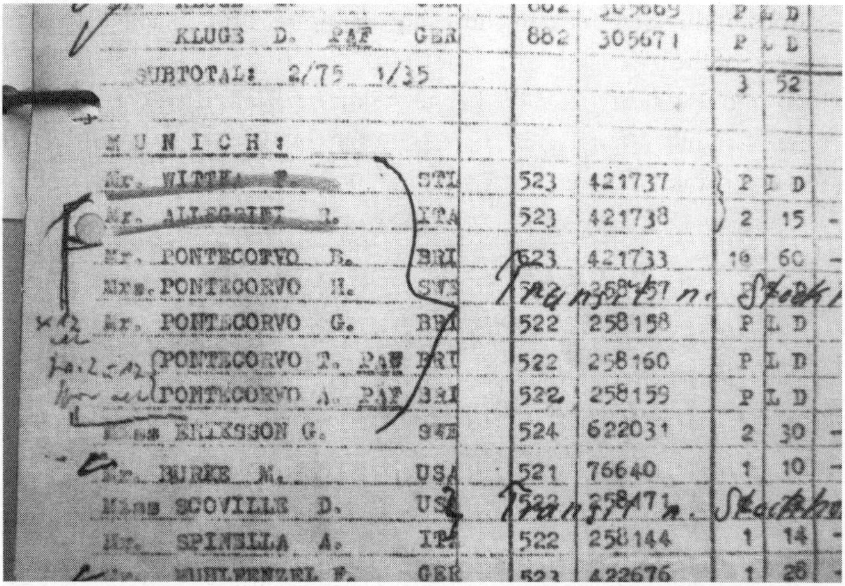

KLUGE D.	PAF	GER	882	305671	P	L	D
SUBTOTAL: 2/75 1/35					3	52	
MUNICH:							
Mr. WITTKA F.		STL	523	421737	P	L	D
Mr. ALLEGRINI R.		ITA	523	421738	2	15	
Mr. PONTECORVO B.		BRI	523	421733	10	60	
Mrs. PONTECORVO M.		SWE	522	258157			
Mr. PONTECORVO G.		BRI	522	258158	P	L	D
PONTECORVO T.	PAF	BRI	522	258160	P	L	D
PONTECORVO A.	PAF	BRI	522	258159	P	L	D
ERIKSSON G.		SWE	524	622031	2	30	
Mr. BURKE M.		USA	521	76640	1	10	
Mrs SCOVILLE D.		USA	522	258171			
Mr. SPINELLA A.		ITA	522	258144	1	14	

IMAGE 11.1. Airline manifest of the Pontecorvos' flight from Rome. This shows that Bruno's ticket number is more akin to those of two accompanying passengers—Wittka and Allegrini—than those of his own family. Ticket numbers are listed individually within a block (e.g., 421733) and by the number of the block (e.g., 523). Names are followed by nationality: STL implies stateless, SWE Swedish, and BRI British. The two columns to the right of the ticket number show the number of pieces of luggage and their total weight in kilograms. Thus the Pontecorvo family took ten pieces totalling 60 kilograms. The shaded highlight against the names of Wittka and Allegrini—who also had minimal baggage, a mere 15 kilograms between them—appear to have been made by MI5. (Author, The National Archives.)

named Wittka and Allegrini, whose tickets came from the same block as Bruno's and are almost in sequence with his. These men took the same flights as the Pontecorvos, from Rome to Stockholm, via Munich. One of them was stateless, with no checked baggage; the other, identified as Italian, checked two pieces totalling a mere fifteen kilograms. MI5 later suspected that these mystery men might have been assigned to watch the Pontecorvos.[68] MI5 made investigations in Rome to ascertain why the family had booked tickets separately, and to identify Messrs. R. Allegrini and F. Wittka. The agency was unsuccessful on both counts. In any case, the presence of the mysterious pair fits with the KGB's standard practice of firm control.[69]

At Munich, the Pontecorvos went to the transit lounge with the other passengers and then flew on to Stockholm, where they arrived late that evening.[70] According to MI5 sources, they had booked reservations for the night at a Salvation Army hostel. The sources also reported that Bruno met with a man in Stockholm late that evening. The Stockholm correspondent of the Italian newspaper *Il Tempo* was more explicit, stating that the Pontecorvos spent the night of September 1 in "a house occupied by the Soviet Embassy" and went to the airport "at about 11 a.m." on September 2 "in a Soviet Embassy car."[71] Gil recalls only that they were taken to a small hotel.

Whatever the details, it is certain that they did not visit Marianne's parents, who lived no more than a couple of miles from the airport. Gil later recalled, "I wanted to go to my grandmother's. I asked—why not?—but I don't recall the answer."[72]

On September 2, the Pontecorvos arrived in Helsinki, Finland. Someone in Stockholm had provided further tickets and money. In Rome Bruno had purchased tickets only as far as Stockholm, and had arrived in Sweden with no US dollars. However, when he reached Helsinki, he had over four hundred dollars in cash.[73]

Just before the plane landed in Helsinki, a man and woman turned up at the airport and told the staff that they had come to meet the Pontecorvo family.[74] The pair asked that the Pontecorvos' luggage be taken to their car, rather than being placed on the airline coach. A customs official told MI5 that these two people had often visited the Helsinki airport, but he didn't know who they were. He had always thought they were officials of the British legation.[75]

The Swedish newspaper *Dagens Nyheter* later learned that there had been several slipups by Swedish officials. For example, Marianne and the children were travelling on a temporary Swedish passport, issued by the Swedish embassy in London. This should have been confiscated upon their arrival in Stockholm. The error was compounded the following day, as they should not have been allowed to leave Sweden for Helsinki on this passport.[76] It is thus ironic that the entry into Finland was straightforward for Marianne and the boys, whereas Bruno ran into trouble.

Bruno was travelling on his British passport, and there was a hitch when Finnish officials demanded to see his visa—there was none. They impounded his passport and deposited it with the Finnish foreign office. As Bruno was a British citizen, this was not a major issue—a visa would be issued overnight and he could recover the passport the next day. He signed various entry papers without hesitation, stating that tourism was the reason for his visit, and that the length of stay would be about a week. He also told officials that they could get in touch with him at his hotel.

But there was no hotel, and he never collected the passport. That was the last confirmed sighting of him in the West. On September 2, 1950, the Pontecorvos vanished, not to be heard from again for the next five years.

TWELVE

THE DEAR DEPARTED
1950

The first hint of trouble came at the end of August, when Bruno's parents turned up at the Hotel Terminus in Chamonix. Arriving on the twenty-fourth, they expected to spend a few days with Bruno and his family, as they had arranged just a fortnight ago. For three days, they searched all the campsites. They telephoned the cosmic ray laboratory on the Pic du Midi, where Bruno's colleagues were working, and eventually met one of them, Dr. D. P. Price. He told them that he too wanted to see Bruno for help with some experiments, but that the Italian scientist had not been seen.

The telegram Bruno had sent to Milan on the evening of August 23 didn't reach his parents in Chamonix until the twenty-sixth. On August 31 he sent the second telegram. The letter, which blamed his absence on the car accident and the children's ailments, took longer. After receiving these communications, his parents waited in Chamonix until September 9, hoping to receive a further letter. Then they spent three days in Annecy, and returned to Italy.

They had received two telegrams and a brief letter, and they naturally expected to hear more from Bruno, who, they believed, should by then have been home in England. On September 11, during their stay in Annecy, his mother sent the first anguished letter to his house in Abingdon:

"Dearest we no longer know what has happened to you and we are very worried. We begged you to send us word as soon as you arrived in England for our peace of mind. You are now in England and you must have got there on the 3rd crossing the sea but still no word. . . . Write IMMEDIATELY to Milan on how the journey went and how the children are: THE WHOLE TRUTH."[1]

While Bruno's parents had been away, they had received a telegram from Egon Bretscher of Harwell, inviting Bruno to Chamonix to help with the cosmic ray experiment. They wrote to his colleague, Dr. Price, whom they had met in Chamonix, and explained why they had been unable to deliver the telegram to Bruno. Their letter shows their worry. They describe themselves as "restless," as they have been without news of "their boy" for some time: "We've had no communication from him since his telegram and letter. We travelled to Savoie and because he knew we had to change our address in Milan, since we left the furnished rooms that we inhabited, he couldn't write to us anymore. We have sent him in Abingdon our address . . . but he would not get the letter before he returned to Abingdon."[2]

On September 15, their daughter Laura phoned. Unknown to them, she was in Rome. She "calmed" them with her news: Bruno had left on August 31, the accident had caused only cosmetic damage to the car, and Bruno and the children were well. The next day, Bruno's father wrote to him in Abingdon. He said that, after all their worries, hearing from Laura had at last put their minds at rest, though a small chastisement was included: "We had no news and we imagined all sorts of mishaps after the car accident and the children's indisposition. But . . . a postcard to say you had arrived safely as we asked you for should have been sent."

Bruno's mother, however, was less happy. Her intuition had told her that something was amiss. Here is her postscript to the letter:

My dear Bruno. I must correct the fact that papa says WE were calmed down by Laura. It would be better if he said _he_ was calmed down because as for _me_ the knowledge that you had taken up your return journey gives me great pleasure but only half sets my mind at ease. To be quite happy about everything we needed just two words (it is not too much to ask and it would be enough) to tell us that you had arrived in England, well, without accident, and without the children being too over-tired! But these words have not arrived up to now. And your silence is inexplicable. Ever since the accident that you had, you should understand that our hearts stood still. And in this state of mind we spent the time at Chamonix and Annecy on the way home. We hope that a line will come from you now and that you will tell us that all is well and the children have got over their indisposition. . . . Never do it again.

Having admonished him, his mother's maternal love took over, with tragic irony: "Poor little Bruno. Besides your misfortunes you have had complaints from us but be sure for another time, if there ever is another time dear, think of us. Kisses and love to all [signed] Mama."[3] Bruno, by then in the USSR, never received this letter. Which is profoundly sad: there never would be another time.

This letter eventually arrived at number 5 Letcombe Avenue, in Abingdon. The postman pushed it through the front door, after which it joined the accumulating mail and newspapers on the floor. Bruno's parents' anxieties erupted again when they received no reply. On September 21, in Glasgow, Guido received a letter from them, asking if Bruno had returned to England as "they had no news of him."[4] Guido did his best to find answers; upon drawing a blank, he sent them a telegram on the twenty-third: "Bruno not back yet. Colleagues enquiring."[5] Giovanni, in Amersham, also received a postcard from his parents, on the twenty-fourth. This too expressed their anxiety.

As they had recently moved to a new apartment in Milan, they worried that news from Bruno might have gone astray. In a desperate hope that he might have written to them at Chamonix, they contacted the Hotel Terminus. On September 26, the manager received their postcard, which pleaded for news of their son. He was unable to help.

ENGLAND

During their summer absence, the Pontecorvos had cancelled the daily deliveries of milk, newspapers, and mail to their Abingdon house. (The deliveries were scheduled to resume again upon their return.) A memory of that long-ago summer has remained with Anthony Gardner ever since:

his mother would periodically come into the living room and tell Paul and Anthony, "The ducks have disappeared again."[6]

The birds needed water. There was no pond in either the Pontecorvos' garden or the Gardners'. Paul and Anthony knew where to look. The River Stert, which was actually more of a stream than a river, flowed near their house, and was the ducks' favourite haunt. When their mother expressed concern, the boys would go to the stream, find the birds, and bring them home.

On Monday, September 4, soon after dawn, the milkman left bottles on the Pontecorvos' doorstep. This was followed by a delivery of letters and cards. The Gardner family, having spent the summer in a pas de deux with the Pontecorvos' ducks, eagerly anticipated their neighbours' return.

At first they thought the Pontecorvos had been delayed. Then they began to worry that something awful had happened. Milk bottles accumulated outside the door on a daily basis. The milk turned sour, and deliveries stopped. The Pontecorvos owed money to the dairy for milk delivered both prior to the their departure, and after their (scheduled) return. The dairy owner, John Candy, later sent a bill to the Soviet embassy, but he "never had a reply and is still waiting for his money."[7]

Although neighbours, local schoolchildren, and various deliverymen were all aware that the Pontecorvos hadn't returned, no one at Harwell initially suspected anything untoward. However, in hindsight, clues were rapidly accumulating.

Joe Hatton, a physicist who worked with Bruno Pontecorvo at Chalk River and Harwell, in his nineties remained a fountain of knowledge. His memories of Bruno are typical of everyone who knew him: "He was one of the most delightful men you could imagine. He had an extraordinary presence. He could charm a bullfinch out of a tree." As for the summer of 1950, when Pontecorvo disappeared: "I can remember my reaction. A postcard came [from] somewhere in Italy."

Joe recalled that it seemed like a typical holiday postcard, except that it read as if Pontecorvo wasn't going to see him again. He ruminated poignantly: "I was surprised. I wondered what it meant. He must have sent it when he was about to run for it. . . . Soon after that, it was announced that he had done a bunk."[8]

Some weeks elapsed before the truth came to light. Hatton received his postcard a few days before the Harwell conference on nuclear physics.

Scheduled to begin on September 7 and last a week, it would be a major international gathering. On opening day, there was no sign of Bruno, but as yet there was no undue concern. Niels Bohr arrived on the eleventh, an event that experts such as Bruno Pontecorvo would not want to miss.[9] Professor Piero Caldirola, whom Bruno had met in Italy and invited to the conference, looked in vain for his new friend. On the final day, Caldirola went to Pontecorvo's house in Abingdon and put a card through the door. It explained that Caldirola had hoped they would meet at the conference, but that this had not been possible as "you had not arrived" by the time it finished. Caldirola departed, puzzled.[10]

It was becoming clear that the Pontecorvos would not return to Abingdon anytime soon—or perhaps at all. The Gardners accepted the inevitable, leading to the first casualties of the "Pontecorvo affair": they ate the ducks.

ROYSSE'S: SEPTEMBER

At Roysse's, during the weekend of September 17, the boys were returning for the new school year.

Among them was Mark Bretscher, whose father, Egon, was a close colleague of Bruno's. Mark was one of the new boys, preparing for his first day at school. As mentioned earlier, the plan was that Gil Pontecorvo, who had been at the school a year already, would be Mark's companion, guiding him through this rite of passage. However, Gil was not home, so Mark had to take the plunge alone. That was when he first realized that "something was up."[11]

The school records show that fees were due two or three weeks in advance of the term. Hitherto, the Pontecorvos had always paid on time, although as the year went along, the payments got nearer to the deadline.[12] If the previous year was any guide, the fees should have been paid toward the end of August 1950. At first glance, the school records appear to indicate that they were: "Pontecorvo—fees paid—for the Michaelmas term, which began on 19 September (boarders return 18 September)."[13]

However, this is a pro forma entry, and, in the "paid" column, instead of a date, there is a note that the anticipated cheque had not arrived. Gil Pontecorvo's name is crossed out in pencil, with a question mark beside it. The administrators at Roysse's didn't know what was happening; Gil hadn't showed up and his fees were overdue.

FIRST WORRIES

A Harwell employee who had returned from the Pic du Midi for the nuclear physics conference told his colleagues how Bruno had cancelled his visit to Chamonix due to an accident and his children's illness. This news, which originated with Bruno's parents, confirmed Bruno's own written message to Harwell, sent on August 31.[14]

The conference ended on September 14, and, given that there was still no sign of Bruno, Egon Bretscher sent telegrams to Edoardo Amaldi and Bruno's parents. He asked them to deliver a message to Bruno, if they happened to be in touch with him: the cosmic ray team at Pic du Midi was having some difficulties with their apparatus and would welcome Bruno's help. It was this telegram that first confirmed Bruno's parents' worst fear: their son had not reached home. This stimulated their letters to Guido and Giovanni. Bruno's parents then contacted Dr. Price, at Chamonix, to tell him that they had heard nothing and were becoming "restless." Price forwarded this news to Bretscher at Harwell; it was at this point that events relating to Bruno's disappearance began to be recorded in the British security files. These provide a crucial record of the unfolding crisis.

It wasn't until September 20 that Cockcroft, the director of Harwell, "became anxious [and] asked the Security Officer [Henry Arnold] to make enquiries about Pontecorvo's whereabouts."[15] These inquiries drew a blank. On the twenty-sixth, Arnold phoned Colonel John Collard, his contact in MI5, to say that Pontecorvo—who had been expected back by the ninth—had not yet returned. Collard advised Arnold to treat this "initially as an administrative matter." He instructed Arnold to make a casual inquiry with Bruno's sister Anna about his movements, but advised that she should not be alerted to the security implications of his disappearance.[16]

This advice seems to suggest a rather relaxed attitude on the part of both Henry Arnold and MI5, especially given the amount of time that had elapsed. Better late than never, someone finally took action. The fact was that Bruno Pontecorvo had vanished, and there was a strong possibility that he had gone to the USSR. Ronnie Reed, the head of counterespionage against the Soviets at MI5, took charge of the case.

Meanwhile agents of MI6 began making inquiries in Italy, and throughout mainland Europe. Edoardo Amaldi had already received the telegram from Bretscher. Now he was called by "someone from British Intelligence," who asked Amaldi whether Bruno, if he was in Italy, would be likely to come to see him. Amaldi said this was likely, as Bruno was a longtime colleague and friend. "If he does, please ring this number," the caller requested, without elaborating further.[17] The significance of the call only became clear later, when news of Bruno's disappearance broke.[18]

On October 5, the following note was added to Pontecorvo's file: "In addition to action taken by MI6, enquiries by Harwell of his Bank Manager and the Automobile Association [confirm] a return passage on the Dunkirk Ferry was booked for PONTECORVO'S car on 4 September but was not used."[19] MI5 now decided to examine the Pontecorvo home in Abingdon. A security officer and a local police constable broke into the house "under the pretext of turning off the water."[20] Forcing the locks was child's play. However, getting inside was another matter: the piles of letters, postcards, and newspapers that littered the hallway kept getting trapped under the door.

Children's toys still lay on the living room floor, and family photos sat in their frames on the sideboard. In the wardrobe upstairs, they found Marianne's fur coat.[21] Ronnie Reed's analysis was that the family had expected to return to Abingdon, and that Bruno's disappearance, far from being preplanned, had come out of the blue: "Much property and many of his belongings that would have been invaluable to him and his family in Russia had been left behind. Their winter clothes are still there."[22] This last point was confirmed by a fellow passenger on the plane from Stockholm to Helsinki, who described the Pontecorvos as "much worse dressed than the normal aircraft passengers. Marianne did not even appear to have a fur coat."[23]

NEWS BREAKS

By the third week of October, classes at Roysse's had been in session for a month, and still there was no sign of Gil. On Saturday, October 21, Paul Gardner went to the Regal Cinema with David Lees.[24] The name of the film they saw has long been forgotten, but their memories of what happened when they returned home are still fresh, half a century later.

Letcombe Avenue is a quiet street even today. In 1950 it was on the outskirts of Abingdon, bordering the countryside. As Paul turned the corner onto the avenue he found it "chock-a-block with people. Men in trilby hats were milling around." To reach his own house he had to walk past the Pontecorvos', and as he did so "one man grabbed me and said, 'do you live here?' I said yes—I was ten years old. 'Do you know the family that lives here?' 'Yes' 'Do you have any photos of them?' I said that I didn't know, and at that moment my mother emerged, and pulled me into our house. She said, 'I don't know what's happening out there, but it's obviously something to do with the Pontecorvos.'"[25]

His older brother, Anthony, recalls reporters knocking at their door, wanting to know about the Pontecorvos: "Dad told us to say nothing other than that we were friends." Hordes of photographers took pictures of the Pontecorvos' house—"a very boring house I might add," made of red brick, a typical dwelling on a government estate.[26]

David Lees too recalled the street outside the Pontecorvo house "swarming with press reporters and photographers." He laughed as he remembered how "us young lads found it rather flattering to be the best sources of information available." His abiding memory is of a London cab, which pulled up on Letcombe Avenue: "Out stepped a very smartly dressed man in a three piece suit, who exclaimed, 'Hold on, cabbie, I won't be long, keep the engine running. Can anybody tell me anything?' he asked. So we boys all chimed in with our little anecdotes and off he went. It all took about ten minutes." Thus many of the "facts" about the Pontecorvos' lifestyle, such as the claim that they were looking for a Russian language teacher, had given hints that they were about to disappear, and other fanciful tales that have been propagated over the decades stemmed from the imaginations of ten-year-old boys.[27]

Then the headlines hit the papers: "Hunt for Missing Atom Scientist: The British Intelligence Service has been brought into the hunt for the missing atom scientist"; Bruno Pontecorvo "skipped just ahead of Italian police and British Intelligence"; "speculation the family may have gone to the Soviet Union."[28]

The story was front-page news around the world. "British atom scientist 'lost' in Europe," wrote the *Sydney Morning Herald*. In the United

States, the *Palm Beach Post* reported that Finland (where Pontecorvo was last sighted) was being "combed for the missing atom scientist [who has] evaporated into thin air." Newspapers throughout the country ran stories speculating on Pontecorvo's whereabouts, next to stories about spy trials, wars against the "Reds" in Asia, and the possibility of an atomic bomb being dropped on the reader's town. By inference, and even by explicit assertion, Pontecorvo was linked to all of these. The Western media had no doubts: Pontecorvo was the third and potentially most dangerous of the "atom spies" who had fled to the Soviet Union as the net was about to close around him. A police guard was placed on his vacant home in Abingdon. The Melbourne *Age* announced that "to aid MI5" the Americans had sent "two G men" who had "moved into Abingdon disguised as farmers."[29]

The media then discovered that a Russian ship had apparently left the Helsinki docks, bound for Leningrad, a few hours after the Pontecorvos' arrival in Finland. With echoes of a spy novel, the media revealed that, in addition to eleven "cabin trunks," Bruno Pontecorvo had with him a "bulging brown briefcase" that he "kept close to his person."[30] This left little doubt in readers' minds that the contents included atomic secrets.

Because the Pontecorvos no longer had passports, and there was no record of them staying at any hotel in Finland, it was assumed that they had escaped to Russia on the ship. More careful reporters, however, found that the vessel had actually departed some hours before their plane had arrived. Nonetheless, the debunked theory has remained part of the folk wisdom about the affair.

Anthony Gardner recalled that "everyone at school who knew Gil was excited. The speculation that he and his family had gone to a communist country behind the Iron Curtain, and indeed that they were now in the USSR, was like reading a spy thriller and taking part in it." Anthony was already interested in languages, and considered Gil to be lucky in a way: "My goodness! He speaks English and Italian, and now he's going to learn Russian."[31]

GILLO PONTECORVO'S SON, LUDO, HIMSELF A PHYSICIST, LATER recalled, "Dad said that when he heard that Bruno had disappeared, 'I knew at once where he'd gone: to USSR.'"[32] Ugo Amaldi told me that his father, Edoardo, said as much also.[33] But none of them could explain why it all happened so suddenly and catastrophically; everyone who had seen

Bruno and Marianne in the dog days of August 1950 remarked how relaxed they had seemed.

Bruno's son Gil told me, "I thought [my aunt] Giuliana was involved. When I stayed with her in Rome in the 1980s, I asked, 'Did you know anything?' She said, 'Of course I didn't.'" Then Gil added, laughing, "But that means nothing."[34] Gil may be perceptive, as Giuliana's responses to inquiries by the Italian police, made in 1950 on behalf of MI5, suggest that she knew more than she admitted.

When the Italian police interviewed Giuliana, early in October, she told them a story whose chronology was demonstrably wrong. She said that Bruno had "put his car in the garage on the 31 August for minor repairs." On the "5 or 6 September" the Pontecorvo family had, in her version, "risen early and said they wanted to return to England in stages." She claimed that "somewhere about the middle of September" she received letters from Bruno "datelined Rome," asking her to pay the garage expenses and send the car on to England. She could not remember the date or the postmark.

MI5 learned from their Italian colleagues that Giuliana had appeared "evasive" while giving answers. The director general of MI5 now asked the Italians to visit her again to clarify this misleading information.[35]

When the inconsistencies in her story were pointed out to her at this second interview, which occurred around October 10,[36] she apparently replied, "Oh, I must have got the dates confused."[37]

Other members of the family share Gil's suspicion that Giuliana was party to the flight, although none admitted that she had ever said as much. In 1950, Guido remarked to Ronnie Reed that "Giuliana and her husband Tabet . . . might have influenced Bruno but could not have organised anything, whereas Emilio Sereni was powerful enough to do so and quite possibly may have done."[38] Ronnie Reed was also convinced that Emilio Sereni, Bruno's well-placed communist cousin, was involved.[39] The possibility that Giuliana, his communist sister, was ignorant of his plans seems most unlikely.

DAMAGE CONTROL

Whereas for Gil's friends in Abingdon this was a spy story come true, for others the implications of Pontecorvo's defection were more serious.

When news of Pontecorvo's disappearance finally broke on the weekend of October 21, and suspicion grew that he had defected to the USSR, the strategy of the British authorities was to minimize the damage. Given Fuchs's exposure earlier that year, and the resulting strain on Anglo-American relations, this was no time to admit that another spy had slipped through the net.

A telegram sent from the Cabinet Office to the British embassy in Washington on October 20 proposed responses to the anticipated press stories. Marked "TOP SECRET," it concludes, "There is no definite proof he was bound for Russia and we shall do everything to play down the fact. Please inform State Department and AEC immediately and ask them to reply to any press enquiries on same lines as we are. It is just possible the press here may refrain from publishing anything through fear of libel action if Pontecorvo should turn up here again."[40]

Meanwhile Guy Liddell, the deputy director general of MI5, continued to write in his diary. It records that on October 21 he received "the latest news about the disappearance of PONTECORVO a scientist at Harwell." At this same time, Liddell apparently first became aware of the fact that in 1943 the FBI had written to the British Security Coordination in New York to report finding communist literature in Pontecorvo's house. Liddell noted, "No one knows what happened to these reports, since the records of BSC have been destroyed."[41] As we shall see, these letters later became central to the affair.

WITH PONTECORVO ALMOST CERTAINLY IN THE USSR, AN URGENT review of his files began. Harwell security had described him in 1948 as "a straightforward fellow with no political leanings." The records stated that he was "Not politically minded. Expresses no political views. Dr Cockcroft has confirmed these opinions." Bruno had been naturalized as a British citizen while in Canada, rather than the United Kingdom. "Had he been in the UK," the record notes, "he would have been submitted to the naturalization enquiries by the police."[42] The British authorities were already manoeuvering to deflect blame toward anyone but themselves.

There was one troubling aspect of the case, however—the mislaid letters, which the FBI had sent to the British Security Coordination Office in 1943. The MI5 minutes state, "FBI had reported to BSC on 2/2/43 and again on 19/2/43 about the search of Pontecorvo's home in Oklahoma. The Royal Canadian Mounted Police were asked for any adverse information. On 31/12/46 they sent three names but not Pontecorvo. Therefore the Canadians had nothing negative on him in Dec 46."[43]

London here is attempting to pass any possible blame over to Canada. Meanwhile the Canadian authorities' strategy was to stress their innocence. They noted that Pontecorvo's name had not been mentioned in any of the documents handed over by Gouzenko in 1946, and that "at no time prior to his disappearance had Canadian authorities received any information [from UK or US sources] indicating him to be a security risk." Then came the coup de grâce: "Pontecorvo was cleared by British security authorities when he came to Canada . . . in accordance with agreed procedure." The fact that Bruno was employed by the British atomic energy team, which "was sent to Canada," meant that he should have been "the responsibility of British authorities for screening."[44]

With Canada off the hook, concern grew in London and Washington that something had gone seriously amiss. The British embassy in Washington sent a secret telegram to London on October 21:

> Patterson who has been in touch with the FBI tells us that there have been communications from the FBI in 1943 and [again] recently asking for information about reports that Pontecorvo had communist sympathies.
>
> If this is so, the FBI may well in response to enquiries from the press here admit that they have been in touch with the British Security Authorities about Pontecorvo. If this happens it will inevitably be concluded that he was under suspicion.
>
> The question will then be asked over here is why he was allowed to leave the UK if he was under suspicion. This will be a very awkward one to deal with. We are not sure about the exact legal position but presumably it is not feasible to detain persons or to prevent their travelling merely on suspicion. At the same time American public opinion puts the protection of atomic secrets as top priority in the national security. Coming after the Fuchs case it will be hard to persuade them that we were not lax in letting Pontecorvo go if we had suspicions about him.[45]

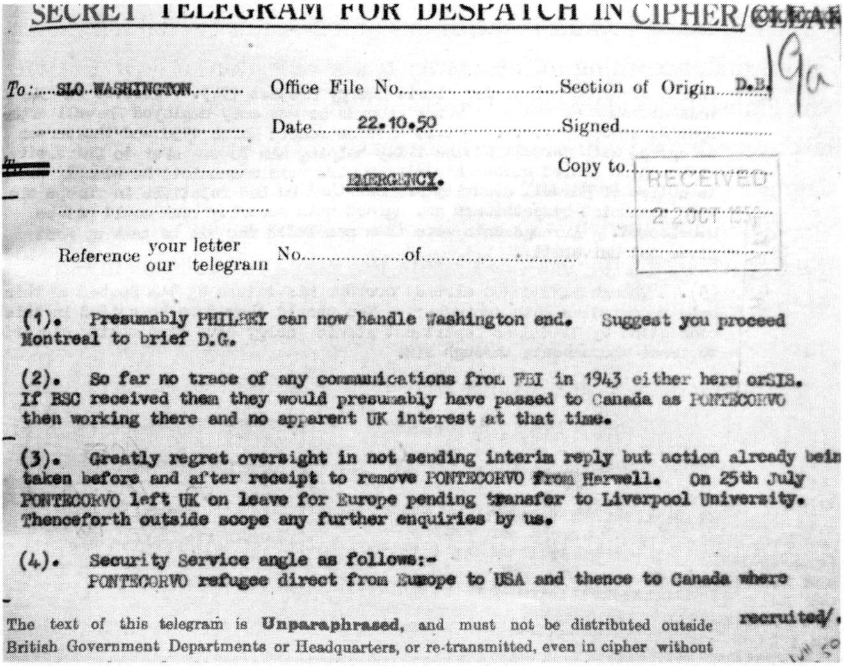

IMAGE 12.1. Secret telegram of October 22, 1950, requesting that "Philbey" [*sic*] spearhead the Washington office's efforts to manage the Pontecorvo fallout. Paragraph (2) again refers to the missing FBI communication of 1943. (AUTHOR, THE NATIONAL ARCHIVES.)

The reply, which came on October 22, is shown in Image 12.1. The request that the matter be handled by Kim Philby, who was later exposed as a Soviet double agent, is ironic.

By October 23, the British and American authorities were both reviewing their records, to see if they had overlooked anything. As the full implications of the mislaid FBI report were realized, MI5 set in motion what amounts to a cover-up. The strategy was to downplay Pontecorvo's significance not only to the public, but also to the British prime minister.

Immediately after Pontecorvo's disappearance was confirmed, Cockcroft reported that "for past several years he has had hardly any contact with secret work and has mainly been concerned with Cosmic Ray studies. In the early period in Canada he had access to information about the Pile there and in so far as he has recently had any secret work it has been on detailed problems."[46] This narrative was used to brief Prime Minister Clement Attlee. Guy Liddell wrote in his diary about his briefing of Attlee:

"The PM asked how far Pontecorvo had had access to vital information. I said that, according to the answer to the Question to be put down in the House today, D At En [Director of Atomic Energy, Cockcroft] had expressed the view that for several years Pontecorvo had hardly had any contact with secret work."[47]

Chapman Pincher, who, as scientific correspondent for the *Daily Express*, covered the Pontecorvo affair in 1950, and has zealously investigated British intelligence for half a century, described Liddell's brief to the prime minister as having "grossly understated" Pontecorvo's access to secrets.[48] The statement that "for several years he had hardly any contact with secret work" is disingenuous, given what we know about his research. Moreover, it utterly contradicts MI5's own statement to Harwell on April 25 that Pontecorvo was a security risk due to his "access to top-secret information"![49] A line in a report to the Foreign Office, which initially stated, "Bruno Pontecorvo had access to top secret information," was changed to "From the security standpoint a potential security risk existed."[50] This duplicity reveals the pressure that MI5 was under to keep a lid on an internal disaster.

On Monday, November 6, the UK House of Commons began a debate about Pontecorvo, "who might have atomic secrets of value to an enemy."[51] On Monday morning, before the debate began, Michael Perrin briefed Sir Roger Makins, deputy undersecretary at the Foreign Office, a "most courteous man, tall and thin with a commanding presence, like a great Norman knight."[52] Perrin's briefing to him went as follows:

> Pontecorvo's main value to the Russian atomic energy project would result from his knowledge of the main NUCLEAR features of the Canadian Heavy Water Research pile. He was unlikely to have expert knowledge of the important technological features of the pile, such as heavy water purification and recombination system; canning procedures etc. He has a good general picture of the possibilities of different types of future reactors likely to be important in a power programme though he would not be able to write out a detailed specification for anyone. He had no contact with atomic weapons work. An outstanding nuclear physicist as he is would, however, be of great general value if he were admitted freely to the project. Previous experience suggests that he is more likely, however, to be interrogated and consulted but not allowed to work on the main project.[53]

While the Foreign Office and prime minister were being presented with political spin, J. Edgar Hoover, head of the FBI in Washington, and his counterpart in London, Sir Percy Sillitoe, director general of MI5, formed their own conspiracy. They agreed to cover up possible failures in their respective agencies relating to the handling of the now-infamous FBI letters.

On the day of the debate, Roger Hollis of MI5 spoke to his senior colleagues about his briefing of the Minister of Supply, George Strauss. Strauss wanted to know how to respond if he was asked questions during the debate about items in his briefing that were marked "not for disclosure." Hollis told his colleagues that Director General Sillitoe had recently seen J. Edgar Hoover and the two had agreed: "Neither [would] make press statements about the other's office without first clearing with the other. It was important that nothing be said that give any indication that the FBI had not passed on their information to the British authorities." This was an outrageous statement by Hollis, implying that the FBI was to blame. A bland response was concocted for Strauss: "The British and American security authorities are constantly exchanging information on matters of security interests."[54]

In the ensuing debate in the House of Commons, Strauss claimed that Pontecorvo had had no direct access to secret subjects for some time, "except in a very limited way." However, he admitted that it was impossible to be sure that he had not obtained information from Harwell or Canada that would be of "value to an enemy." He added, "I have no conclusive evidence of his present whereabouts but I am sure that he is in Russia."[55]

Although the FBI and MI5 were "constantly exchanging information" with each other, they were misleading officials in other branches of government. For example, a letter sent on November 29 from Sir Oliver Franks (the British ambassador in Washington) to Roger Makins referred to a memo that had been shared with the FBI. The version of Franks's letter that was sent to Gordon Arneson in the US State Department, however, omitted a crucial paragraph. The missing passage mentioned that in February 1943 the FBI had sent the damaging memo about Pontecorvo to the British security services, and that the two nations' intelligence agencies had been discussing possible reasons that the memo had been discounted or overlooked. An annotation in the minutes attached to the letter notes, "It is naturally desirable that these facts should not become

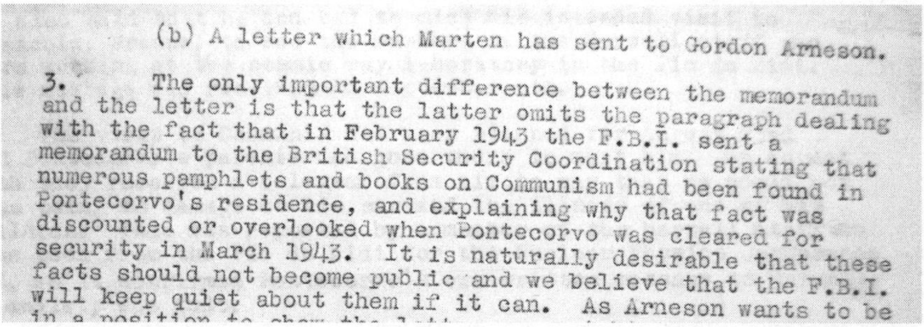

(b) A letter which Marten has sent to Gordon Arneson.

3. The only important difference between the memorandum and the letter is that the latter omits the paragraph dealing with the fact that in February 1943 the F.B.I. sent a memorandum to the British Security Coordination stating that numerous pamphlets and books on Communism had been found in Pontecorvo's residence, and explaining why that fact was discounted or overlooked when Pontecorvo was cleared for security in March 1943. It is naturally desirable that these facts should not become public and we believe that the F.B.I. will keep quiet about them if it can. As Arneson wants to be

IMAGE 12.2. Part of a letter showing attempts to cover up alleged failings on the part of MI5 and the FBI, linked to the loss of information about Pontecorvo's communist background. (AUTHOR, THE NATIONAL ARCHIVES.)

public."[56] In this the agencies succeeded. These machinations remained secret for nearly half a century.

There would be an ironic postscript to this fiasco. In 1951, two senior Italian intelligence officers visited Liddell. They confirmed that Bruno Pontecorvo's activities with Emilio Sereni and other prominent communists in 1930s Paris had not gone unnoticed. The Italians had a file on Bruno, which recorded his communist affiliations. Why had no one consulted them in 1943 when Bruno's security was vetted? As Liddell remarked, with a sigh: in 1943 Britain had been at war with Italy, so "consultation would have been rather difficult."[57]

FALLOUT

On October 23, Guido Pontecorvo wrote to his parents about his suspicions that Bruno had gone to the USSR. They replied, "The whole thing seems mad to us and we cannot believe that our Bruno could have gone entirely of his own free will. Whatever theory one produces comes up against insuperable objections, especially for those, like us, who know him. We seem to be living in a dream."

The press had besieged them; their letter to Guido refers to the "loathsome press campaign." His father wrote, "I will not tell you about the siege of journalists but we were unable to get away from flash photographs." Bruno's mother added, in anguish, "What has happened to our boy and his dear ones?"[58]

Guido responded to his parents with a sober assessment of the implications of Bruno's disappearance. They wrote back to him, in turn: "We do hope they are treated well and that Bruno is not too unhappy. It is all a great and most painful mystery. The fact that we cannot know or receive even one word is insupportable. But you say that with this we must be prepared and we must do everything to assure ourselves that one day everything will be cleared up in the best possible way even though today it is not understandable."

And finally his mother added, "I think so much about the three children of Bruno with anguish. They were so happy and it was such a serene little family."[59]

Marianne's family was devastated as well, and deeply hurt once it became clear that their daughter and grandchildren had passed through Stockholm without contacting them. Marianne was the favourite sibling of her older brother. Nothing was known of her fate for several years. In 1954 her brother and his wife had a daughter. They baptized her Ann Mari Helen. Her missing aunt's forenames were Helene Marianne.

THE SECURITY SERVICES CONTINUED TO KEEP A CLOSE WATCH ON members of Bruno's extended family. On April 6, 1952, when Guido and his wife returned to Glasgow from Copenhagen, their arrival at Prestwick Airport was reported to the intelligence authorities.[60]

On September 6 of that year, Bruno's parents, Massimo and Maria, arrived in Folkestone, England, having crossed the English Channel from Calais. MI5 were informed that "the examinations showed they were parents of the missing scientist. They said there had been no news of him despite attempts they had made to trace him through the Italian Communist Party. They were here to spend 4 to 6 weeks with David Guido and Anna." The record then bears the cozy addition: "They appeared to be a decent old couple, and were therefore landed conditionally as above." Nonetheless, MI5 was informed.[61]

At least these actions were discreet and had no effect on the couple's lives. However, for Bruno's brother Paolo, who was living in the United States under the anglicized name of Paul, the fallout was severe.

Paul was working for Raytheon, an electronics company involved in national defense. He was on track to become a senior executive in the company, but, after Bruno's disappearance, Paul's promising career mysteriously stalled.[62] This was not his imagination, but the result of government interference. The British had learned that Paul Pontecorvo was employed in the US "on research work connected with radar," and MI5's director general decided that this "should be reported to the FBI." The FBI took note. An FBI file dated December 11, 1950, remarks that Paul Jacob Pontecorvo, a radio engineer at Raytheon with "access to restricted information," is the brother of Bruno, who "allegedly fled to the USSR on or about 2 September."[63] The remainder of the file is blacked out.

Bruno's brother Giovanni also suffered. He not only missed out on Bruno's promised financial help with his poultry farm; his business was threatened with collapse when the media hounded him during an agricultural exhibition in London, and clients were put off as a result. He later changed his name from Giovanni David Pontecorvo to David Maroni, adopting his mother's maiden name. Only then was he able to carry on and avoid persecution.

THE PATENT SAGA

The patent for the discovery of the slow-neutron phenomenon also became caught up in the Pontecorvo affair, and the Via Panisperna Boys lost their chance to make a fortune. Emilio Segrè was a very astute businessman, and for him, more than perhaps any of the others, the promise of riches from the patents was very powerful. Whether Segrè was right or wrong about Pontecorvo's motives for choosing Harwell in 1949, the consequences of his intervention with the FBI would prove disastrous—for Bruno as well as for Segrè himself.

The slow-neutron phenomenon that had been discovered by the Via Panisperna Boys was crucial for both nuclear reactors and nuclear weapons. The scientists and engineers of the Manhattan Project had exploited it. Now, postwar, the ownership of the rights to the discovery became mired in dispute.

In 1935, the team had received an Italian patent for their discovery. They were not businessmen, and had no idea how to register patents in other countries. Slow neutrons were a good thing, but slowing their research in order to deal with lawyers was not. A lucky break came their way in the guise of Gabriello Giannini, one of Fermi's first students. Ambitious and eager to make his fortune, Giannini had immigrated in 1930 to the United States, where his quick mind and self-confidence enabled him to succeed, despite the Great Depression. Soon he had gained some legal experience. His former colleagues decided that he was their man, and made a deal with him: if Giannini could register patents in Europe and America for the slow-neutron process, he could become the eighth member of the consortium and receive an eighth of any eventual profits.

Giannini first won some patents in Europe, and then turned to the United States. In October 1935, "G. M. Giannini & Co." filed an application for a patent with the US Patent Office. Five years passed before the authorities agreed that the applicants were the inventors of the slow-neutron process. On July 2, 1940, they received their American patent. Within months, their breakthrough had become the heart of the Manhattan Project, and the patent gave the team a financial stake in the technology the US government was using to make plutonium.[64] After the war ended, the slow-neutron method became central to another new enterprise—the development of nuclear power for peaceful purposes. Suddenly, all sorts of difficulties erupted around the patents.

During the war, the Manhattan Project had been a military enterprise, a closed secret. Postwar, the Manhattan Project ended and the civilian Atomic Energy Commission was born. In 1946, the Atomic Energy Act authorized payments to be made for patents that had been used during the war. It looked as if the Via Panisperna Boys were about to become rich, until they learned of a hiccup: Enrico Fermi was an adviser to the AEC. He received no salary, but the government lawyers argued that nonetheless he was a government employee, which meant that he and his coinventors could pursue no claim against the United States.

There matters rested for four years. Meanwhile, in February 1949, Pontecorvo moved to Harwell. In August of that year, the Soviet Union detonated its first atomic bomb. It soon became obvious that atomic secrets had been passed to the Soviet Union not just by Nunn May in Canada, but also from Los Alamos, and paranoia grew that a communist fifth column was at work in the United States. Against a rising clamour of anticommunist persecution, Giannini resumed his efforts to reach an agreement with the AEC over payment for the use of the patent. On August 21, 1950, without having consulted the "Boys," Giannini filed a lawsuit against the government for nonpayment of patent royalties and patent infringement.[65] The claim, for around $100 million in modern values, seemed outrageous to the scientists; the fact that *l'Unità* accused the US government of "defrauding Enrico Fermi" inflamed passions even more.[66]

Giannini tried to calm his partners. He explained that this was how the game worked; the vast amount was really an imaginary figure used to get the ball rolling, whereas the actual sum would turn out to be very different. But then an astonishing development occurred. On October 21, before any progress had been made on the lawsuit, the news broke that Pontecorvo—one of the inventors and claimants in the lawsuit—had vanished without a trace, probably having fled behind the Iron Curtain. In those feverish times, it hardly mattered whether Pontecorvo was a spy, as the media speculated, or simply a communist "fellow traveller." Giannini got cold feet. He didn't want to lead a lawsuit against the US government on behalf of a group containing a communist, who had in all probability defected to the Soviet Union. Giannini soon withdrew the suit.

Giannini gave up because, according to him, the group didn't "wish to be associated even remotely with anyone involved or reputed to be involved in any sort of international mystery."[67] Later, in 1953, Giannini reached a settlement with the Atomic Energy Commission for a fraction of the claim. After legal expenses had been paid, each inventor received much less than they had hoped for originally. The British *News Chronicle* reported in November 1953 that the US government was holding $18,750 for Bruno. The article, whose headline read, "Pontecorvo—here it is," noted that the money was waiting for the missing scientist, but that "he has to collect it himself."[68]

The Pontecorvo affair created unexpected fallout, whose influence was felt far beyond Bruno, Marianne, and the boys. As for Bruno, he had reached his half life, slipping from one world to another. It is perhaps appropriate that the term *half life* is taken from the realm of nuclear decays; with Bruno's flight to the USSR, his chances of winning a Nobel Prize decayed also.

THIRTEEN

THE MI5 LETTERS

"Did MI5 get back to you after I forwarded them your letter?" The neat, handwritten note, on House of Lords stationery, was brief and to the point.[1] When I received it, about two years into my research, I had no idea that it would lead me to solve the mystery of Bruno Pontecorvo's sudden disappearance.

MI5 did get back to me, and confirmed what my correspondent had hinted at: that a file of "lost" papers regarding Pontecorvo had been "found." The documents in question turned out to contain the history of MI5's interest in Pontecorvo during the months that led up to his defection.[2]

The final entry in the MI5 record before Bruno Pontecorvo's disappearance was a letter received on July 19, 1950, from the British embassy in Washington. The document, which is marked "SECRET," appears to have had little impact. No action was taken based on its transmission. Years later, however, its contents would embarrass the entire British security apparatus. The question of whether it had truly been lost before my interest was brought to MI5's attention, or whether it had been "lost" out of convenience, is for conspiracy theorists to debate.[3] The lack of action should also be intriguing to those who have argued that Roger Hollis, the director general of MI5 from 1956 to 1965, was in reality a double agent working for the Soviet Union.[4] Chapman Pincher, the journalist and veteran spy-catcher, has remarked that Hollis was so reluctant to take action on various occasions that he was either incompetent or deliberately duplicitous. The Pontecorvo file is a notable case in point. Hollis was fully aware of the serious nature of the Pontecorvo case, as he added written comments to the file in May 1950. Yet when the letter from Washington arrived in July, there was no action within MI5. It would, however, lead to action in the Soviet Union. In any event, once I saw the letter, the kaleidoscope of facts began to settle into a clearer picture.

The significance of the letter will become apparent once we understand certain events that had been taking place in the United States over the preceding months.

A SECRET WAR

In the years immediately after World War II, the British embassy in Washington, DC, was the weak point of British and Allied security. Unknown to the authorities, it played host to three members of the infamous Cambridge Five spy ring. From 1944 to 1948, one of these spies, Donald Maclean, exploited his position as the British representative on the American-British-Canadian council on the sharing of atomic secrets. He was, of course, privy to these secrets, and passed news about development of the atomic bomb and nuclear power to the Soviets. Meanwhile, another member of the group, Guy Burgess, was based in the Foreign Office in London until late 1950. For a period in 1949 and 1950, Burgess forwarded to the KGB information that had originated with Kim Philby in Washington. This continued until Burgess too moved to the Washington embassy.[5]

Kim Philby had arrived at the embassy in September 1949. He formal title was First Secretary but his specific (and covert) role was as a representative of the Secret Intelligence Service (SIS), or MI6. Ever since World War II, the United Kingdom and the United States have shared intelligence. Thus, one of Philby's duties was to liaise with the CIA, which meant that he was aware of some American operations, in addition to British ones. At this stage of his outstanding career, many saw him as a future "C"—Chief of SIS.

In reality, Philby was a traitor throughout his career, from 1934 until his exposure in 1963.[6] His autobiography admits that he was a double agent, who worked for the Soviet Union but was paid by the British. Philby himself wrote of his "total commitment to the Soviet Union." He regarded his "SIS appointments purely [as] cover-jobs" to be carried out only well enough to enable his "service to the Soviet Union to be most effective."[7] Like Maclean and Burgess, Philby was a member of the infamous Cambridge Five ring of traitors, who were groomed at Cambridge University in the 1930s, and who rose to senior positions in the British civil service. His résumé of duplicity includes giving alerts to the Soviets when their atomic spies came under suspicion. In 1945, as head of the SIS's Soviet counterintelligence section, Philby kept the Soviets aware of developments in the case of Alan Nunn May. As the net closed around the physicist, Philby warned Moscow that MI5 had caught wind of a meeting planned in London between Nunn May and his Soviet contact. As we saw earlier, Philby's intervention caused the meeting to be aborted.[8] Philby used his position to alert the Soviets not only about Nunn May but also about Fuchs. It now seems that he tipped them off about Pontecorvo too.

Philby was the centre of the Cambridge spy ring. Suspicion about him grew after Burgess and Maclean defected to the USSR in 1951, but Philby himself managed to maintain his double life for another twelve years. The fact that Philby evaded detection for so long was due to a combination of skill and good fortune, as well as his powerful position at the heart of British intelligence operations. This privileged role gave him early access to critical information.

Most significantly, Philby was one of a handful of people who were party to the biggest diplomatic secret in the postwar West: the VENONA project, an American programme to intercept and decrypt Soviet intelligence traffic. In the summer of 1949, Meredith Gardner, a lean and gangly American linguist, cracked the Soviet diplomatic codes. Radio messages between Moscow and its Soviet embassies in North America were now open to the West. Philby was briefed about VENONA in September 1949, soon after the Soviets detonated their first atomic bomb. He immediately told the Soviets that their codes had been cracked.

The decrypts contained references to three scientists who had been working on the Manhattan Project. The message revealed that the trio, code-named CHARLZ, QUANTUM and MLAD, had passed atomic secrets to the Soviet Union. Although we now know that MLAD was Ted Hall, his identity and that of QUANTUM were still a mystery in 1949. Within weeks of his arrival in Washington, however, Philby learned that CHARLZ had been identified as Klaus Fuchs.[9] As we have seen, Fuchs was placed under surveillance by MI5 in October 1949. His arrest and imprisonment occurred despite Philby's best efforts.

Philby was being ultracareful. The Soviets' intelligence operation at their embassy in Washington was in a turbulent state, two of their residents having been recalled to the USSR in the months prior to Philby's arrival in the city. He therefore refused to deal with any Soviet intelligence officers in the US, and for about a year his only contact with Moscow headquarters was via messages sent through Burgess in London.[10] Through this circuitous route, Philby alerted Moscow that Fuchs had been exposed, and warned that any Soviet agents who had dealt with Fuchs might be compromised. Philby's warning was right: Fuchs confessed and was arrested; next, Fuchs's courier Harry Gold was arrested in the United States in May 1950, and within weeks the US government was pursuing communists—real or imaginary—with a zeal reminiscent of 1930s Nazi Germany. Two weeks later the US invaded Korea, and up to a dozen "atom bomb spies," as the headlines described them, were arrested.

When Julius and Ethel Rosenberg, a husband-and-wife spy team, were arrested in the summer of 1950, the Soviet network in North America was decimated. Kim Philby's invaluable work as a double agent in the British embassy in Washington continued undetected, but he was an exception. The Soviets now cut their losses and extricated their agents from North America. In late June or early July, for example, Lona and Morris Cohen were rescued. They were first smuggled out of the US to Central America, and then flown to Moscow later that summer.[11] They arrived in the Soviet capital a few weeks before Bruno Pontecorvo.

Meanwhile, back in the United States, Senator Joseph McCarthy continued to rant about a "Red threat." This Cold War hysteria fanned the flames of political intimidation. In New York, mounted police broke up a protest meeting that was calling for a cease-fire in Korea. Hundreds were injured or arrested. Bruno Pontecorvo feared that right-wing extremism, which he had fled in Europe, was now reappearing in the United States; with the downfall of Joliot-Curie, it even seemed to be spreading to France, and was threatening to come to the United Kingdom.[12]

The pace of events quickened that summer. In June, Philby learned of a new breakthrough in the VENONA decryptions: Soviet telegrams from 1945 had revealed the existence of a ring of spies at the heart of British intelligence. However, the information from VENONA was fragmentary. At this stage, they had only partial decrypts. This was just as well for Philby, who immediately understood that this information referred to himself and his colleagues in the Cambridge Five.

In order to keep abreast of developments, Philby manoeuvered to increase his access to VENONA, and arranged for SIS to provide him immediately with copies of any new VENONA material. The official reason for this arrangement, as stated in a letter sent to the director general of SIS on July 18, 1950, by Geoffrey Patterson, the Washington embassy's liaison with US security, was that it would enable Philby to absorb and analyze new information before he met with the FBI.[13] In reality, of course, Philby was simply trying to protect himself. As a result of these manoeuvres, Philby learned that VENONA had identified a code name, HOMER, which he recognized as referring to Maclean. This information was passed on to Moscow. The following year, as the net closed, Maclean—along with Burgess, who had also been compromised—would defect to the USSR.

Tim Marten worked in the British embassy in Washington at this time, and one of his responsibilities involved communications on atomic energy. Tim recalled how these messages "were ultra-secret and therefore went through the MI6 communication channel. . . . So Philby, as head of MI6 in Washington, had direct access to every telegram that I sent or received." At this memory, Tim gave an ironic laugh. He then continued: "But of course I thought Philby was rather a good egg at the time. He appeared to be . . . quite a wheel, in constant touch with the CIA and state department. He was a very highly regarded person all round and obviously very competent. What we didn't know at that moment was that he was passing everything on to the Russians."[14]

Philby's unique access to American and British intelligence enabled him to conduct the espionage orchestra during the critical months of 1950, when the Soviet networks in North America were in danger. In addition to information that affected him directly, or references to the Cambridge spy ring, he kept a careful watch for anything that would interest his real employer, the Soviet Union. In the middle of July, he saw another letter that Patterson had drafted.

Written on July 13, 1950, and received by the director general of MI5 in London on July 19 (and most probably by Philby's contacts in Moscow

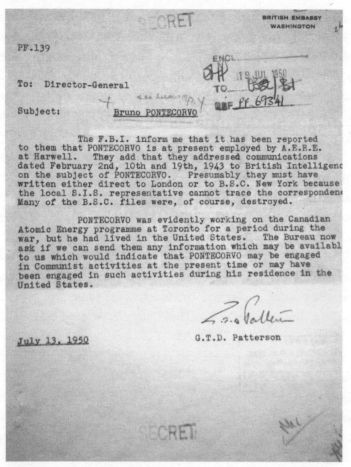

IMAGE 13.1. Letter from Geoffrey Patterson to MI5, sent July 13, 1950, received in London July 19, alerting them to the FBI's interest in Bruno Pontecorvo. Note also the reference to Philby on line seven. This was the final entry added to MI5's file on Pontecorvo before he fled to the USSR. (AUTHOR, THE NATIONAL ARCHIVES.)

soon after), Patterson's letter concerned a subject of particular interest for readers of this book: "The FBI inform me that it has been reported to them that PONTECORVO is at present employed by AERE at Harwell. They add that they addressed communications dated February 2nd, 10th and 19th, 1943 to British Intelligence on the subject of PONTECORVO. Presumably they must have written either direct to London or to BSC New York because the local SIS representative cannot trace the correspondence. Many of the BSC files were . . . destroyed [at the end of the War]."

Patterson then pointed out that Pontecorvo had worked on the Anglo-Canadian atomic energy project during the war, and had also lived in the United States. "The [FBI] now ask if we can send them any information which may be available to us which would indicate that PONTECORVO may be engaged in Communist activities at the present time or may have been engaged in such activities during his residence in the United States."[15]

Patterson's letter, which explicitly mentions the "local SIS representative" in Washington, shows that Philby was fully aware of these developments.

Philby was the overall SIS chief in Washington. He discussed atomic affairs on a regular basis with the embassy's specialist on atomic matters, Dr. Wilfrid Basil Mann. Their offices in the embassy were next to each other. Judging from Tim Marten's testimony on Philby's access to material, and the assessment of others who were familiar with his style and the workings of the Washington embassy at the time, it is most improbable that Philby was unaware of Patterson's letter. As he had done with Alan Nunn May, and then Klaus Fuchs, he now had to tell Moscow of the West's interest in another atomic scientist: Bruno Pontecorvo.[16]

As stated by Patterson, the British intelligence team in Washington was unable to locate the 1943 letters from the FBI, even when the letters' existence was brought to their attention. Given Philby's reputation, one might imagine that the failure to find the letters occurred because he had destroyed the evidence.[17] However, it seems more likely that, on this occasion, Philby was acting in good faith: the 1943 correspondence was indeed lost, possibly when the British Security Coordination closed at the end of World War II and many files were destroyed.

The FBI subsequently forwarded copies of the letters to MI5.[18] They showed evidence only of Bruno's communist associations. They did not show evidence that he was a spy. Their resurrection in July 1950 suggests that they were part of a fishing expedition conducted by the Americans, inspired by McCarthyism and a desire to undermine the Via Panisperna Boys' lawsuit against the US government.[19] If Philby had seen these letters, they would have raised little alarm. However, it seems he did not. All he knew was that the FBI was interested in an atomic scientist named Bruno Pontecorvo, that they had written not just one but *three* letters about him within seventeen days in 1943, and that VENONA had revealed the existence of two still-unidentified spies at the heart of the atomic project, code-named MLAD and QUANTUM.

As the reader knows, MLAD would eventually be identified as Ted Hall, a brilliant young physicist who was arguably the most successful of the "atom spies." QUANTUM remained an enigma until 2009, when KGB files identified him as Boris Podolsky, a US-born Russian physicist.[20] None of this was known to Philby in 1950. We have no hard evidence that Philby warned Moscow about the FBI's interest in Pontecorvo, or what the warning might have consisted of, but it is most improbable that a warning was not transmitted.

BUT WHAT ABOUT LONDON? WHAT REACTION DID PATTERSON'S LETTER cause there?

A deafening silence, it would seem. The receipt date on the letter is Wednesday, July 19, less than a week before Pontecorvo left the United Kingdom, never to return. There is a pencilled note reading "See Lumes" (although the script is hard to decipher), but no further mention of the letter in Pontecorvo's file, nor any record of action being taken. Indeed, this is the final entry in the file before news of Pontecorvo's disappearance erupted.

The personal diary of Guy Liddell records the inside story of MI5 that summer. At the time, there was a lot of concern about the possibility of the British getting involved in a war; the Korean War had begun in June, and the recent news that the Soviets had the atomic bomb worried everyone. His diary also reveals tensions between MI5 and the FBI, which have relevance to the Pontecorvo case.

On July 29, just ten days after Patterson's letter arrived from Washington, Liddell recorded, "Hollis is worried about the nature of enquiries we are receiving from the FBI." The FBI had recently asked MI5 to place two people under surveillance, and had followed up by seeking information about a third person, who McCarthy had asserted was a spy, without any evidence. Even President Truman denounced McCarthy for this behaviour, and MI5 was concerned that if they started making inquiries on behalf of the Americans, they would get bogged down in "this mud surrounding Senator McCarthy." Liddell continued: "Unfortunately [J. Edgar] Hoover is taking a personal interest, since he doubtless wishes to have a dig at the State Department."

It is clear that there was a history of tension between the two intelligence agencies. Liddell said his primary aim was "not to exacerbate the rather strained relations between [MI5] and the FBI." Ever since Fuchs's arrest, the FBI and Hoover had put pressure on MI5, given the British media information that was detrimental to MI5, and blamed the UK authorities for having been lax. In London patience had worn very thin. Liddell recorded this sorry saga in his diary throughout the spring and early summer. He also suggested that MI5 should "politely" point out to the Americans that "we are not one of their field offices."[21]

Under these circumstances, it is possible that the UK's lack of interest in Bruno Pontecorvo's possible "Communist activities" was a case of the FBI having cried wolf too often. Even so, given that MI5 had been interested in Pontecorvo for several months, one would expect some response to Patterson's letter. Although Liddell's diary mentions three subjects of particular interest to the FBI, none of these relate to Pontecorvo. Given Liddell's suggestion that MI5 was being treated as a "field office" of the FBI, it's possible that the British agency could have regarded the wording of Patterson's letter as presumptuous and thus not given it high priority. Hollis conspiracy theorists, on the other hand, might add this to their list of his "unfortunate oversights."

Unaware of Philby's duplicity, the security chiefs in England took their holidays, or passed afternoons at the Oval cricket ground that summer, watching the magical West Indian spinners, Sonny Ramadhin and Alf Valentine. It was a more relaxed era than today, the Cold War notwithstanding; a fortnight-long holiday was standard. Liddell, for example, went to Ireland for the last two weeks of August.

Even if MI5 had taken an interest in the FBI's letters, it could not have told Pontecorvo's superiors anything more damning than "we have reports that Pontecorvo and his wife have expressed communist opinions." Although this would be a serious matter for an intellectual in the United States, where anticommunist paranoia was rife, it was no crime in the United Kingdom. For Pontecorvo it would have simply meant that he was a security risk. But that was old news, of course: the transfer to Liverpool had already provided a pragmatic solution to that problem. Even taking into account the new information, MI5 had no evidence that would have justified a "purge" (prosecution) of Bruno Pontecorvo.

So it is unlikely that the inquiry by the FBI would have radically altered Pontecorvo's prospects in the United Kingdom. However, that conclusion could only be drawn by someone who was conversant with the state of knowledge about Pontecorvo in MI5. For Philby in Washington, deeply connected to VENONA but remote from the minutiae of MI5's London office, let alone Harwell security, the reference to the "Communist activities" of a nuclear physicist, who might have been MLAD or QUANTUM, demanded action.

Assuming that Philby passed on this information to Burgess for transmission to Moscow, which was his normal modus operandi at the time, there would have been little time for them to act before Bruno left on holiday. The story about the man who interrupted the tennis match might have some sinister significance, but it is hard to square this with Bruno's relaxed attitude, which continued until his final days in Italy.[22] There was a story in the media that two men had contacted Bruno in the Alps, during August, but no source was ever provided.[23] Bruno was camping and on the move; thus the earliest opportunity for the Soviets to make contact with him would have been late in his holiday, through mutual friends in Italy. This fits with his sudden change of behaviour during the last week of August. If the Soviet embassy in Rome gave Bruno money to pay for his airline tickets, as was suspected by MI5, this would have been a logical response to Philby's alert. Emilio Sereni, his cousin, the good communist and activist, would have been a convenient liaison in the Soviets' quest to make contact with Bruno.

As for the decision to flee, Bruno's son Gil told me, "It looked to me like a sudden decision. [Whatever] the reason, it was [made] quite late."[24] Today it seems probable that the letter from Washington was a key—perhaps *the* key—to his unpremeditated flight.

EARLY IN SEPTEMBER 1950, BRUNO WAS CROSS-EXAMINED IN MOSCOW by the KGB.[25] According to one account, his interrogators included Stalin's enforcer, Lavrenti Beria himself.[26] We do not know the actual date of this interrogation, but there would be no advantage to delaying it. It probably occurred before the following event.

On September 12, when no one in the West yet realized that Pontecorvo had disappeared, Philby—who happened to be in London—dropped by to see Guy Liddell, the deputy director general of MI5. Philby, of course, was a high-ranking member of MI6, also known as SIS. Liddell's diary records what happened next.

"I had a long talk with Kim Philby. . . . I thought I discerned a fly thrown over me in the form of a suggestion that it was really unnecessary for us [MI5] to have a Washington representative [Patterson], and that he [Philby] could carry the whole business. . . . I told him that whatever the flow of information I was quite convinced that [MI5] ought to have a man in [Washington]."[27]

The reference to a "fly being thrown" is an idiom from fishing. The fly is a lure, typically when fishing for trout. The moment the fish takes the fly into its mouth, the angler jerks the line so that the hook penetrates the fish's mouth and captures it. In this scenario, Liddell is the fish; Philby the fisherman.

Philby's suggestion could be perfectly innocent but this is unlikely: Philby calculated everything with a view toward his personal safety. His colleague Donald Maclean was already under suspicion, and, in the memory of Lorna Arnold, "looked like he had ants in his pants."[28] If, as seems probable, Philby's actions had led to Pontecorvo's flight, there was a potential danger for Philby that he needed to guard against. Given his role in MI6, Philby could be reasonably certain there was no British double agent planted within the KGB, but he could not be sure the same was true of the Americans. If Philby's role became apparent during Bruno's cross-examination by the Soviets, this hypothetical double agent might hear of it. Philby wanted total control over the flow of information between Washington and MI5.

As it turned out, Philby had no need to worry about any such double agent. No one in the West knew what had become of Bruno Pontecorvo and his family for five more years. By that time, Philby's partners in crime, Guy Burgess and Donald Maclean, had both defected to the USSR, and suspicions about Philby had also come to a boil. On September 12, 1950, however, Philby had little to worry about. He had done his job well; no one suspected that he had played any role in the Pontecorvo affair.

SECOND HALF

"Midway on our life's journey, **I found myself in dark woods**, the right road lost."

—*Dante's Inferno*

FOURTEEN

IN DARK WOODS

Even today, Finland is a land dominated by forests, which stretch for hundreds of miles.

In 1950 these dark woods extended from the outskirts of Helsinki into the Soviet Union. That September, two cars sped along empty forest tracks toward the Russian border. One contained Marianne and her three sons; the other contained Bruno hidden in its trunk.[1] Gil Pontecorvo, who was twelve at the time, did not know who the drivers were, but he laughed wryly as he told me, "They were Russians for sure."

Gil had no idea where they were. After the idyllic weeks at camps in the mountains, and the long, sunny days by the warm Mediterranean, in the company of grandparents, uncles, aunts, and cousins, everything had changed.

When the family left Italy for Stockholm, Gil "thought these journeys were all part of their holiday. I had *no idea* that we were not going back. We didn't have much luggage [and] had left lots at home." But he realized something strange was going on when the family failed to visit his maternal grandmother, who lived near the airport, and instead flew on to Helsinki, to be met by strangers. "We spent one or two nights there," he recalled. So it would have been on the third or fourth of September that they made their drive through the woods. Six decades later, Gil remembers this episode clearly: the forest "went on for mile after mile, seemingly without end." For entertainment, he "had a book by Jack London to read." As for his father being in the trunk of one of the cars: "I knew something was up."[2]

At the time, some Western reporters imagined that the Pontecorvos had been taken to Porkkala, twenty miles from Helsinki, where a long bridge spanned a creek separating Finland from an area that in 1950 belonged to Russia. There were no guards on the Finnish side, as no Finn "in their right senses" would want to cross; the media saw this as an explanation for the fact that the Pontecorvo family had vanished without trace. At the Russian end of the bridge, sentries from the Red Army supposedly welcomed the party. This, however, is one of the many myths about their flight. In reality, they had driven eastward, eventually reaching the USSR at Vyborg, which had been part of Finland until it was lost to the Soviets in the Winter War of 1939–1940. Only now, inside the Soviet Union after more than one hundred miles on the road, was Bruno able to come out into the open.[3]

Another eighty miles brought them to Leningrad (formerly, and again today, known as Saint Petersburg).

Their KGB guards politely refused Bruno's request to look around the city, explaining that there would be plenty of opportunities to see it once he was settled in the USSR.

During his time in the car trunk, Bruno had been preparing a statement. Years later, he claimed that, for him, the die was already cast and the future lay in the USSR. He wanted to send a message to family and colleagues in the West, explaining his actions: the Soviet Union was a peace-loving nation whose ideals he believed in. Exhausted, the family sat down to their first meal in Russia, all the while under guard, and spent a night in an "anonymous apartment but no one slept very much."[4]

THE NEXT MORNING A MAN AND WOMAN CAME TO THE HOTEL TO accompany the family to the railway station and onto the train to Moscow. The man had short hair and was about Bruno's age; the woman younger. She carried a bunch of flowers, which she presented to Marianne.

Upon arriving in the capital, the Pontecorvos were met by a small delegation. A "tall and elegant man," who stood out as the leader, greeted them with pleasantries. He asked Bruno in English if they'd had a good journey and whether they needed anything. Bruno replied that he had prepared a statement that he wanted to read out on the radio, directed to his colleagues in the West, explaining his support for the Soviet Union. The first hint of the restrictive nature of life in Stalin's fiefdom now showed itself, with the man's polite but firm refusal: "There will be plenty of time to do so in due course." Bruno acquiesced.[5]

The party got into a pair of black cars, with grey curtains drawn across the windows. Bruno rode in one car, with the senior KGB man, while Marianne and the children rode in another. Separated from Bruno for a while, they tried to sneak peeks at Moscow.

During his time at school in Abingdon, Gil had got the impression that Moscow was essentially a Wild West town made of concrete, where the rule of law was marginal and unwary travellers were at risk. However, when he saw the city in person for the first time, it appeared no different from the other major cities he had passed through that summer: "My first shock in Moscow was to discover that it was Moscow!"[6] The one exception was the traffic: there were very few cars, many old buses, and lots of open trucks containing men dressed in army fatigues, standing shoulder to shoulder. For Bruno, his arrival in Moscow induced profound emotion: "I felt like the Jew who found the Promised Land."[7]

The car stopped in front of a huge apartment block on a wide boulevard—Gorky Street—not far from the Kremlin. The building had been built just three years before, and one of its luxury apartments had been prepared for the Pontecorvos' arrival; it would be Bruno's Moscow base for the rest of his life.

Their apartment was large, and located on the seventh floor. There were three bedrooms, a living room, a kitchen, a dining room, and a study, all filled with traditional furniture. The ceilings were high, embossed with stucco, and the living room had large windows, draped with white curtains that reached the ceiling. Bizarrely, the apartment contained a set of classic French books.

Bruno spent many hours reading the French classics to fill the time between Russian language lessons, as the family was effectively "locked in the house" for several weeks.[8] Their entire existence was now under state control.

A young KGB officer had been assigned to watch over them. To Bruno's surprise, the man asked him for news of the war in Korea. In the afternoon a copy of the state newspaper, *Pravda*, arrived, along with an interpreter, who was also eager to hear the headline news from Bruno personally.[9] Already Bruno was starting to realize how news was controlled under Stalin's rule.

On the day after their arrival, the KGB officer explained that it would be best if Bruno postponed the release of the public statement he had prepared. Bruno was surprised, as he felt that it would benefit the Soviet Union if he explained his decision to his friends and colleagues in the West. The Soviet leaders, he was informed, viewed things differently. He could release his statement "all in good time." In reality five years were to elapse before the time was deemed appropriate, two years after Stalin's death. The officer explained that it was necessary for the Pontecorvos to stay indoors "for vigilance," as "no one must know" they were in the USSR; for some unexplained reason, this secrecy was deemed vitally important.

The most urgent task for the Pontecorvos was to learn Russian. They were provided with a teacher, who spoke English perfectly. First they learned to read the Cyrillic alphabet; then they learned a few basic words, spelling them out in the strange script. This required concentration, discipline, and time. There was no shortage of the latter, and Bruno and Gil set about the task enthusiastically. Marianne, however, did not. Despite the insistence of the teacher, she gave up after a few lessons, complaining that the Russian language was too difficult, and that studying it gave her headaches. That, at least, is what Bruno recalled years later in interviews with Miriam Mafai. Gil's memory of that time differs, however. He recalls his mother as a good linguist.[10] In any event, cut off from her family in Sweden, and from her friends in the West, Marianne was now isolated in a strange land.

Her isolation soon began to deepen, even within her own home. Every morning a fat maid with bleached hair and a gregarious personality came to take care of the family's needs. She cleaned the children's rooms and the kitchen, and prepared food. That much was fine. The problem was that she was very voluble, and tried to talk with Marianne in Russian, asking what she wanted to eat. Marianne generally opted out, wanting to let the maid decide, but the maid kept on asking during every visit. After a few weeks, Marianne found this unbearable, and began locking herself in her room when the woman arrived.[11]

Bruno arranged for the maid to be replaced by a younger woman, who didn't talk so much and made her own decisions. Nonetheless, Marianne continued to stay in her room for long periods, incommunicado. Like the rest of the family, she was restricted to staying indoors, based on the KGB's insistence that their whereabouts remain secret, as well as the fact that she did not understand enough Russian to go to the grocery store anyway.

The difficulty of keeping three young boys occupied added to the family problems. One day, Gil left the apartment and crossed the street. Years later, he still recalled what happened next: "I bumped into a lady who had a bag full of eggs. I think one of the eggs must have smashed. I didn't understand any Russian and spoke in English." One can imagine the scene from the Muscovites' perspective. In 1950, the USSR was tightly controlled, foreigners were regarded with suspicion, and contact with them was restricted. So the appearance on a busy street of a strange boy speaking a foreign language must have created quite a stir. Gil continued: "A crowd gathered. Someone started shouting. I didn't know what they were saying. A policeman came and dispersed them."[12] The incident was so singular that it stayed in Gil's memory, but he has no recollection of his parents' reaction. In any case, the incident can hardly have improved the family's chances of liberation.

Such events spurred Gil and his younger brothers to ask more and more questions: "What is happening? Why are we here? When are we going home?" The family had left Abingdon on July 25, at the end of the summer term at Roysse's. The new school year was due to begin on Tuesday, September 19. Gil remembers being "promised I would be at school in England by the 16th."[13] This deadline was fast approaching, and Gil was getting anxious. After two weeks in Moscow, the date finally came and his fears were confirmed: "I was hysterical because I was sure I was not going back." He threw a fit.

It is intriguing that, forty-three years later, when father and son were reminiscing, Bruno had no recollection of this event, which to Gil had been one of the singular moments of his life. Gil recalled, "He seemed well, and I reminisced with him, asking if he remembered my tantrum. I am a placid person and that is the only time in my entire life that I made such a scene. It is surprising how different people remember events. Bruno did not even remember my hysteria."[14]

Initially Bruno and Marianne may have seen their move to the USSR as simply another stop on the peripatetic path their lives had taken up until then. After all, they had already left their homelands for France, escaped the Nazis by fleeing to the United States, spent three years in Canada, then moved back to Europe when Bruno joined the team at Harwell.

Their relocation to the USSR, the mecca of communism and the home of a new atomic laboratory at Dubna, made some logical sense. Bruno's cousin Emilio Sereni travelled to and from the Soviet Union regularly, so there was initially no reason for Bruno to suppose that he would be treated differently. It seems Bruno thought he could work at Dubna without a major disruption to his life. Once the family was settled in their new home, their friends and relatives in the West could be contacted, Gil's schooling could be dealt with, and life would carry on as before. Unfortunately, life in the USSR of the 1950s was not so simple.[15]

As we have seen, the first signs of the restrictive environment that awaited the Pontecorvos came soon after their arrival in Moscow, when the authorities refused to allow Bruno to make a statement. He was stranded, unable to explain his reasons for fleeing to the Soviet Union, or even to let his family know that he was safe. This incident, along with a summons to the Kremlin during the family's first few days in Moscow, changed Bruno's view of his situation.

IN 1950, BORIS IOFFE WAS A YOUNG THEORETICAL PHYSICIST WORKING in Moscow at "Laboratory Number Three." Today this lab is home to the Institute of Theoretical Physics, but at the time it was dedicated to the development of nuclear reactors—and thus, indirectly, to the physics of atomic and hydrogen bombs.[16]

Ioffe had joined the lab on January 1, and for a few months he worked purely on theoretical problems. In May, however, an order suddenly came "from the highest level"—in Ioffe's opinion, from Beria or "probably even Stalin himself." The young scientist was to help design a heavy-water nuclear reactor, using enriched uranium, for the purpose of producing tritium in the shortest possible time. "All theoreticians were mobilized to make the physical design," he recalled later. "From that time, for years, I worked on pure science in parallel with the physics of nuclear reactors."[17]

A hydrogen bomb's explosive power comes from the fusion of tritium and deuterium, two isotopes of hydrogen. At the start of the 1950s, when the possibility of creating this weapon arose, there was practically no tritium in the USSR. The isotope is unstable, with a half-life of twelve years. Only trifling amounts of tritium are found in nature, but it can be made in nuclear reactors, using heavy water and enriched uranium. However, there were no such reactors in the USSR at the time, and design plans had barely begun. The government hoped to produce enough tritium for a weapon in two to three years; it was clear to scientists that this was out of the question, but Stalin insisted that they succeed.[18] Thus there was an urgent push in the Soviet Union to build nuclear reactors, or to find some alternative to tritium. The arrival of Bruno Pontecorvo could hardly have been more opportune.

The Soviets already had some clues to help them design a heavy-water reactor, including "blueprints of the Canadian heavy water research reactor."[19] Ioffe doesn't know exactly when this information arrived, but he believes it was shortly before he started working at the laboratory at the start of 1950.[20]

How did the Soviets get hold of these design plans? Nunn May was not the source—he had been jailed in 1946, and his deathbed confession made no mention of the plans. The available facts are consistent with the theory that Pontecorvo was the source, although there is no proof. All we can be sure of is that there was some well-placed collaborator, in addition to Nunn May, who passed information about the Canadian reactor to the USSR.

The senior theoreticians in the Soviet reactor programme included Isaak Pomeranchuk, who was Ioffe's supervisor, and A. D. Galanin, a reactor expert. Ioffe later recalled how, from the middle of 1950 until early 1951, Pomeranchuk was away from the lab, working at Arzamas-16, the Soviet equivalent to Los Alamos. During his absence, one day in the fall of 1950, Galanin was "summoned to the Kremlin." This was a singular occurrence, as, in Ioffe's words, "People were summoned to various places, but never the Kremlin."[21]

When Galanin returned to the lab, he said nothing. Ioffe and his colleagues followed the standard rule of life in the USSR: don't ask; if you need to know, you will be told. It wasn't until several years later, after Pontecorvo's presence in the country had become public knowledge, that Galanin revealed what had taken place.

Sometime in mid-September, Bruno Pontecorvo's KGB minders ushered him into one of the ubiquitous black sedans with curtained windows, and took him to the Kremlin. There he met with a group of physicists, which included Galanin. Their goal was to find out what Pontecorvo knew about the "atomic problem" and Western nuclear technology in general.

The USSR had no shortage of first-rate scientists. The central problem for the Soviet nuclear programme in the postwar years was not a lack of technical know-how, but a lack of access to uranium. The nation's first nuclear reactor, built in 1946, only succeeded due to the Soviet army's chance discovery of one hundred tons of uranium in a German repository during the war. Although the meeting with Pontecorvo taught Soviet scientists nothing significant about nuclear technology (they already had the blueprints of the Canadian reactor), it confirmed their hopes that Pontecorvo's unique expertise in uranium prospecting could be of considerable value to Stalin's nuclear strategy.

It is interesting, therefore, that during the next five years stories appeared periodically in the Western media claiming that Pontecorvo had been seen at various uranium mining sites in Eastern Europe. At the time, the significance of uranium for the Soviet programme was not generally known in the West, nor did anyone know of the Kremlin debriefing, which suggests that these rumours may have had a basis in fact.

WHAT IS HARDER TO EVALUATE IS THE CLAIM THAT PONTECORVO was cross-examined at the Kremlin meeting about matters unrelated to science—including security issues and his reasons for coming to the USSR. The accuracy of this claim, which originates with an anonymous former KGB source, is impossible to assess.[22] However it seems very plausible under the circumstances.

It is obvious that the KGB would need answers to some big questions. They knew that Pontecorvo had been interviewed by MI5 in April; they needed to find out what he had been asked, and what he had said in response. If he had been spying for the Soviets, they would want to know if he had confessed to the UK. In any case, the KGB would be very interested in finding out what MI5 knew about the spy trail linking Canada to Moscow—and whether the West knew anything about the Cohens, whom the Soviets had rescued from the United States just a few weeks earlier.

Bruno had expected to be debriefed about his atomic research in the West, but he was not prepared for this aggressive inquisition into his motives. In a featureless room, lit by low-wattage bulbs, with walls painted dull grey and mustard, the emotionless interrogators pointed out that a spy for the Soviet Union, whose colleagues have been exposed, might obtain immunity from prosecution by agreeing to work for the other side.

Whether or not Pontecorvo was a spy, it is probable that the KGB also had another worry—this one involving the letter composed by Geoffrey Patterson in July, which alerted Philby to the FBI's interest in Pontecorvo. Was this a genuine document, showing that the game was up for Pontecorvo in the West? Or was it an elaborate trap devised by the British, using Pontecorvo as bait to expose Philby and other Soviet agents? After all, if the Soviets were to express a sudden interest in Pontecorvo, this could confirm that Patterson's letter was known in Moscow, allowing the British to reel in Philby.[23]

If the Soviets truly questioned Bruno about the Cohens, there was no possibility of letting him go. He now knew too much, even if he hadn't before. Furthermore, it's doubtful that Bruno would have been welcome in the West. By this point, Philby's news that the FBI was investigating Bruno and his communist associations had presumably reached him. Having been confronted with this news, Pontecorvo could hardly expect the British to take him back, except perhaps in order to put him in a noose.

This mention of the death penalty may sound like the stuff of a spy novel, but it is sadly based in reality. Bruno Pontecorvo had worked closely with Alan Nunn May in Canada, and had even helped him pry classified information from the US team in Chicago—some of which had ended up in the USSR. This could easily constitute grounds for a capital charge in the United States. Klaus Fuchs had prepared himself mentally for execution by the British, only to discover that, because he had not passed secrets to an enemy (the USSR was an ally at the time), he was not guilty of a capital offense in the UK. The possibility that he might be extradited to the US, where he could face execution in the electric chair, was a serious concern. This encouraged Fuchs to plead guilty in the British court. While researching this book, I interviewed the immediate relatives of two other confirmed atomic spies from that era—Alan Nunn May and Ted Hall. In both cases, the relatives confirmed that the fear of execution had been very real. Nunn May's stepson stated that the possibility of extradition to (and execution by) the US is what led Nunn May to cooperate so readily with British prosecutors. Ted Hall's wife, Joan, confirmed that he too had feared execution for treason. I asked, "If Ted had been threatened with exposure in 1950, would you have gone to the USSR?" This elicited an instant response: "For sure we would!"[24]

Bruno would always claim that his decision to enter the USSR was made purely for idealistic reasons, based on his profound belief in communism and a wish to use his scientific knowledge for peace, away from the perceived persecution of the West. However, none of this explains why he left on a whim, in the middle of his summer holiday, rather than making an organized transition to the USSR, which would not have been difficult for a man of Bruno's intelligence.[25] In 1946, Bruno had refused to join Harwell unless he was given the freedom to travel; now, by going to the Soviet Union, he had lost all freedom—not only to travel but also to communicate with his parents, siblings, friends, and scientific colleagues. Parting the Iron Curtain in 1950 was like entering a black hole, where people and information could be lost forever. Gil could not return to school in England; Bruno and Marianne could not contact their families in the West; the very fact of the Pontecorvos' presence in the Soviet Union would remain a closely guarded secret for five years. During this period, they were referred to by their relatives as "the dear departed."[26]

BY THE END OF SEPTEMBER, BRUNO, MARIANNE, AND THE CHILDREN were allowed to leave the apartment, as long as they were accompanied by bodyguards. It was not yet winter, but the weather was very cold and the Pontecorvos had only summer clothing. One morning, their Russian teacher arrived with a girl carrying packages of fur-lined coats, hats, gloves, and boots. No wonder Bruno, years later, recalled that they felt like "privileged guests, protected" and that these early days in Moscow were "very peaceful." However, it would seem that Marianne might not have shared this positive impression, if the family's first trip out is any guide.

The Pontecorvos' apartment was close to several of Moscow's major stores, whose windows displayed a range of goods available for the home. On this initial excursion, the family entered one of the shops, where Marianne was "discontented" by the sparse offering of products, the long lines, and the sullen attitude of the staff. According to Bruno's recollections, he and Gil scolded Marianne. They told her that in Paris or London the stores were less crowded "because only the rich could buy and the staff were forced to smile in order not to be made redundant." Bruno admitted that the goods were of lesser quality than those in the West, and were poorly packaged and presented, but he argued that this drawback was outweighed by the fact that "everyone could buy everything," and the goods were not merely available to the well-off.

Bruno later recalled that Marianne "did not seem convinced," but chose not to discuss the episode further.[27]

AT THE END OF OCTOBER, AFTER SPENDING NEARLY TWO MONTHS marooned in their Moscow apartment, the Pontecorvos were on the move again, being driven for miles on unpaved roads through dark woods of fir trees. They passed a few villages filled with rustic houses, some abandoned and decaying at the side of the road. After two hours they reached Dubna, a village some seventy miles north of Moscow, on the banks of the Volga—the home of a secret nuclear physics research centre.

In former times, Dubna had been a peasant village, far away from the cities, an ideal spot for rest and relaxation. Even after the laboratory took over, it retained a rustic charm, with its quaint streets cut through the forest. Gil recalled, "In 1950 when we arrived, Dubna was little more than a Russian village, with two or three gravel roads and log cabins. These were similar to what I remembered in Canada." He added, with a laugh, "All governments put nuclear research laboratories in the backwoods."

One of the first things Gil noticed about Dubna was the mosquitoes: "I had grown up in Canada where they were everywhere, so I was used to them, but Bruno hated them."[28] Mosquitoes thrived at Dubna because the village and laboratory were built on reclaimed swampland, formerly part of the Gulag Archipelago. The village itself is situated on an island at the junction of the Volga River, the Dubna River, and a canal that links the Volga to the Moskva. This canal had been dug in the 1930s by inmates from prison camps, who provided forced labour for the laboratory after the war. They dug the foundations, broke rocks, and built the entire edifice of the lab.

In addition to the town's basic bungalows, there were some more substantial homes where senior scientists lived. The Pontecorvos were presented with an elegant detached house, painted in ocher. It was two stories high, with a third-floor gable at one end. The front door opened on to a stone patio, fringed by an elegant low wall. The property was surrounded by a green wicker fence, which enclosed numerous tall fir trees and a garden. Compared to the brick estate house in Abingdon, this was a land of enchantment. In Dubna the Pontecorvos had space, and a house such as one might find in the forests of North America. After the privations of Moscow, Marianne felt very positive about her new home. Its surroundings reminded her of her native Sweden.

No one could enter Dubna without special permission, and its entire population was specially selected. One might hope to enjoy a certain amount of freedom within such a protected environment, yet Bruno could not leave his home, even for the short walk to the laboratory, without being accompanied. His protests were met with the explanation that the physicists employed in important research work "need protection."[29]

Bruno was beginning to discover the strange reality of his new life in Dubna. Even today, visiting the area can feel like passing through a time warp. Two security fences reminiscent of the Berlin Wall still surround parts of the site. After you pass through the first fence, the door to the outside world closes behind you. Armed guards examine your papers before allowing you to continue. Beyond the second fence, there is nothing other than the laboratory, hidden in the woods, and the mosquitoes.

In Moscow, Gil had "hated the loneliness." Now he attended the local school, where he enjoyed the company of other children. On his first day of school in Dubna, "all problems vanished."[30] Over sixty years later he still lives there.

FIFTEEN

EXILE

THE ATOMIC NUCLEUS HAD REVEALED ITS AWESOME POWER IN THE explosions at Hiroshima and Nagasaki; exploring its deepest structure was the obvious next step for the world's postwar governments, and an intellectual challenge for scientists.

In 1944 Soviet scientist Vladimir Veksler had shown that it was possible to create stable beams of high-energy particles, which could be used to bombard atomic nuclei and shatter them. The following year, in the United States, Edwin McMillan independently discovered the technique. This breakthrough raised the possibility of revealing the deep secrets of the nucleus by bombarding it with pions produced by a high-energy particle accelerator. This line of research soon became a top priority in the West, and Igor Kurchatov, the father of the Soviet atomic bomb, urged his superiors to make it a priority in the USSR as well. As a result, the Soviet government decided in August 1946 to build a special laboratory—Dubna—that would contain the world's largest high-energy particle accelerator, known as a synchrocyclotron.

Up to that time, energy had been extracted from the nucleus under a limited set of circumstances, which required either a rare isotope of uranium or artificially created plutonium. Even though the results of these methods could be explosively dramatic, the amount of energy they liberated was still less than 1 percent of what was locked inside the nucleus by the powerful nuclear forces.

The discovery of pions in 1947 gave scientists a new understanding of those forces. Just as photons are the material embodiment of electromagnetic fields, so are pions the material embodiment of the much-stronger nuclear fields. Scientists initially studied pions out of curiosity; their research had no immediate military significance, and high-energy particle physics in the West was developed in the open. Even so, some thought that pions might be able to unleash nuclear energy in quantities that would make all previous methods pale in comparison, which was one reason for Western governments to support this new field.

The motivations of the scientists at Dubna were probably no different than those of their counterparts in the West; the Soviet government, however, mindful of the strategic possibilities, kept the existence of the Dubna accelerator a secret. Stalin's agenda was to create atomic and hydrogen bombs, for military purposes. He distrusted intellectuals, but realized that he needed physicists to do the job. The advice of Lavrenti Beria, Stalin's security chief, was characteristically direct: "Let them get on with it; we can always shoot them later."[1]

Once the Soviets decided to build the accelerator, a site had to be chosen. Beria set up a meeting, where three possible locations were discussed. The choices didn't include Dubna, however. Beria, who enjoyed hunting near Dubna, then pointed at the map and announced that the laboratory would be built—right there!

The Dubna area was hardly an ideal place to construct a particle accelerator, as it was full of swamps. To this objection, Beria announced, "We will drain them." As for the lack of roads: "We will build them."[2] And, as for the workforce, Beria had an answer for that too: the area was full of forced labour camps, part of the Gulag. Throughout the 1950s, in Dubna's early years, scientists travelling from Moscow would routinely see prisoners with shaved heads building the roads.

And so Dubna was born. The project's purpose was disguised by the name *Hydro Technical Laboratory*. The construction was completed in December 1949, and in January 1950 physicists from Moscow began conducting experiments at what was then the world's highest-energy accelerator.[3] It retained this honour until 1953, when the Cosmotron accelerator at New York's Brookhaven National Laboratory took the blue ribbon.

THERE WAS GREAT EXCITEMENT AT DUBNA WHEN, IN THE FALL OF 1950, the senior management learned that Bruno Pontecorvo—"student of the famous Enrico Fermi"—had arrived at the laboratory. According to Venedict Dzhelepov, who later became Dubna's director, the fact "that such a talented and well-known scientist was to work in the then small scientific community of our laboratory was very valuable."[4]

Only a select few knew of Bruno Pontecorvo's presence in Dubna. Irina Pokrovskaya, who served as Bruno's secretary at the laboratory for forty years, initially knew him only as "the professor," a man with no name.

Nonetheless, the Western media was sure that Pontecorvo was in the USSR, and over the next five years reporters made some fanciful claims.

Often, half-truths and rumours were elevated to the level of supposedly factual stories. One notable example occurred in November 1951, when newspapers in Rome claimed that Pontecorvo had been arrested by the Russians in an effort to stop their atomic secrets from being leaked to the United States; Pontecorvo, apparently, was suspected of being a double agent. The article quoted unnamed Russian sources. The story's genesis was apparently President Truman's announcement that atomic explosions had taken place in the USSR; as these explosions were meant to be secret, the Soviets thought Truman must have a source, a Western spy among their top-ranking scientists. A Harwell spokesman commented that if the news of the explosions was true, it was intriguing that it had percolated through the Iron Curtain.[5] A closer analysis of the facts, however, reveals the Pontecorvo rumours to be nonsensical. The White House had indeed made the announcement regarding atomic explosions in Russia—but this had happened in the fall of 1949, nearly a year before Pontecorvo disappeared.

Another story, popular in North America, was that Pontecorvo was in China's Xinjiang Province, working at a "huge atomic stronghold" that Russia was setting up there. Such stories were accepted unquestioningly by the *Los Angeles Times*, *Chicago Tribune*, and *Christian Science Monitor*. The *Glasgow Bulletin* was more skeptical, announcing that Swedish and Finnish sources had poured cold water on the story. The reports of Pontecorvo's presence in China had been ascribed to refugees who had escaped from the USSR to Helsinki and Stockholm. However, inquiries showed that no Soviet refugees had actually reached those cities since

Pontecorvo's disappearance. Even so, in 1951 a US congressional committee declared Pontecorvo to be the "second deadliest" spy in history (Klaus
Fuchs being the first). The committee also claimed that Fuchs and Pontecorvo had advanced the Soviet weapons programme by eighteen months.[6]

The bizarre suggestions that Pontecorvo was working on helium weapons or atomic fogs are best seen as science fiction, and never had any
scientific credibility. Reports that he was used in the Soviet quest for uranium are harder to dismiss, as his expertise in this area meshed so well
with the USSR's needs at the time. Indeed, uranium mining soon became
one of the jobs performed by forced labourers in the Gulag.[7]

Bruno always denied having worked on atomic weapons at any stage. Although this may be literally true, Isaak Pomeranchuk, head of the nuclear-
reactor research programme in Moscow, consulted Bruno frequently
during the latter's first five years in Russia. Boris Ioffe recalls that Pomeranchuk "often visited Dubna at that time and many times said after returning
that he discussed such and such a question with 'a professor,' or 'a professor said this.'"[8] Samoil Bilenky, who would later work closely with Bruno,
was a young scientist at that time, and a student of Pomeranchuk. He later
recalled a car journey he took with Pomeranchuk and another senior scientist. Pomeranchuk kept repeating, "Professor said this; professor said that."
Bilenky remembered the incident because it had seemed so strange. "Why
did he not say the name of the professor? Naturally I knew not to ask."[9]

Bilenky and Ioffe both stressed the fact that Pomeranchuk "never said
who the professor was." Ioffe added that "mentioning Pontecorvo's name
was taboo" until 1955.[10] Only later did Pomeranchuk confirm what Ioffe
already suspected: "the professor" was Bruno Pontecorvo.

POMERANCHUK'S QUEST

What did Pomeranchuk need from Pontecorvo? Why did he consult him
so frequently?

In 1945 Pomeranchuk and three colleagues had worked on mathematical problems relating to "the tube"—a conceptual method in which
deuterium and tritium could be used to make a thermonuclear weapon.
A conventional atomic explosion would heat the tritium, which would
then provide the spark to ignite the rest of the bomb, which consisted of
a tube full of deuterium. Only a small amount of tritium was needed, and

because deuterium was cheap, the tube could be made as long as necessary. The plan was for a shock wave to pass down its length and cause the nuclei of deuterium to fuse explosively.[11]

Until 1949, all the physics research in the USSR had been geared toward making a traditional, fission-based atomic bomb. This was because the Soviets were eager to demonstrate their power to the West, and because a fission explosion is needed to ignite the tritium in a hydrogen bomb—so mastering fission explosions was a necessary first step on the path toward their ultimate goal of a hydrogen bomb. After 1949 the Soviet quest for a hydrogen bomb began in earnest. The basic physical principles behind such a weapon were clear; what was uncertain was whether the reaction would explode or fizzle. One of the unknowns involved how energy would spread through the device. If too much escaped, there would be no explosion. During 1949 and 1950, a group at Arzamas-16, the "Soviet Los Alamos," investigated this intensely. They focused on how energy, carried by gamma-ray photons, would dissipate as the photons bounced off of electrons in the device—a phenomenon known as Compton scattering.

Compton scattering was one of many processes that could be studied using the new breakthroughs in quantum electrodynamics (QED), the quantum theory of light and matter. The theory of QED had been successfully completed in 1947 by theorists in Japan and the US, and subsequently published in the literature. Around the world, physicists investigated its implications. In fact, this is exactly what Boris Ioffe was studying for his PhD thesis. Individual electrons and photons can spin in flight. In 1950 Ioffe was told to calculate how certain properties of Compton scattering depended on the relative orientations of the particles' intrinsic spins. This was, apparently, a question of academic interest, an application of a new theory, which could be used to test its limitations. However, it also had considerable relevance to the innards of a thermonuclear weapon. In order to maintain secrecy, the examining committee for Ioffe's thesis was carefully chosen. At the end of the examination, one of the members, L. V. Groshev, agreed that the thesis was sound but didn't understand one point: Why it was so secret? The chairman, Lev Artsimovich, replied that it was "very good that you didn't understand it."[12]

Ioffe was part of Pomeranchuk's team. In the summer of 1950, Pomeranchuk had been sent to Arzamas on a "long assignment." However, he wanted to discuss the revolutionary discoveries in QED with his colleagues in Moscow. He managed to convince the authorities that it would be best if his group returned to Moscow, where he could work on both QED and "the problem," as the bomb project was known. The team's expertise was mainly in reactors, and their research was subject to the highest level of security—"Top Secret Special Folder." At this level, the protocol was so restrictive that reports were not typed by the carefully vetted special secretarial staff, but were written longhand by the scientists themselves.

In Moscow, Pomeranchuk's group took on the task of assessing how much energy in "the tube" would be lost due to Compton scattering. By 1952 they had the answer: so much energy would be lost that the bomb would not work.

Thus, at the start of the 1950s, Pomeranchuk's interests included QED as it applied to the H-bomb, and the design of nuclear reactors for making tritium. In the realm of pure physics, he was also interested in how particles scatter off one another at high energy. Which brings us back to our question: What did Pomeranchuk need from Pontecorvo?

In general, Bruno's focus was different from that of Pomeranchuk. His interests lay in the neutrino, whose existence had yet to be proven; in cosmic rays and the "strange" particles that had been discovered within them; and in the relationship between electrons, muons, and the weak force of radioactivity. Moreover, Bruno was primarily an experimentalist, with little to offer a first-rank theorist like Pomeranchuk on subjects like QED or high-energy scattering. If Pomeranchuk needed advice on theoretical physics, he had considerable talent at his disposal in Moscow. It is also unlikely that he made the visits to Dubna to discuss basic physics for the purpose of pure research: Pontecorvo and Pomeranchuk never published a joint paper on basic physics during their lifetimes, nor did their primary interests overlap strongly. It is known that Pomeranchuk asked Bruno about the strange particles, which had interested physicists in Moscow, but this was a transient interest and Pomeranchuk played no role in this field. However, the two men did have a mutual interest in nuclear reactors.

Pomeranchuk was, in Ioffe's opinion, the "main contributor" in the Soviet Union to the theory of nuclear reactors. In 1947 Pomeranchuk wrote the first text in the world on the principles of nuclear reactors.[13] He was a theoretical physicist, and the reigning expert on the underlying concepts. However, as we saw during the building of the reactor at Chalk River, there is no substitute for hands-on experience. In a way, building a nuclear reactor is like learning to drive: you can read books on the subject, but until you actually go out on the highway, you won't get the experience necessary to pass the test. For the Soviets, it was imperative to build a heavy-water reactor that actually worked. Pomeranchuk had read the books, but he'd never been on the highway. In this regard, the presence of Bruno Pontecorvo, who had already passed the driving test, was invaluable.

Indeed, considered from this angle, the reasons for Pomeranchuk's visits become obvious. When the first reactors were built in the West, a new problem emerged, known as "creep." The intense neutron bombardment, combined with the heat produced by the reactor, deformed the metal in the cooling system and threatened to kill the reactor entirely. By 1946 Kurchatov had four institutions working on ways to seal the uranium rods and avoid this problem.[14] For these purposes, information from agents in the West was useful, but limited. Much of the research at Chalk River and Harwell was not formally recorded, but "done on chalkboards and by coffee-housing" and would have been "taken to the USSR in Bruno's head."[15] Indeed, in 1955, when Pontecorvo's presence in Russia was finally acknowledged, he would admit, "A few years ago I had occasion to discuss with Soviet colleagues some problems regarding radiation protection for nuclear power plants intended for peaceful purposes."[16]

Given the paramount secrecy of the atomic project, it is obvious that Bruno would not have been told the real agenda behind Pomeranchuk's questions. And when we consider the fact that the Soviets' top priority was to construct reactors that could breed tritium, it is naive to suppose that none of Pomeranchuk's discussions with Pontecorvo had any relevance to the "atomic problem." It is also naive to assume that Bruno Pontecorvo would not have deduced what was going on.

THE SECRET NOTEBOOKS

For Bruno's first five years in the USSR, he recorded his work at Dubna in classified logbooks. At the end of each day, a letter-sized journal, with секрет ("secret") stamped on its maroon cover, would be deposited in the laboratory safe. There it would remain until Bruno returned.[17]

Each journal consisted of two hundred pages, with every page numbered so that it would be apparent if any were removed. The date of

IMAGE 15.1. Cover of Bruno Pontecorvo's first secret
logbook in the USSR, 1950. (Courtesy of Gil Pontecorvo and
the Pontecorvo Centenary Exposition, University of Pisa.)

Bruno's arrival in Dubna is established by his first journal entry. The front
cover of his first logbook declares, "начато 1950" (started 1950); the handwritten date on the first page is "1 ноябрь" (November 1). The dates are
the only entries in Russian, and appear to have been written by someone
other than Bruno. His personal log from each day is written in English.

The first logbook begins with a brief entry on how to measure the energies of neutrons very precisely. The application of this question becomes
apparent in the next entry, on page 2: "Fission from highly excited states."

By this point in his life, Bruno Pontecorvo was a world-leading authority in this field of instrumentation. In Oklahoma, he'd designed a neutron
detector for oil prospectors; in Canada, this device had been used to find
uranium; at Harwell, he'd developed more sophisticated detectors. Normal
fission happens when high-energy neutrons hit atomic nuclei, which are
in their most stable state. In the logbook, Bruno comments, "As the fission
of medium A [nuclei in the middle of the periodic table] shows, there must
[occasionally] be fissions arising from *very highly excited states*"—states in

which one of the constituent neutrons or protons has been raised up the energy ladder temporarily. (Italics added.) He then notes, "These fissions must . . . release plenty of energy in uranium and thorium."

Halfway down the second page we come to another question discussed on Bruno's first day at Dubna: "Is it possible to detect H4 particles inside the chamber?"

In 1955, H4, or quadium, would play a prominent role in the satirical novel *The Mouse That Roared* as the isotope powering the "Q-bomb."[18] Today we know that quadium is so highly unstable as to be effectively nonexistent and useless for military purposes. However, the Soviets' attempt to isolate this exotic isotope was reasonable, given the tritium shortage that hindered their production of thermonuclear bombs. The scientists at Dubna hoped that when beams of deuterons, or alpha particles, smashed into suitable targets, H4 particles might be produced.[19] If they were, this might be a fast-track solution to Stalin's challenge. But first the scientists would have to successfully detect H4 particles, and for this they turned to Bruno Pontecorvo.

In his logbook, Bruno suggests using the magnetic fields of the cyclotron to curve the path of "the electrons" (the beta particles produced when the H4 isotope decays), and from this deduce the transient presence of quadium.[20] If its existence was established, methods of producing the isotope in greater quantities could be developed. The discussion continued on November 3, when Bruno noted the possibility of performing the experiment with an arrangement of electronic counters.

Pontecorvo also evaluated ways of detecting pions. On this same date, he had some ideas on "Fission [caused] by mu meson." However, this last phrase is crossed out and not developed further. In addition, he recorded some thoughts about the strange particles—thoughts that would mature a couple of years later. And, tantalizingly, he considered the possibility that the decay of a muon produces two neutrinos of different characters. This set of ideas is a rough outline of Bruno's initial hopes for his particle physics experiments at Dubna, which would have no immediate military significance.

After three days, Bruno's flow of ideas is interrupted, as the November entries suddenly end halfway down page 9 of the logbook. The writing on the lower half of the page is inverted. The explanation for this is given below.

The above burst of activity seems to reflect an initial period of brain-storming, after Bruno's arrival at Dubna. At the end of this period, he was apparently assigned another task: nothing further is recorded, and with ninety pages still empty he is given a new logbook dated November 30. The first few pages have sporadic entries, until April 1951 when there begins a detailed record of particle physics research at the Dubna cyclotron. This logbook was filled by September 1951, and only then did he return to the original logbook, which had remained fallow since November 3, 1950.[21] When he resumed writing in this logbook, he turned it upside down so that the final page became the first. He then maintained his daily research log for Dubna until, by March 1952, he had worked his way "forward" to page 9. His particle physics research, which the authorities would later claim to be his sole activity in the USSR, occupied him full-time from April 1951 onward.

Bruno's decision to record these data in the rear of the original logbook, separate from the initial inquiries, is a deliberate act. It would be natural for him to retain space in the logbook for any further work on the original questions; hence the new material is recorded from the rear. However, it is clear that by April 1951 he had given his employers all the information they required on the question of fission and H4 particles.

The first notebooks having been filled, Bruno began a new one: "начато 1952" (started 1952). This and subsequent logbooks record his ongoing research interests, which from then on appear exclusively to involve experiments at the Dubna cyclotron. Other pages constitute lesson plans or drafts of research papers. This continues until his "coming out" in 1955.

So what can we conclude about Bruno Pontecorvo's first year in the Soviet Union? For one thing, the questions involving H4 and fission, which occupied Bruno during his first days at Dubna, are in marked contrast to his subsequent work there.

The entries from those first days of November appear to be responses to problems that Dubna had grappled with before Bruno's arrival.[22] It is possible to view them as genuinely "pure" physics questions, but their nature and scope make them more obviously applicable to strategic issues related to the release of energy from atomic nuclei. Specifically, the first two entries suggest that the Soviet scientists are looking for ways to increase the energy released by the fission of strategically important elements (that is, elements of relevance to energy release in weapons or reactors, including, but not limited to, uranium and thorium). Such an interest would be esoteric in the context of pure nuclear physics. And to identify the phenomenon with certainty, very precise measurements of the energies of neutrons would be needed; hence a reason for the initial entry. The logbooks suggest that, upon arriving at Dubna, Bruno was consulted on questions relating to the production of energy by fission or fusion, with potential applications to power production in reactors or explosively in atomic and thermonuclear bombs.

Once this particular task was completed, he was consulted on other secret matters, which explains the gaps in the record.[23] We know that, in general, Bruno discussed aspects of nuclear reactors with Pomeranchuk and others. If he performed any detailed work for the Soviets on reactor physics or uranium, it must have taken place during this period, between November 1950 and March 1951. After that, he then took up full-time work on particle physics at Dubna.[24]

STRANGE PARTICLES

In 1951, Bruno was thirty-eight years old, and in the prime of life. Over and above his initial significance for the Soviet government's nuclear programme, he played an active role in the work at Dubna. He both inspired new lines of research, and helped drive existing lines forward. The most significant new line of inquiry he was involved in dealt with the so-called strange particles.

Three years before Bruno fled to the USSR, physicists had discovered "strange" particles in cosmic rays. They were dubbed strange because they lived for about a hundred-millionth of a second, which, although short by everyday standards, is about a million billion times longer than expected. To illustrate the unexpected duration of these particles, one scientist said, "It's as if Cleopatra fell off her barge in 40 BC and hasn't hit the water yet."[25]

As we saw earlier, one goal of the research at Dubna was to understand pions—the particles that are the embodiment of nuclear energy. A single pion contains about one-seventh of the energy normally locked within a proton or neutron, similar to the amount liberated in a single fission of uranium. So the discovery of a "strange" sibling, about three times heavier (and thus possessing three times the energy of a pion), was tantalizing. This new particle became known as the kaon, or K-meson.

The Dubna synchrocyclotron was powerful enough to make pions, but in 1950 did not have enough energy to make the more massive kaon, and solve the mystery of strangeness. The enigma began to be unravelled after another strange particle turned up in the debris from cosmic collisions in 1951: the Lambda. A Lambda is like a neutron that carries this mysterious strangeness. And, in the Soviet Union, it was Bruno who made the breakthrough that helped scientists understand it.

In 1951, soon after the discovery of the Lambda, Bruno wrote a classified paper in which he drew attention to the anomalous properties of the strange particles. This articulated the possibility that "the process of formation of these particles is not the reverse of their decay."[26] In other words, even though strange particles are produced by the strong force in pairs, they decay individually due to the weak force—the same universal weak force that Bruno helped identify in 1947. A kaon and a Lambda are born together but die alone. Bruno proposed the idea of what is now called "associated production": this states that strange particles are born in pairs, but then part company. The strong interaction only operates when the two of them are close together, after which they are freed from its snare and the weak force takes over. The weak force is very feeble, however, compared to the strong, which is why an isolated strange particle can survive for an unexpectedly long time.

Bruno may have been the first scientist to make that insight, but he remained flummoxed as to what made particles such as the kaon and Lambda "strange," whereas the proton, neutron, and pion were "normal." Other scientists would solve that conundrum.

The answer is that, just as some particles carry electric charge, while others do not, some particles carry this attribute of strangeness. You cannot create a positive or negative electric charge in isolation; every positive charge must be balanced by a negative charge, and vice versa. A similar idea applies to strangeness. The kaon has, let's say, one unit of positive strangeness, and the Lambda has a negative unit of the same value. The protons and neutrons in the nucleus have no strangeness, so when a proton in a cosmic ray or an accelerator collides with a nucleus, the creation of a Lambda (negative strangeness) must be counterbalanced by the simultaneous appearance of a kaon (positive strangeness). The total strangeness remains zero.

Bruno, however, had not conceived of positive and negative strangeness, so the system he developed only required that strange particles appear in pairs. For example, we now know that a collision between two neutrons cannot spawn two Lambdas, as the latter pair carries two negative units of strangeness, whereas Bruno mistakenly thought that this reaction was possible.

Although Dubna lacked the energy necessary to make kaons, it did have enough to produce Lambdas. In a Dubna report from 1953, which was classified as secret, he proposed a search for the reaction described above, in which two colliding neutrons spawn two Lambdas. However, when Bruno and his colleagues carried out the experiment in 1954, there was no sign of Lambda particles. Indeed, no such reaction has been seen to this day.

By 1954 the Cosmotron had begun operating in the United States, smashing protons into targets at energies far in excess of what Dubna could achieve. The Cosmotron could even make beams of pions, and direct them at atomic nuclei. The result: a kaon emerged along with a Lambda. Thus Bruno Pontecorvo had correctly identified the phenomenon of associated production, but had failed to realize its deeper significance.[27]

In any case, he received no credit for his insight. Like the tree that falls in the forest with no one to hear, Pontecorvo's work was unknown outside of Dubna. That same year, in the US, theorist Abraham Pais had independently come up with the idea, and many textbooks credit him alone with the discovery.[28]

The hypothesis had far-reaching consequences. In 1961, the concept of strangeness inspired American physicist Murray Gell Mann to propose the "Eightfold Way" scheme for classifying strongly interacting particles. This in turn led to discovery of a deeper layer of reality that exists within these particles: they are made of smaller particles called quarks.[29]

LIFE IN THE USSR

When the Pontecorvos arrived in Dubna in 1950, barrack dormitories surrounded the town. Beria's Gulag prisoners, with uniforms and shaved heads, were building the roads. Bruno saw these workers, but gave them no special attention. Years later he said that he didn't know who they were or why they had been imprisoned, and that "there are laws in every country and when broken this can lead to imprisonment." Yet he was also aware that "there had been great trials and death sentences. I knew people were detained in camps and prisons. I thought, as a million other communists, that this was the inevitable consequence of the class struggle still in process."[30]

Bruno too was still a prisoner, albeit in a gilded cage. On one hand, living in Dubna gave him plentiful access to food and conveniences of a higher quality than those available to the general Soviet populace at that time. However, if anyone wanted to leave the limited area around the laboratory encampments, they had to obtain special permission. For Bruno the restrictions were even tighter. No documents or books could be taken from the lab for work at home. His logbooks were locked in the laboratory safe overnight. And there were other, more serious, constraints as well.

Whenever he left his house, even for the short walk to the laboratory, two guards accompanied him, supposedly for protection. Given that Dubna was a restricted area whose inmates—for at times that is how Bruno saw them—were specially selected, the concept of protection seemed absurd. The guards did play an important role, however: they prevented Bruno from speaking to strangers as he walked. One of the guards had a habit of whistling, which annoyed Bruno. When this man was assigned to accompany him, Bruno would walk fast and get to his destination quickly to minimize the time spent in his presence.

This enforced isolation even extended to his social life.

In the West, it had been customary for scientists to visit one another's homes, for their families to have dinner together and take communal trips on the weekends. Bruno experienced none of this during his five secret years at Dubna. He worked closely with colleagues at the laboratory, had lunch with them in the canteen, and then returned to his home in the woods, always accompanied by his KGB minders. Social contact with his fellow scientists, whose houses were in glades among the trees near his own, was nonexistent. No one ever invited him to their home. Bruno, the gregarious extrovert who thrived in company, was trapped by a cordon sanitaire.

BRUNO MAXIMOVITCH PONTECORVO

We have seen how Pomeranchuk would refer to Bruno only as "the professor." Even in Dubna itself, this denial of Pontecorvo's identity was the norm. He was simply the professor, whose first name was Bruno.

IMAGE 15.2. Bruno and Marianne at Dubna. (Courtesy
Gil Pontecorvo; Pontecorvo family archives.)

Russians traditionally address one another by their given name and
their patronymic—this second name essentially meaning "son or daughter of X." Bruno, however, was called simply "Bruno," the only identity
allowed for the professor. This caused embarrassment. The formalities of
the Soviet Union in the 1950s meant that colleagues could no more address him as "Bruno" than a member of the general public in England
could have greeted the prime minister as "Winston." The problem was
solved when one of Bruno's senior colleagues asked him for the first name
of his father. On learning that it was Massimo, his colleagues agreed to
call him "Bruno Maximovitch." Once his presence was officially acknowledged in 1955, he would be known socially and professionally as B. M.
Pontecorvo.

Even his children had lost their family name. At school they were
known as Gil, Tito, and Antonio Ivanov. Gil confirmed that Bruno's presence in Dubna was a state secret, necessitating the peculiar name change:
"Being called Ivanov was strange, but we didn't care much."[31] Bruno's assessment was more blunt: "Some things had to be kept secret."[32]

Bruno and his family were effectively in exile. In the judgment of a
former head of MI5, it sounded like the Soviets didn't trust him.[33] There
are similarities here to the cases of Guy Burgess and Donald Maclean, of

the infamous Cambridge Five spy ring. Following their defection to the USSR in 1951, Burgess and Maclean were exiled to the industrial city of Kuibishev for several months, while "a thorough check of their credentials was made to ensure that they had not been turned." Their presence in the Soviet Union was also initially kept secret; Philby's role in both their defection and Bruno's had to be protected. According to Chapman Pincher, "The KGB always put the safety of its agents first, [and preferred] silence to any short-term propaganda gain."[34]

Whereas Bruno suffered, Marianne considered the isolation "not a problem," as she had been almost pathologically averse to social life to begin with.[35] In the United States, when she had balked at meeting the Fermis, Bruno had admonished her. In Dubna, there were no social appointments to avoid. Marianne, who was "always silent and a little distant," was beginning to "forget names, appointments and things." She would "read for hours, or remain lying on the bed, gazing out of the window at the trees."[36] In these descriptions, given to Miriam Mafai years later, Bruno reveals Marianne in a state of depression, en route to the psychological breakdown that would eventually lead to long periods in a sanatorium.

BRUNO AND POLITICS

For Bruno's first five years in the Soviet Union, he was unknown to all but a handful of people there. To the Western media, his disappearance remained an enigma, as it did to his family and friends.

Bruno recalled, "I read *Pravda* each day, and occasionally saw *L'Unità*. Even so, it was difficult if not impossible to get a rounded view of world events."[37] An example of this difficulty occurred in 1953, shortly before Stalin died, during the coverage of the so-called Doctors' Plot, in which several Jewish doctors were arrested "for aiming to remove by harmful treatment the lives of the active leaders of the USSR." Bruno read this coverage in the cloistered confines of Dubna. He decided that the claims might be true: "I believed it. Not the actual words, as they were simply propaganda, but the substance could have been true. In history there are plenty of examples of political assassinations around the world. It was 1953, the height of the Cold War [and] of rebuilding the Soviet economy. Killing Stalin could have been a real objective."[38]

One of the doctors died in prison, and the others would undoubtedly have been shot if Stalin himself had not died on March 5, 1953.[39] Within weeks, the state news agency announced that the whole case had been a "misunderstanding" and that the surviving doctors had been released. Speaking in 1990, Bruno interpreted this as a sign of a "political struggle taking place within the Kremlin. I did not know much about this fight. It might seem inexplicable now, but [at the time] I believed [the reports]."[40]

Stalin maintained control by the periodic use of terror. Bruno was not overly concerned about this. He defended Stalin's harshness on the grounds that it had saved the October Revolution and defeated Nazism, drawing analogies to the Jacobin Reign of Terror that had saved the French Revolution.

For many in the Soviet Union, Stalin was a hero, who had saved them from the Nazis. Nonetheless, he feared the power of the United States, as well as the other Western nations, who had attempted to crush the Bolshevik Revolution in its infancy, and then developed atomic weapons without informing their Soviet ally. One reason Fuchs, Nunn May, and other sympathetic Western scientists had passed atomic secrets to the Soviets was because they believed it was vital for the West's power to be balanced by that of the USSR, which would require both sides to possess nuclear weapons.

At the time of Stalin's death, Bruno, like millions of others in the USSR, still had great faith in the communist state and its leaders. The entire nation went into mourning. Hundreds of thousands filed past Stalin's coffin, and many died of suffocation within the crush of people in Red Square. Bruno, however, was not among them; he was still trapped in Dubna, listening to the radio's coverage of the events. Gil, by then fourteen years old, took note of the occasion: there was "a week or ten days of solemn music on the radio. It was the first time I appreciated classical music."[41]

In 1954 Bruno joined the Communist Party of the Soviet Union. Whereas this normally involved a lengthy application process, in which two existing members served as guarantors and the applicant was required to surmount various hurdles, Bruno simply told the local party secretary that he wanted to register—"and my request was automatically accepted." He soon found that Soviet party meetings were a far cry from his prewar experiences in Paris, however. In those earlier meetings, "the leaders had the last word but the debate took place. Here, there was none." At Dubna, to Bruno's surprise, "the meetings discussed more personal matters than politics. Much of the time was taken up with examination of the individual conduct of members of the cell. I found it somewhat medieval, talking about this in public and justifying one's behaviour. I felt it an undue interference in personal life." Typically the cell would hear reports of a man who was frequently drunk, or a woman who was betraying her husband. Bruno later confessed, "I was never able to understand the true meaning of this ceremony, which was a kind of confession of sins where punishment or absolution was administered not by the priest but the secretary of the cell."[42]

SIXTEEN

RESURRECTION

WITHOUT WARNING, IN EARLY 1955, BRUNO SUDDENLY SURFACED. AT the time, there was a global campaign by the World Peace Council, calling for the destruction of all atomic weapons and a ban on building new ones. In February, an article in the Soviet newspaper *Isvestia* associated Bruno's name with this campaign. Bruno was indeed interviewed by the paper, but he had no idea why the authorities had allowed this to happen: "No one asked my opinion. A mid-level party official told me that a decision had been taken [that I should do this]. I had already realized that a person at my level does not decide anything."[1]

In the article, he also explained the reasons for his defection. He revealed that in 1936 he had become a committed antifascist, and had learned "undeniable facts" about the USSR's leading role in the struggle against war. Later, after the explosion of the atomic bombs over Hiroshima and Nagasaki, he had experienced "moral suffering" as a physicist. He also claimed that, in 1950, while still in Britain, he had been subjected to intolerable "direct questioning and systematic blackmail by the police authorities."[2] He had quit the West in order to maintain his personal dignity because preparations for the military use of atomic energy made him "ashamed of my profession."[3]

In 1950, Guido Pontecorvo told MI5 that, according to Gillo, Bruno had become apolitical at Chalk River and Harwell because it "suited his circumstances."[4] We might similarly conclude that, in 1955, Bruno's professed devotion to peace, and his criticism of Western scientists for their moral failure, suited his circumstances in that it undoubtedly pleased his new masters.

Certainly Bruno's statement ignores the fact that there had been many peace-loving physicists in the UK, who had preached their message throughout his time at Harwell. Given his close relationship with these scientists, and the media attention they received, it is inconceivable that he could have been unaware of their initiative. For example, Herbert Skinner, his former colleague at Harwell, was one of a growing group of British physicists concerned about nuclear weapons. Another member of this group was Rudolf Peierls, who had initiated the atomic project in 1941, was a close friend of Klaus Fuchs, and served as a consultant at Harwell. Peierls had taken part in the Manhattan Project because, as a Jewish émigré from Nazi Germany, he feared that the Nazis would create a nuclear weapon. When Germany surrendered, the moral justification for working on the atomic bomb ended, as far as he was concerned.

The same was true for many scientists. Peierls led a significant number of UK physicists in a campaign against the proliferation of nuclear weapons. In 1947 this group wrote to the *Times*, making the case for international arms control. MI5 duly took note of this initiative, as did Senator Joseph McCarthy in the US. In July Skinner was part of a study group that produced a widely publicized paper on the subject, which was even noted by highly placed Soviet politicians.[5] When President Truman revealed the US government's interest in the hydrogen bomb at the start of 1950, the news was so upsetting to many British scientists, including Nobel laureate George Thomson, that they joined with American physicists in urging the president to publicly refuse to use the weapon. Peierls and other scientists published letters in major newspapers. For the rest of the year, there were articles in the media, describing the scientists' concerns about the bomb. The *Daily Mirror* even ran the story on its front page in March.[6] Throughout Pontecorvo's time at Harwell, such issues were debated widely among physicists, but there was no sign that he himself was ever active in the movement.

The ultimate irony, perhaps, is that the position offered to Pontecorvo in Liverpool had opened up due to the departure of Joseph Rotblat, who left the Manhattan Project on grounds of conscience, became one of the most vociferous critics of the arms race, and later won the Nobel Peace

Prize for his work. In his statement of 1955, Bruno Pontecorvo speaks as if none of this ever happened.

The timing of Bruno's coming out has some significance. The USSR had by then built an arsenal of atomic weapons and detonated its first hydrogen bomb, and its nuclear-reactor programme was well advanced. Stalin had died and his successor, Nikita Khrushchev, was preparing to embrace a new strategy called "Atoms for Peace" at the Geneva Summit. This strategy promoted the idea that nuclear energy could be used for peaceful purposes, such as generating electricity or powering ships, rather than for purposes of destruction, and was intended to show that the Soviet Union wanted peaceful coexistence. It therefore suited Soviet propaganda to present Bruno Pontecorvo as a peace lover who had never worked on the atomic bomb.

Also, two of the key reasons for keeping Pontecorvo's presence in the USSR secret had disappeared a few months earlier. The KGB was always careful to maintain firewalls around its agents, even once they were back in the Soviet Union. In 1954, the Cohens, who had been extracted from the United States in 1950 and brought to Moscow just weeks before Bruno arrived, were sent to England to run a new spy ring under the aliases Helen and Peter Kroger.[7] By this time MI5 also knew of Kim Philby's treachery and had removed him from security work, so in 1955 there were fewer reasons for the KGB to be concerned about Pontecorvo.

ALTHOUGH WESTERN GOVERNMENTS HAD LONG SUSPECTED THAT Bruno was in the USSR, this confirmation nonetheless created a sensation in the global media. In response, the Soviets arranged a press conference for March 4, during which selected journalists would speak to Bruno at the Soviet Academy of Sciences in Moscow. Bruno later admitted, "The prospect of appearing at a press conference bothered me a little. I didn't know what questions I would be asked and I prepared for a long time."[8]

At four o'clock in the afternoon, Bruno arrived in the hall of what used to be the czar's summer palace. He wore a dapper grey suit, with the tip of a white handkerchief poking out from its breast pocket, and sported the gold medal and red ribbon of the Stalin Prize on his right lapel.[9] Two interpreters accompanied him, and he addressed the hall in Italian.

He began: "Journalists, friends, companions, I knew that after my appeal many journalists would want to meet me so I have come here for a frank and friendly exchange of views and to have a chat with the world's press. The Atlantic powers want war, [whereas] my time in the USSR has convinced me that the Soviet people want peace. The government of the USSR is taking all possible measures to prevent war. I appeal to all honest men, and in particular scientists, to take a stand. Today one cannot simply remain a spectator."

He concluded his statement by asking the press to send greetings from himself, Marianne, and the children to all their relatives, and in particular to his and Marianne's parents, who for nearly five years had received no news of them. With that, reporters thrust their hands in the air, anxious to ask questions.

What followed was a tightly controlled question-and-answer session. When one journalist asked about the circumstances of Bruno's flight to the USSR, he refused to answer. The reporters learned something of his life since his arrival in the country, however. He revealed that he split his time between his Moscow apartment and his house at the Dubna site, near the laboratory, where he conducted experiments with the giant cyclotron, and that he had received the Stalin Prize "for work on high energy physics." In answer to a question about his citizenship, he held up an identity document and announced that he had been a Soviet citizen since 1952.

The audience wanted to know what his work consisted of. He offered to explain what "high-energy particle physics" was, but added, "I doubt you would understand." When asked if this work had any military applications, he replied, "Absolutely not. I have never worked in areas that could impact the military." To the follow-up question ("Do the results have no applications to atomic weapons?"), he maintained his position: "The USSR government has a tireless campaign for the prohibition of atomic weapons, which cannot be said for the USA." He added, "When I came to the USSR I made several proposals in atomic energy, all of a peaceful nature," and explained that he had also discussed issues of radiation-protection in nuclear power plants.[10]

After two hours, the conference came to an end. One result was that on May 24, 1955, the United Kingdom revoked his citizenship. This is an extremely rare occurrence.[11]

As far as the Soviets were concerned, Bruno must have performed satisfactorily, as the next day he and Marianne were allowed to write letters

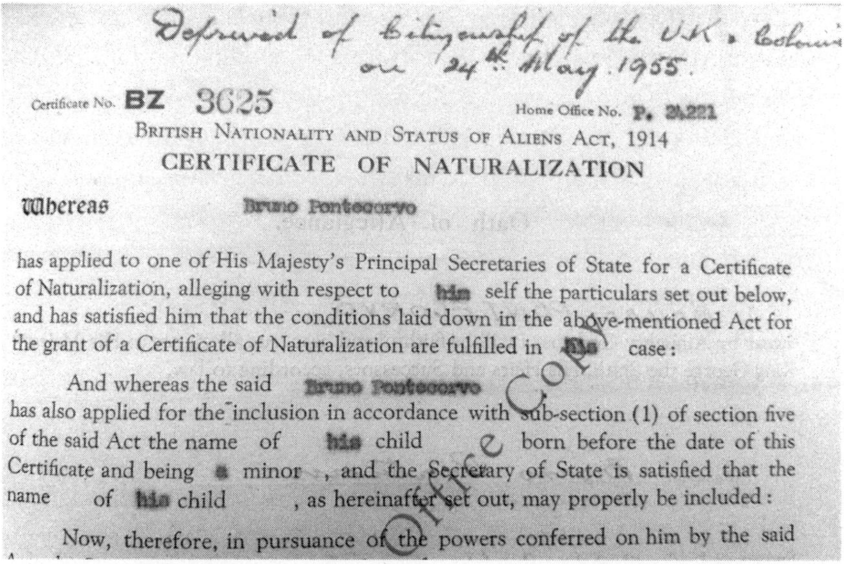

Deprived of Citizenship of the UK + Colonies on 24th May 1955.

Certificate No. **BZ 3625**   Home Office No. **P. 24221**

BRITISH NATIONALITY AND STATUS OF ALIENS ACT, 1914

CERTIFICATE OF NATURALIZATION

𝔚hereas   Bruno Pontecorvo

has applied to one of His Majesty's Principal Secretaries of State for a Certificate of Naturalization, alleging with respect to   him self the particulars set out below, and has satisfied him that the conditions laid down in the above-mentioned Act for the grant of a Certificate of Naturalization are fulfilled in   his   case :

And whereas the said   Bruno Pontecorvo
has also applied for the inclusion in accordance with sub-section (1) of section five of the said Act the name   of   his child   born before the date of this Certificate and being a minor , and the Secretary of State is satisfied that the name   of his child   , as hereinafter set out, may properly be included :

Now, therefore, in pursuance of the powers conferred on him by the said

IMAGE 16.1. Bruno Pontecorvo's British naturalization certificate, overwritten with removal of citizenship on May 24, 1955. (AUTHOR, THE NATIONAL ARCHIVES.)

to their parents, their first communication with them—or indeed with anyone outside the USSR—in five years. Speaking thirty-five years later to Miriam Mafai, Pontecorvo claimed that communications during the preceding years had not technically been forbidden: "No one said we could not write. But we did not write." Then he added enigmatically, "There are things that you cannot understand." When I asked Mafai for her interpretation of Bruno's statement, she said, "There are things you can only understand if you are communist."

With respect to Bruno and Marianne's parents, letters were as good as it would ever get. Travel outside the USSR remained off limits to the couple. Marianne was refused an exit visa to attend her mother's funeral in 1967. When Bruno's mother died in 1958, and his father in 1975, the Soviets similarly denied him permission to attend the funerals—despite the fact that Gil was granted an exit visa to attend a conference in Italy at the end of 1974. Bruno did invite his parents to the USSR, but they declined.[12] And although though they could write to their son, they could not know where he lived: Bruno's official address in the Soviet Union was a post office box.

AFTER STALIN DIED, NIKITA KHRUSHCHEV OVERSAW SIGNIFICANT changes in the Soviet state and its relations with the West. In 1956, the Hungarian Revolution against Soviet control led to the collapse of the Hungarian government, along with a decision to hold free elections and withdraw from the Warsaw Pact. Moscow could not tolerate this affront to its authority. On November 1, Russian troops entered Hungary and put down the uprising.

For supporters of communism in the West, this was probably the most serious crisis of conscience since the Soviet pact with the Nazis in 1939. The explanation put out by the Soviet authorities was that dissatisfaction in Hungary had been exploited by fascists and Western powers, which had led Hungarian patriots to call for aid from the USSR in smashing the insurrection. Even within the Soviet Union, a number of intellectuals found this story implausible, but Pontecorvo was prepared to accept the official version of events, and rationalize the invasion as a necessary act.

For Bruno, loyalty to the USSR was one of the first responsibilities of anyone who believed in communism. Mafai, after interviewing him in 1990, described the focus of a committed communist of the Cold War era as follows: "Loyalty to the USSR is the essential core of his identity. Every decision, every act of their life is dedicated to support the USSR."[13] Bruno was devastated a few months later when he learned that his favourite brother, Gillo, like many European intellectuals, saw the invasion of Hungary as an act of Soviet aggression, and had quit the Italian Communist Party. Bruno could not understand Gillo's lack of commitment to the cause. Nevertheless, in his later years, Bruno seemed to want to distance himself from this loyalist stance. In 1990 he described himself coyly as having been interested in physics and tennis, and not in politics.[14]

PROFESSOR PONTECORVO

During his time in the West, Pontecorvo had begun to consider how one might detect neutrinos emitted by nuclear reactors, or by the sun. By 1955, Bruno's existence had been confirmed, but his beloved neutrino still remained hypothetical.

It was around that time that V. P. Peshkov, one of the top experimentalists at the Institute of Nuclear Problems in Moscow, told one of his graduate students, I. I. Medvedev, to think about how one might detect neutrinos, which, if the prevailing theory was correct, were pouring from nuclear reactors in great numbers. Peshkov was a senior member of the State Committee for Science and Technology, and well connected politically. One consequence of Peshkov's position was that he had access to operating nuclear reactors. It was this that led him to muse about their possible role in revealing the neutrino.

Medvedev knew that Semen Gershtein, one of his fellow students, was interested in the weak interaction (the fundamental force responsible for beta decay) as well as the idea of the neutrino. So he made contact with Gershtein to discuss the problem. Hardly anyone took the pursuit of neutrinos seriously at the time, so when Gershtein and Medvedev learned that Bruno Pontecorvo—one of the world's foremost experts in neutrinos—was in residence at Dubna, with an apartment in Moscow, they decided to consult him about the feasibility of conducting an experiment to detect them. To their surprise, they soon realized that Pontecorvo had the answers ready almost before they had asked the questions.

One problem in detecting a ghostly neutrino is that, once you have captured a faint signal, how can you be sure it is from a neutrino and not from some other source? Cosmic rays contribute to this problem. Although the atmosphere acts like an umbrella and cuts out or disintegrates the powerful primary cosmic-ray particles, there is still a gentle rain of secondary particles at ground level, enough to mask the faint trace of a neutrino in a detector. Pontecorvo recommended that the students put their detector underneath the reactor so as to increase the protection against cosmic rays. He even estimated their chances of success, and provided details on the necessary equipment.

It was obvious that he had already given much thought to the subject. At the end of the conversation, Gershtein asked why he had not performed the experiment himself. Bruno blushed, as he frequently did when he felt uncertain, and avoided answering. During further discussions, Gershtein repeated the question, and Pontecorvo reluctantly, and "with embarrassment," revealed that he had wanted to perform the experiment soon after his arrival but was forbidden access "to any reactor."

Igor Kurchatov had been an admirer of Bruno's work for several years, but even he—the father of the Soviet atomic bomb—was unable to convince the authorities to give Pontecorvo access to a reactor.[15] Gershtein was astonished when he learned this, and years later expressed his belief that "without doubt, if investigations had started in 1950, when many industrial reactors were already operating and new ones under construction, Bruno Pontecorvo could have been the first to detect neutrinos."[16]

This folk wisdom is hard to sustain, however, as Pontecorvo's Dubna logbooks tell a different story. One journal contains notes, written toward the end of 1951, on the possibility of detecting neutrinos by means of the chlorine method, which he had advocated in 1946. He seems to have envisioned an experiment at a nuclear reactor. However, even if he had been given access to a reactor, he would not have succeeded in detecting neutrinos, for the same reason that Ray Davis failed to detect the neutrino in this way at an American reactor in 1955—namely, that a reactor produces antineutrinos, which cannot be detected by this method. In 1956, Clyde Cowan and Frederick Reines, working at the Savannah River nuclear reactor in South Carolina, succeeded in confirming the existence of these antineutrinos using a different approach. I found no evidence in Bruno's logbooks that he considered their method.

FORMER FRIENDS

We saw how, in August 1950, days before Bruno left on his long journey from Rome to the Soviet Union, he had hoped to visit Edoardo Amaldi, but missed him because Amaldi was in the United States. During his time in North America, Amaldi witnessed the construction of the Cosmotron accelerator at Brookhaven, which gave him the idea that a similar accelerator should be constructed in Europe. This was the genesis of CERN, the European Council for Nuclear Research, which was formally established in 1954, as a joint venture among twelve Western European countries. CERN's first accelerator was a relatively modest synchrocyclotron, completed in 1957, which was less powerful than that at Dubna.[17] However, by 1959 CERN had finished construction on the Proton Synchrotron, which produced protons with about forty times more energy than Dubna's, and rivaled the Cosmotron for several years as the world's leader.

Following CERN's example, the nations of the Warsaw Pact decided to launch a similar venture, based at Dubna. In 1956 the Joint Institute for Nuclear Research (JINR) was inaugurated, and construction began on the Synchrophasotron, an accelerator with five times the power of Dubna's existing one. As before, most of the construction work was done by prisoners from the Gulag, and the entire site was hidden in the woods behind two fences and a high embankment. The tilled soil between the two rings of fences gave the area the appearance of the border between two rival states, the earth ready to reveal the footprints of anyone trying to sneak across.

The Synchrophasotron began operation in 1957, allowing scientists to explore new areas in nuclear and high-energy particle physics. However, it was unable to compete against the even-higher energies that would soon be produced by CERN's Proton Synchrotron, or against the sophisticated electronics available to researchers in the West.

Bruno's role in invigorating the experimental physics programme at Dubna culminated in 1958, with his election to the Soviet Academy of Sciences. Members of the Academy and their families enjoyed privileges that were utterly exceptional. In a country where quality goods were in short supply, an Academician had access to special stores that sold Western furniture, perfumes, and wine. Holidays in exclusive hotels also became possible, along with access to first-rate medical care and other benefits shared by those in the highest echelons of the party.

In terms of privilege and status, this lifestyle far exceeded that of most Western scientists. However, in one area Bruno still lost out: Western scientists were free to take international holidays, and to perform experiments anywhere they liked. In the USSR, foreign travel was all but forbidden, even to celebrated Academicians. Thus the occasion of a major international conference on high-energy particle physics, to be held in Kiev in 1959, gave Bruno a rare opportunity to meet colleagues from around the world.

BY THIS POINT, THE EXISTENCE OF THE NEUTRINO HAD BEEN ESTABLISHED. However, Bruno still had questions about the mysterious particle. For example, under what circumstances do neutrinos maintain an identity, and what determines it?

At Chalk River, Bruno had pondered the relationship between the electron and the muon. By 1959 he was beginning to extend this line of thinking to neutrinos. He asked the following question: Are neutrinos that are produced along with electrons or positrons in beta decay the same as those produced along with a muon in the pion decay? In other words: Are "electron-neutrinos" the same as "muon-neutrinos"? At the conference,

Bruno proposed that there might indeed be more than one variety of neutrino, and suggested ways of testing this idea in experiments, which will be described in the next chapter.

Bruno's presentation at the conference was a triumph; his interactions with his Western colleagues, however, were mixed. Nino Zichichi, then a young Italian researcher, met Bruno at the conference for the first time. He told me of his excitement at meeting Fermi's former student, who had done such important work in his youth and then disappeared.[18] Those who had known Bruno in his former life were less welcoming.

Edoardo Amaldi, who had developed the slow-neutron method with Bruno in 1934, acknowledged his former colleague only with a nod, according to Bruno's great friend from his Via Panisperna days, Gian Carlo Wick.[19] Wick himself, according to Bruno, was "very cool" toward him. Luis Alvarez, a leading physicist from Berkeley, who had worked with Fermi and was a colleague of Segrè, had "on many occasions expressed his suspicions" about Bruno, following his defection. When Wick acknowledged Bruno's presence, Alvarez gave him an "evil look" for doing so—or at least this is how Wick remembered the occasion later. Emilio Segrè—in the opinion of Bruno and others who were present—was especially harsh, snubbing his old friend entirely. Bruno's impression, as he told Miriam Mafai years later, was that people he'd regarded as real friends ten years before, were now "very icy" and had "never forgiven him" for his move to the Soviet Union.[20]

IMAGE 16.2. Bruno Pontecorvo, c. 1980, with signature in Cyrillic script. (COURTESY GIL PONTECORVO; PONTECORVO FAMILY ARCHIVES.)

SEVENTEEN

MR. NEUTRINO

When the mysteries of neutrinos were finally solved, it enabled other puzzles in particle physics and cosmology to be solved as well. Bruno Pontecorvo made several seminal contributions in this field. During his time in the West he identified the weak force, and during the second half of his life he realized that neutrinos were the key to learning more about this fundamental force, which is the key to the production of elements in stars. Thanks to neutrinos emitted from the heart of the sun, and to techniques inspired by Bruno Pontecorvo, we have established how the sun creates those elements. Neutrinos emitted by a supernova have even revealed what happens when a star collapses.

Bruno Pontecorvo not only inspired the use of neutrinos as a tool in cosmology; he also thought deeply about the nature of the neutrino itself. He was fascinated by the mystery of how a particle that is as close to nothing as anything we know—with no electric charge, a mass so small that it has yet to be determined, and an extraordinary aversion to being detected—could nonetheless come in two distinct varieties: those made of matter and those made of antimatter. How does a neutrino know its identity?

Jack Steinberger, who won the Nobel Prize in 1988 for his own work with neutrinos, summarized Pontecorvo's contributions as follows: "There are few of us who can boast of a single original and important idea. Bruno's wealth of seminal suggestions establish him as a truly unique

contributor to the remarkable advances of high energy physics in the latter half of the twentieth century."[1]

Of all Bruno's ideas, perhaps the most famous is his insight that neutrinos exist in more than one variety. This great contribution to physics is
forever recorded on his memorial at the Campo Cestio in Rome, with an
equation that declares the separate identities of the electron-neutrino and
the muon-neutrino.[2]

Such work is why Bruno Pontecorvo has been given the sobriquet
"Mr. Neutrino." The reasons he never won a Nobel Prize for any of these
contributions are secreted in the closed archives of the Nobel Foundation
in Stockholm. However, there is a consensus that this may be the price he
paid for his flight to the USSR. Once in the Soviet Union, he was forced to
publish in Russian journals, which meant that his work only appeared in
English after a gap of about two years—a disastrous delay in a competitive
and fast-moving field. Also, because Bruno was not free to travel outside
the USSR, he was unable to perform various crucial experiments. These
restrictions limited his ability to test his ideas about the enigmatic neutrinos, and other scientists ended up gaining the spoils.

ANTIMATTER NEUTRINOS

In 1956, American physicists Fred Reines and Clyde Cowan confirmed
the existence of the neutrino. The discovery owed nothing to Pontecorvo,
but it would stimulate him to come up with a series of ideas.

Pontecorvo's 1946 paper advocated the use of chlorine as a target, and
predicted that the impact of a neutrino would convert a neutron into a
proton, thus changing the chlorine into a radioactive form of argon, which
lies immediately next to chlorine in the periodic table of the elements. The
neutrino, meanwhile, would turn into a negatively charged electron to con-
serve the total amount of electric charge throughout the process. This was
not the approach that Cowan and Reines used. And, in fact, the neutrinos
produced by a reactor do not generate this sequence of events. Instead, they
convert a proton into a neutron, which would change chlorine into sul-
phur. And instead of turning into an electron, the neutrino becomes a positron—the positively charged antiparticle of an electron.[3] The conventional
way to differentiate between these two alternatives is to say that, in the

former case (where an electron emerges), a neutrino has struck, whereas in the latter case (where a positron emerges), it was an *antineutrino* that made the impact. This makes sense if the terms *matter* and *antimatter* have some intrinsic meaning: a neutrino (matter) turns into an electron; an antineutrino (antimatter) turns into an antielectron, or positron. However, this raises the question of what it is about the antineutrino, as it flies through space, that identifies it as such, and differentiates it from a neutrino.

Pontecorvo began to ponder this question, and in 1957 gave a talk at the Dubna laboratory in which he suggested that a neutrino might transform into an antineutrino, or vice versa, in midflight. His idea stemmed from two circumstances. The first was a rumour, which turned out to be false, about Ray Davis's quest for neutrinos; the second was a discovery relating to a specific strange particle: the neutral kaon, or K-zero.

By this time it was clear that there were two electrically neutral strange particles, one with positive strangeness and the other with negative, known as the K-zero and the anti-K-zero, respectively. As their names suggest, one is the antiparticle of the other. Experiments had shown that when either of them decays, the debris can be a pair of pions. Quantum theory implied that a pair of pions from a K-zero decay could then come back together—this time forming an anti-K-zero. Through this two-step process, a piece of electrically neutral matter could turn into antimatter. This possibility fascinated high-energy physicists. In 1957, Bruno wondered if this idea could also apply to neutrinos and antineutrinos.[4]

The motivation for his conjecture seems to have been a game of "Chinese whispers". In 1957, a rumour arrived in Dubna to the effect that the American experimentalist Ray Davis, who had been using Bruno's chlorine method, had detected neutrinos from a nuclear reactor. Bruno deduced that if the report was correct, the antineutrino that left the reactor must have have changed into a neutrino capable of triggering Davis's chlorine detector. Unaware that the rumour was false, he proposed that neutrinos and antineutrinos could "oscillate" back and forth, shifting their identities from one to the other.

The idea was extremely audacious, and was regarded by many as the "fantasy of a prominent physicist."[5] Scientists were skeptical because, according to quantum theory, such a transmutation could occur only if the neutrino had mass, and the received wisdom at the time was that it had none. This point became moot when the rumour about Davis's experiment was found to be false. In reality, Davis had not recorded any neutrinos at the reactor. This seemed to confirm the emerging conventional picture: a reactor produces antineutrinos, and there was no reason to suspect that an antineutrino can switch to become a neutrino.[6]

Thus, in 1957, Pontecorvo's idea about oscillation was forgotten, though not by him. A few years later, he would resurrect it in a new guise— one that we now know to be correct. For neutrinos are not massless, and although their mass is so trifling that it has not yet been measured, Pontecorvo's theory of "neutrino oscillations" has today become the focus of a whole branch of science. I shall return to this point later.

STELLAR NEUTRINOS

During his time at Chalk River Bruno Pontecorvo had been intrigued by the muon. Indeed, it was he who had established that it was a relative of the electron—and also that there was some unique character, other than its mass, that made it distinct from an electron. This enigmatic quality troubled Bruno, as it did many others, and continues to be troubling even today. The image of the muon invaded his mind and would not go away. He did not determine what makes a muon different from an electron, other than its mass, but the kaleidoscope of confusion settled into an unexpected picture: Bruno realized that neutrinos might play an important role in astrophysics and cosmology.

I can only guess how Bruno reached this conclusion, but we do know the basic pieces of his puzzle, and from these a possible path emerges. It is as follows:

Muons are heavier than electrons. Due to Einstein's relation between mass and energy, a muon therefore has more energy locked within its mass than an electron. A muon can convert this energy into an electron, a neutrino, and an antineutrino. This is the traditional way that a muon decays. Bruno now imagined what might happen if an electron had a lot of energy: Could it shed that energy by "radiating" a neutrino and an antineutrino, analogous to the decay of a muon? There was no reason why not. How would an energetic electron do this, while maintaining the sacrosanct laws of energy- and momentum-balance? Bruno knew the answer:

when an electron passes near an atomic nucleus, it picks up energy from the atom's electric field. It can then shed that energy by transmuting into a lower-energy electron, neutrino, and antineutrino.

Bruno's insight was truly novel, and momentous not just for particle physics but for cosmology as well.

The link lies inside of stars. The production of neutrinos involves the weak force. As its name suggests, the effects of this force are feeble relative to those of the electromagnetic force. This is why the production of neutrinos is normally so rare compared to the radiation of light. However, at high energies the relative power of the weak force grows. Bruno realized that inside stars that are much hotter than the sun, the weak force becomes more powerful, potentially comparable to the electromagnetic force, in which case the production of neutrinos could, in theory, occur as easily as the radiation of photons.

Neutrinos penetrate matter much more easily than photons do. This leads to a startling consequence. Whereas most photons are absorbed within the stellar mass, neutrinos produced in the heart of a star can escape. The implication, which Bruno pointed out in his paper of April 1958, is this: "At some stage in the evolution of a star, it may well be that the energies radiated into space in the form of neutrinos and photons become comparable."[7] Thirty years later this insight would be confirmed when a burst of neutrinos was detected coming from a distant supernova—SN1987A.

THE FLAVOURED NEUTRINO

Bruno's work on neutrinos became vital in saving a developing theory of the weak force.

According to Fermi's theory of beta decay, the interactions of neutrinos (and weak-force interactions in general) are not always feeble. Instead, the chance of a reaction depends sensitively on the energies of the particles involved. Double their energy and the chance increases fourfold; triple their energy and the chance increases by a factor of nine. In general, the growth is proportional to the square of the energy.

However, this chance cannot grow indefinitely in reality, for if it did, at a certain point it would imply that some processes occur with greater than 100 percent probability, even infinite probability, which is nonsense. So Fermi's theory can only be an approximation of some more complete explanation.

In 1956, Julian Schwinger, the American theorist who shared a Nobel Prize for his work on the electromagnetic force, made a crucial proposal about the weak force, which went beyond Fermi's theory.[8] In a nutshell, Schwinger suggested that the weak force and the electromagnetic force have something in common.

Electromagnetic radiation comes in particle bundles—photons. In quantum theory, the electromagnetic force between two particles arises when they exchange photons. Schwinger suggested that a similar process occurs for the weak force. He predicted the existence of the W ("weak") boson, analogous to the photon. In Schwinger's hypothesis, particles experience the weak force when they exchange W bosons.

This was confirmed in 1983 with the discovery of the W boson, but in 1956 it was just a hypothesis. Nonetheless, the idea was compelling. In addition to its seductive implication that two fundamental forces (electromagnetic and weak) were analogous and could perhaps be united theoretically, it also avoided the nonsensical probabilities implicit in Fermi's original model. When particles interact at high energies, W bosons can be produced. When this new process is included in the quantum accounts, the troubles with Fermi's model disappear.[9]

Almost immediately there was a problem. Soon after Schwinger had made his suggestion, another American theorist, Gerald Feinberg, noticed that if the weak force is indeed due to the action of W bosons, there is an unwanted implication for the decay of the muon. By this stage it had been established that a muon decays into an electron and two neutrinos. If this occurs through the intermediate action of a W boson, as Schwinger proposed, the laws of quantum mechanics imply that you can do away with the neutrinos and have the muon decay into an electron and a photon. However, no example of such behaviour had ever been seen. Feinberg calculated that one in every ten thousand decays should result in an electron and photon. This is a small percentage, but the experimental data already showed that if there were any such decays at all, they at most accounted for one in a hundred million![10] Schwinger's theory looked to be in trouble.

Feinberg did point out, however, that there was a loophole in his argument: he had assumed that the neutrino associated with the muon and the neutrino paired with the electron were the same. If a "muon-neutrino" differs from an "electron-neutrino," there is no problem. Having made this observation, Feinberg took it no further.

BRUNO NOW MADE HIS FIRST INTERVENTION. HIS PAPER ON "ELECTRON and Muon Neutrinos" was written on June 29, 1959.[11] He was not actually the first to have pursued the implications of Feinberg's observation— Jogesh Pati and Sadao Oneda in the US had written a paper earlier that year, in which they "deliberately" denoted a distinct "neutral counterpart of the muon" (the muon-neutrino) as well as a neutral counterpart of the electron (the electron-neutrino).[12] They pointed out that if the two were actually identical, then a muon could decay into an electron and a photon. If they were not identical, then this could not occur. Bruno also understood this. However, his efforts in Dubna would go further toward finding a definitive answer to the problem.

Initially, the primary goal of the scientists at Dubna was to investigate how the strong forces of the atomic nucleus produce pions, which are the material embodiment of the energy latent in the nuclear field. A pion is not stable. It self-destructs, leaving either a muon or an electron, accompanied by a neutrino.[13] The traditional decays of nuclear particles produce electrons (or positrons) and neutrinos. Once in every ten thousand decays, the pion does also, but most of the time it decays into a muon and a neutrino. In 1959 Bruno Pontecorvo started wondering: Are the neutrinos produced when a pion decays into a muon the same as those emitted in conventional beta decays?

He began by systematically listing all processes in which neutrinos occur, then moved the subject forward by identifying practical experimental ways of identifying a neutrino's character. He showed that although neutrinos and antineutrinos could be identified as distinct, the question remained open empirically as to whether muon- and electron-neutrinos are different or identical.

In Bruno's paper, one section was titled "Are muon-neutrino and electron-neutrino identical particles?" He acknowledged that the possibility that they are distinct would be "attractive from the point of view of symmetry and the classification of particles." His point is that the muon and electron have distinct "flavours," so if their neutral counterparts occur in two flavours also, there would be symmetry among these particles.

Today this is a basic plank of the Standard Model, which classifies the fundamental forces and the interactions among them.

Bruno evidently suspected that the two neutrinos were indeed different, because in his paper he introduced the nomenclature and notation that is universally used today. The particles became known as "muon-neutrinos" and "electron-neutrinos" forever onward. In the shorthand notation of particle physicists they are written ν_μ and ν_e, respectively.

IT IS NOT UNUSUAL IN SCIENCE FOR A GREAT IDEA TO OCCUR TO MORE than one person, independently. The fact that one person is remembered and another forgotten can be due to many factors: chance, opportunity, or the confidence to follow through on what others might regard as crazy. And when that "crazy" idea turns out to be sensible after all, and winner and loser have talked to each other along the way, versions of history can diverge, as memories differ of who did what.

The saga of how the idea of two neutrinos matured into established lore is a paradigm of such divergence. This tale begins in Moscow, in 1957, with a colleague of Bruno's named Moisy Markov.

Markov was interested in the neutrinos that are produced when cosmic rays hit the upper atmosphere. The collisions produce pions and muons, which in turn shed neutrinos. These can have considerable energy, far more than the neutrinos from nuclear reactors, the main source in the 1950s. Markov wondered how scientists could detect these "atmospheric neutrinos." He decided this might be a good project for his student Igor Zheleznykh to investigate.

Zheleznykh designed a detector. It contained one cubic metre of lead, and was placed deep underground, where it would be shielded from other cosmic ray particles. By 1958, Zheleznykh had shown that the chance of neutrinos interacting with the target grows considerably with their energy. He also remarked, "Different numbers of electrons and muons induced by neutrinos in a detector could give evidence of the existence of two types of neutrino." Like Feinberg, he did not pursue this any further.

During his research, Zheleznykh had stumbled onto a second profound question: Why make calculations only for high-energy atmospheric neutrinos? Why not consider performing neutrino experiments at high-energy accelerators? One evening, late in 1957, Zheleznykh visited Markov at home and asked him.

It turns out that Markov had already asked himself this very question, in connection with the Synchrophasotron, the new higher-energy accelerator planned for Dubna, but had dismissed the idea as impractical. Impractical at Dubna, maybe, but it was an interesting challenge in principle: the experiment might be feasible somewhere, someday. Zheleznykh's question led Markov to reconsider the problem. He went to discuss the idea with Bruno Pontecorvo: "I told [Pontecorvo] that I would like to suggest neutrino experiments at accelerators. [He] liked such an idea very much."

Inspired by Pontecorvo's enthusiasm, Markov gave the problem to another student, Docho Fakirov, who included it in his thesis at Moscow State University in 1958. Markov decided to write a report, titled "On High Energy Neutrino Physics," which he planned to present in 1959 at the conference on high-energy physics in Kiev. However, several colleagues were skeptical, with one influential physicist asking him, "Are you serious?" in a manner that clearly suggested the answer to be no.[14] Markov withdrew the paper. It was a big mistake: Bruno Pontecorvo would be braver, and as a result it was he who was remembered for this advance.

FINDING THE NEUTRINO HAD BEEN HARD ENOUGH; PROVING experimentally that there was more than one variety of the phantom particle would present a new level of difficulty. Bruno's idea was to replicate the Cowan-and-Reines discovery of the antineutrino, using a source of muon-antineutrinos instead of a nuclear reactor, which produces the electron variety.

In Cowan and Reines's experiment, the collision of an antineutrino with a proton converted the proton into a neutron and a positron. This led to the discovery of the electron-antineutrino. Bruno wanted to initiate this same process using antineutrinos produced in association with a muon. If neutrinos have distinct flavours, as he suspected, the subsequent collision with a proton should release a positive muon, not a positron; the production of a positron, on the other hand, would demonstrate that the two neutrinos are identical.

The next question was how best to conduct such an experiment. He noted that using antineutrinos of high energy would be advantageous, as the chance of interaction increases with energy. Thus a high-energy accelerator was needed. The basic idea was to smash a beam of high-energy protons into a target, which would liberate large numbers of positively charged pions. These decay into muons and antineutrinos. A steel shield would absorb the muons, but would be almost transparent to the antineutrinos. Several metres away, another large target would serve as an antineutrino detector. The antineutrinos would have high energy, and hence there would be a reasonable chance that occasionally one would hit an atom in the detector, pick up electric charge, and reveal itself.

He calculated that, with a detector similar to the one used by Cowan and Reines, one collision an hour might be detected "at new accelerators now being discussed in which the intensity of the protons may be [a thousand times larger than at previous accelerators]." He insisted that experiments to test the identity of muon- and electron-neutrinos "must be seriously thought over" when the new accelerators—at CERN in Europe, and Brookhaven in the US—became available.[15]

Bruno mentioned some of these ideas at the end of July 1959, during the International Conference on High Energy Physics, in Kiev. He wrote a report under the auspices of the Joint Institute for Nuclear Research (JINR), which was formally published in the *Soviet Journal of Experimental and Theoretical Physics* later that year. This article was written in Russian, of course, which limited its international reach. Bruno's remarks at the conference would have been translated, but there is no record that they had any memorable effect.[16]

The following year Bruno wrote a second paper, in which he developed his earlier ideas. The first paper had dealt with antineutrinos; the 1960 paper considered how to test specifically for neutrinos of different flavours. He advocated using a high-energy accelerator to produce a beam of neutrinos, pointed at a lump of carbon. Any collisions would convert an atom of carbon into nitrogen.[17]

The beams in an accelerator come in pulses, so the neutrinos they produce also arrive in distinct bursts, separated by a fraction of a second. Bruno realized that with modern electronics it would be possible to record the instant when a burst of neutrinos arrived at the carbon target and see if this coincided with the appearance of an electron. In this way, one could determine if the electron was a genuine signal produced by the collision, or if instead it had strayed out of an atom in the detector itself, and as such was merely background. The nitrogen produced in such a collision would be radioactive and would decay by emitting a positron. This would occur marginally later than the actual collision, and the time delay between the electron and the positron could be used as another check. Bruno had done everything possible to design a realistic experiment; now all that was required was an opportunity to perform it.

Unfortunately, it would be impossible to do the experiment at Dubna. In November 1959, CERN's powerful Proton Synchrotron began operation, and could have served Bruno's purposes, but the Soviet authorities refused to allow him to leave the USSR. Two years would pass before Bruno's ideas were made available in English. By this time, it was too late: he had been scooped.

IN THOSE DAYS, SOVIET IDEAS WERE LARGELY UNKNOWN IN THE US and Europe until they appeared in translation. This meant that ideas often developed independently in the two hemispheres. In New York, during November 1959, Chinese-American theorist T. D. Lee was pondering Fermi's theory on the behaviour of weak interactions. Lee was unaware of Bruno Pontecorvo's ideas, let alone Markov's.

As we have seen, Fermi's theory was only an approximation of some complete explanation, since it gave impossible outcomes for the behaviour of neutrinos at very high energy levels. Schwinger's hypothesis of the W boson solved that problem but ran into its own difficulties regarding the decays of muons. Lee therefore wanted to reveal the solutions to these problems experimentally. The challenge was to find a way to probe the weak interaction in high-energy experiments.

While leading a discussion on this subject, Lee realized that such an experiment would be hard to perform because when particles collide at high energies, the effects of the electromagnetic and strong forces tend to obscure those of the weak force. Melvin Schwartz, a squat experimentalist with a bubbly personality, was one of those present. The lunchtime discussion must have entered his subconscious mind, as in bed that night Schwartz suddenly had the answer: "It was incredibly simple. All one had to do was to use neutrinos."[18] Neutrinos aren't affected by the strong force, and, being electrically neutral, they aren't affected by electromagnetic forces either. As such, Schwartz realized, they are ideal for probing the weak force. His idea was that the production of pions, and their subsequent decays, might produce neutrinos in sufficient numbers that they could be used in experiments.

He wrote a short paper outlining his ideas, which was published in 1960. Pontecorvo's paper had just appeared in English translation, and Schwartz included a comment at the end of his paper noting the "related paper which has just appeared" by Pontecorvo. He also thanked Lee, and Lee's collaborator C. N. Yang, for emphasizing the importance of high-energy neutrino interactions.[19]

Pontecorvo's ideas about the two distinct flavours of neutrinos were not included in Schwartz's paper. However, in the meantime Lee and Yang had been thinking about what might be learned from these experiments. By the summer of 1960 they had reached the same conclusion as Pontecorvo: the absence of muon decay into electron and photon could be the smoking gun proving that the muon-neutrino and electron-neutrino differed. This became the quarry to chase.

In July 1960 the Alternating Gradient Synchrotron (AGS) began operations at Brookhaven. Slightly more powerful than CERN's Proton Synchrotron, the AGS was for the next eight years the world's highest-energy machine. One of the first experiments at the new accelerator led to Nobel Prizes for the three leaders—Melvin Schwartz, Jack Steinberger, and Leon Lederman. In essence they performed the experiment that, unknown to them, Bruno had proposed, and found the result that he was hoping for: muon-neutrinos and electron-neutrinos are distinct. In addition they demonstrated that, at the new high-energy accelerators, the neutrinos are less shy, as Zheleznykh had deduced in 1958, and thereby become useful tools for science. During the next four decades, beams of high-energy neutrinos at accelerators throughout the world opened up new vistas in our understanding of the structure of matter, and the profound patterns at work in the fundamental laws of nature.

The idea that neutrinos have distinct flavours, which parallel those of their electrically charged siblings, electrons and muons, is one of the basic ingredients in the modern theory of fundamental particles, and today Bruno Pontecorvo is recognized as its parent. The American team had the grace to mention his independent insight when they accepted the Nobel Prize in 1988, with Schwartz remarking that Pontecorvo's "overall contribution to the field of neutrino physics was certainly major."[20]

THE TEAM'S NOBEL PRIZE WAS AWARDED FOR THE EXPERIMENTAL "demonstration" of distinct types of neutrino and for the "neutrino beam method." Although Bruno's insights were central to the former, his claims on high-energy neutrino beams are peripheral. As we have seen, others in Moscow, whose work he must have been aware of, had taken the idea somewhat further, but had lacked confidence in their conclusions. Furthermore, as we shall now see, Bruno did not really advocate high-energy beams either.

Although the experiment he proposed could have been done at CERN as well as at the AGS in Brookhaven, it is not clear whether Bruno would have chosen to perform it himself, even if he had been allowed to leave the USSR. His 1960 paper suggests that he was rather pessimistic about the practicality of the enterprise. He noted that the chance of a neutrino interacting grows with its energy, but he worried that, at very high energies, the quantity of neutrinos produced would be smaller. This is due to the time-dilation effect of relativity, which causes fast-moving particles to experience time slowed down, which in turn causes them to live longer. Thus high-energy pions, which are faster, live longer than lower-energy ones. It is the decays of pions that give rise to the beams of neutrinos, so if fewer pions decay, the number of neutrinos falls as well. As a result, Pontecorvo focused on neutrinos with moderate energies, rather than on the very high energies available at CERN.

In fact, he was overly pessimistic. At these high energy levels, the increased chance of interaction more than compensates for the reduction in the number of neutrinos. Also, the effects of collisions at high energy are easier to diagnose. Overall, the rule is: the higher the energy, the better.[21]

BRUNO CONCEIVES THE STANDARD MODEL

Even after the American trio confirmed Bruno's theory that there is more than one variety of neutrino, there remained a further question: Do the two varieties respond to the weak force in precisely the same way? This would be the case if the weak interaction were truly a fundamental force of nature. In 1962 Bruno came up with a way to test the question.

Bruno's idea grew out of work that he and Frédéric Joliot-Curie had done twenty-five years before.

In 1937, Bruno and Joliot-Curie had tried to prove that the beta particles that emerge during radioactive decay are indeed electrons, and not just similar to electrons. They did this by firing electrons at atoms in the hope of inducing the transmutation in reverse—what has become known as the inverse beta process. Their results were inconclusive because there are so many electrons in matter that it was hard to distinguish between signals and background. However, twenty-five years later Bruno remembered their attempt, and now realized that it should be possible to perform the experiment with muons, rather than electrons, and initiate the inverse beta process that way. When it hit the target, the muon would turn into a muon-neutrino, and the experiment could show whether the probability of a muon-induced reaction was the same as that of an electron-induced reaction. Because the neutrino is electrically neutral and would leave no trace as it escaped, it would be important to keep track of the corresponding change in the target to make sure that the reaction had occurred. To do so, he proposed using nuclei of helium-3, in which two protons are bound to a single neutron. If the inverse beta reaction occurred, the helium would convert to tritium—made of one proton and two neutrons.

If a cloud chamber was filled with the helium-3 gas, any tritium formed by the reaction would recoil and leave a wispy trail. So the chamber would act as both a target and a track detector. The problem was that helium-3 was not easy to come by, especially in the Soviet Union. Fortunately, Igor Kurchatov was interested in Pontecorvo's ideas and provided him with sufficient helium-3 (a product of Kurchatov's nuclear weapon programme) for the experiment to succeed.

The result was everything that Bruno had hoped for. He found that the weak interaction acts on muons and muon-neutrinos with the same strength as it does on electrons and electron-neutrinos. Pontecorvo had established the universal nature of the weak force: it acts on muon and electron flavours impartially.

This breakthrough, along with Bruno's earlier idea of distinct neutrino flavours, forms the core of today's Standard Model of particles and forces.[22]

SOLAR NEUTRINOS

Bruno Pontecorvo's 1946 paper, which had inspired Ray Davis's unsuccessful attempt to catch neutrinos coming from a reactor, only mentioned the sun in two sentences. However, the importance of the underlying concept is immense. Ten million solar neutrinos pass through your eyeballs every second, unseen. This is indeed a lot, but still not enough to be easily detected. Furthermore, the chlorine detector is blind to almost all of them.

To be seen, a neutrino first has to hit the nucleus of a chlorine atom. That is a rare enough occurrence, but it's still not sufficient to guarantee that the chlorine will convert into argon. The reason is that the argon nucleus has more mass than chlorine, and this extra mass has to be created from the energy of the incident neutrino. The catch is that the neutrinos produced by the main part of the solar fusion engine (the conversion of hydrogen into helium) have too little energy to do the task.

However, our sun makes more than just helium; other light elements are created in its nuclear furnace. After hydrogen has been converted into helium, the helium nuclei can fuse and, though a series of processes, produce elements such as boron and beryllium. This also liberates neutrinos, some of which have enough energy to activate a chlorine detector. Although these neutrinos are far outnumbered by those created in the hydrogen-to-helium process, the good news is that the chance of a neutrino being captured grows with energy, as we have seen, and so these higher-energy neutrinos are easier to capture.

As the 1950s came to a close, Ray Davis began his quest to capture neutrinos from the sun. This story would continue for forty years, with many twists and turns along the way, before its final chapter was written.[23]

In his first attempt in 1959, Davis used 4,000 litres of cleaning fluid, but he failed to detect anything. He realized that to have any chance he would need to be even more ambitious, so he set out to build a target containing 400,000 litres of the liquid, enough to fill a swimming pool. In 1964 Bruno Pontecorvo held a special seminar in Leningrad to report on Davis's quest. There was a lot of interest in the seminar, but Pontecorvo later said that he was the only person present who believed that the experiment would be successful. Whether this belief was based on insight or simple optimism is hard to say.

By the end of summer 1966—twenty years after Pontecorvo first advertised chlorine as a way to detect neutrinos—Davis's enlarged experiment was ready to begin.

With his mammoth detector, located in a South Dakota gold mine more than a mile underground, where it was shielded from cosmic rays, Davis managed to capture only one or two neutrinos a month. Everything known about the sun implied that he should be detecting one solar neutrino every few days. While this might seem like a small detail, it was in fact a serious worry. Many were convinced that Davis was trying the impossible—and proving them right.

By 1972 Davis had improved his experiment further, and at last managed to convince other scientists that he was indeed seeing solar neutrinos. However, there were still too few of them. Several theorists suggested that our understanding of the sun might be at fault. Bruno Pontecorvo wrote to Davis: "It starts to be really interesting! It would be nice if all this ends with something unexpected from the point of view of [neutrinos]. Unfortunately it will not be easy to demonstrate this even if nature works that way."[24]

Pontecorvo was right. Years went by, and Davis accumulated more data, which consistently disagreed with the predictions of the standard solar model. The quest had reached an impasse. The arguments went on for twenty years, with theorists blaming the experiment, while experimentalists (and, to be fair, several theorists) doubted the accuracy of the solar models. Relatively little attention was paid to the possibility that the sun might be innocent and that the neutrinos were to blame.

Ironically, Bruno Pontecorvo had proposed the solution to the solar neutrino mystery in 1968, as soon as Davis's first results appeared—and been ignored. Of course, he was not primarily interested in the sun. Instead he was focused on the nature of neutrinos, and his own insight that they come in different flavours.

OSCILLATING NEUTRINOS

One reason that it took so long to solve the solar neutrino mystery is that neutrinos can change their properties during the flight from the sun to the earth. Only those that retained the attributes they were born with showed up in Davis's experiment, while those that had changed form escaped detection.

In 1968 Bruno Pontecorvo and a colleague, Vladimir Gribov, resurrected his idea of neutrino oscillations—the possibility that neutrinos can metamorphose from one variety to another. Whereas in 1957 Bruno had proposed that this might occur between neutrinos and antineutrinos, in 1968 he applied the idea to the electron and muon flavours of neutrino.[25]

The sun produces electron-neutrinos and Davis's detector was sensitive to them, but it was blind to any muon-neutrinos. Thus any electron-neutrinos that metamorphosed into muon-neutrinos en route from the sun would be lost. This would explain why Davis detected fewer neutrinos than he expected.

The idea that neutrinos produced in the sun could change identity during their journey ran counter to everything in the textbooks. According to the standard theory of particle physics, this was impossible. Or at least it was impossible if, as scientists then believed, neutrinos were massless and travelled through space at the speed of light. However, Pontecorvo and Gribov realized that the laws of quantum mechanics allowed neutrinos to switch from electron to muon variety and back again, "oscillating" between one state and another—but only if neutrinos had some mass. It didn't need to be large; in fact it could be trifling, thousands of times smaller than the mass of an electron. All that the theory required was the existence of two varieties of neutrinos, with slightly different masses.

In quantum mechanics, certainty is replaced by probability, which rises and falls like a wave. If two particles have the same energy, but slightly different masses, the associated waves have slightly different wavelengths, and interfere with each other as they move through space. Two sound waves whose frequencies differ slightly tend to mingle and produce a pulsing beat. The quantum waves associated with two neutrinos of slightly differing masses produce an analogous rise and fall in intensity. Only occasionally along the journey do the two waves match up in the precise form they started out in. At other points along the route, the initial electron-neutrino is a chimera, appearing sometimes as a muon-neutrino, and sometimes in its original form.

When the flow of these waves is interrupted—for example, by an atom of chlorine in a tank 4,500 feet beneath the hills of South Dakota—it is impossible to know which variety of neutrino will be revealed. If it is an electron-neutrino, Davis's detector records the fact; if it is a muon-neutrino, it is invisible. All quantum theory can tell us is the probability of finding one or the other. In effect, if enough collisions occur, they will be divided roughly fifty-fifty between electron-neutrinos and muon-neutrinos.

Such quantum behaviour is unfamiliar in the everyday world, but is illustrated in M. C. Escher's *Metamorphosis* drawings. On the left side, a picture shows a collection of objects, which gradually change into a new form as you traverse the image. In the spirit of Escher, suppose there were some weird hybrid animal that could metamorphose between a cat and a dog. The dog transforms into a cat as it walks along the street. Halfway along the block the transformation is complete. The former dog (now a cat) continues on its way. By the end of the block it will be a dog once more. When you look at the dog-cat, what you see will depend on how far along the block you are.

Now suppose that you are not receptive to dog-cats, only to things that are one or the other: a dog or a cat. If you are near the start or the end of the block, you will most likely interpret the animal as a dog. If you are near the midpoint, you are more likely to interpret it as a cat. However, if your eyes are only capable of seeing dogs, and are blind to cats, you might conclude that the dog had disappeared.

So it is with neutrinos. In this analogy, the electron-neutrino is the dog and the muon-neutrino is the cat. The sun emitted a dog, and Davis's detector was a dog-catcher. It recorded no reaction when a cat came along. According to this theory, there was nothing wrong with the sun; it was our understanding of neutrinos that was at fault.[26]

If one accepted Pontecorvo and Gribov's theory of oscillating neutrinos, the apparent shortfall of solar neutrinos in Davis's experiment could be understood. All that was required was to give up the myth that neutrinos were massless and travelled at the speed of light. However, few were prepared to do so at the time. The Russian duo's idea was regarded as little more than a mathematical curiosity. Only after a variety of further experiments would Gribov and Pontecorvo be vindicated. That would take three decades and much would happen in the interim.

"TEN TO THE FIFTY-EIGHTH NEUTRINOS! ALL IN ONE GO!"

Physics seems to be a tradition in the Pontecorvo family. In addition to Bruno's son Gil, two of his nephews have chosen the profession: Giuliana's son Eugenio Tabet, and Gillo's son Ludovico. Born in 1964, Ludovico, or Ludo, is an experimentalist, part of the ATLAS collaboration at CERN, which discovered the Higgs boson in 2012. This is the final piece in the Standard Model of particles and forces, whose origins can be traced back to Bruno Pontecorvo's insights about muons and neutrinos in the 1940s.

In 1987, Ludo was a graduate student at university in Rome. By this stage the Soviet policy of glasnost allowed Bruno to visit Italy for extended periods, and he spent a lot of time in Ludo's office during his visit that year. Ludo recalled Bruno's intense excitement when his beloved neutrinos made one of the most remarkable contributions in the history of astronomy: "I arrived at my office where I was doing my thesis. Bruno was already there. I noticed a strange light in his eyes. He said, 'Did you hear what happened today? Ten to the fifty-eighth neutrinos! All in one go!'[27] I said, 'What are you talking about?' and Bruno replied, 'There was a supernova.' I felt he was living for that. [The first thing he said] was not 'supernova'; he said, "'Ten to the fifty-eighth all in one go!'"[28]

Bruno had suggested in 1958 that supernovas would produce neutrinos in vast amounts, capable of being detected on Earth. A supernova explosion releases more than a hundred times as much energy as the sun has put out in its entire life. Thus the breeze of neutrinos from the sun would be dwarfed by the hurricane from a supernova. The sun, however, is close by, whereas supernovas, thankfully, are not. The one that excited Bruno in 1987 had actually occurred some 160,000 years earlier in the Large Magellanic Cloud, and it had taken that long for the shell of neutrinos to reach us. By this stage they were spread over a huge region, larger than the diameter of our galaxy.[29] The blast wave passed through the earth and continued onward into deep space. A handful of these neutrinos were detected by experiments in underground caverns. From this sparse number, it is possible to deduce how much total energy the supernova released. The results confirmed what astrophysics already believed: a supernova occurs when a star implodes and becomes a neutron star. For the first (and, so far, only) time, humans had witnessed a star collapse by observing its neutrinos. Along with the study of solar neutrinos, this marked the beginning of a new science: neutrino astronomy.

SOLAR NEUTRINOS OBSERVED

In 1991 Bruno went to Paris again, and stayed at the Hôtel du Panthéon, located in the square by the mausoleum, near his old residence, the Hôtel des Grands Hommes. He visited the Curie Institute and discussed the neutrino experiments that French scientists were planning to conduct at CERN. By this stage the mystery of solar neutrinos was beginning to be resolved, and Pontecorvo's ideas about oscillations were taken seriously enough that expensive experiments to test them were being designed. Bruno naturally was much in demand.

Meanwhile, after thirty years of pursuing solar neutrinos, Ray Davis was convincing the world that he was seeing some, but not enough. Gribov and Pontecorvo's suggestion that electron-neutrinos were changing form en route from the sun was beginning to be viewed as the most likely explanation. One reason for this shift in thinking was that the scientific community's understanding of neutrinos and their flavours had dramatically evolved in the interim.

Bruno's original theory of neutrino flavour had matched two neutrinos, the electron and muon varieties, with the electron and muon themselves. Meanwhile, starting in the 1960s, the study of nuclear particles had revealed a deeper layer of reality—the quarks. Quarks, as it turns out, also come in pairs. The up and down varieties of quark are the constituents of protons and neutrons; strange and charm quarks form a second pair, which are found within strange and other exotic particles. These four types of quarks, when paired off this way, mirror the four types of leptons—the generic name for the electron, electron-neutrino, muon, and muon-neutrino, which are not affected by the strong nuclear force. Bruno's proposal that there were two pairs of leptons reflected the existence of the two pairs of quarks.

In 1976, a third electrically charged analogue of the electron and muon turned up, known as the tau (the name comes from t as in third). This was soon followed by the discovery of another variety of quark, known as a bottom quark. Theorists anticipated that each of these would be revealed to be part of a pair, completed by a third neutrino (the tau-neutrino, confirmed in 2000) and a third quark (the top quark, discovered in 1995).

Although the third variety of neutrino was not discovered until after Bruno's death, theorists were convinced of its existence from the 1970s onward. One reason was that cosmologists were best able to describe the formation of elements in the early universe if there were three varieties of neutrino. The other reason touched directly on the solar neutrino problem: Davis was seeing only about one-third as many neutrinos as astrophysicists expected. Pontecorvo and Gribov's idea implied that if there were three rather than two alternate guises for an electron-neutrino to take on, then after its 150-million-kilometer trip from the sun, it would be an electron-neutrino one-third of the time, and one of the other varieties two-thirds of the time. Because Davis's experiment was only sensitive to the electron type, he measured one-third as many neutrinos as had set out from the sun.

Davis's shortfall, and the Gribov-Pontecorvo theory, began to appear more and more compelling.

THE FINAL PIECE IN THE SOLAR NEUTRINO ODYSSEY WOULD COME FROM an experiment conducted at the Sudbury Neutrino Observatory (SNO) in Ontario, less than two hundred miles from Chalk River, the laboratory where forty years earlier Bruno had conceived his ideas about neutrinos. The SNO team would eventually confirm his theory that solar neutrinos change form in their journey across space. Sadly, Bruno would not live to see this.

The main innovation at SNO was the use of heavy water, the very material that had been key to the nuclear reactor all those years before. However, this time the liquid was used as a detector of neutrinos rather than as a component of a reactor.

In Davis's experiment, neutrinos that had changed to the muon or tau variety escaped detection. But when any variety of neutrino hits an atom of deuterium, it can bounce off the target and give enough of a kick that the deuterium nucleus breaks up. The recoil of its constituent proton and neutron can then be detected, and the collision of the neutrino can be inferred. The chance that a single neutrino will cause this to happen is given by Fermi's theory, and the number of detected reactions can thus be used to calculate the total number of neutrinos that passed through the target. SNO's results agreed with what the standard theory of the sun had predicted. The team was also able to count the number of collisions caused specifically by electron-neutrinos, as these both split the target atom and convert its neutron into a proton as well.[30] The number of such events was one-third of the total, as suggested by Davis's results. This confirmed that neutrinos are indeed changing form, from the electron type to the other two varieties, on their journey from the sun.

Although Bruno never returned to Canada after his move to Dubna, he was able to see the SNO exhibition at Expo '92 in Seville, Spain. He had been nearby in Granada, at the Neutrino 92 physics conference, and had joined an excursion to the exposition in Seville. This was two years after

IMAGE 17.1. Bruno balancing a stick, in
defiance of Parkinson's disease, c. 1990.
(COURTESY GIL PONTECORVO; PONTECORVO
FAMILY ARCHIVES.)

construction had begun on the SNO observatory, but its completion was
still several years in the future. Art McDonald, the leader of the project,
escorted Bruno around the exhibition. He recalled that the occasion "was
very nice for me," because Bruno had been a prominent supporter of the
project and was profoundly interested in its aims. For Bruno, "It was an
opportunity to ask about Chalk River."

One memory from that day has stayed with Art ever since, as it has for
others. This was a trick that Bruno performed. Always athletic and keen on
sports, Bruno still enjoyed playing to the crowd, for example by balancing a
stick on his foot or his nose. By this stage, Bruno was in the advanced stages
of Parkinson's disease, which caused him to tremble in his movements.
Nonetheless, he got up to his old tricks. "Strange this disease," he mused, as
he balanced his walking stick on his foot without any trembling. The moment he took the stick back in his hand, however, the shaking returned.[31]

EIGHTEEN

PRIVATE BRUNO

THAT LIFE IN THE SOVIET UNION WOULD PROVE DISASTROUS FOR Marianne was obvious from the start; that Bruno too would suffer only became clear later. Cut off from friends and family, having passed through Stockholm without seeing her mother and siblings, Marianne's only role in Russia was as an appendage to Bruno. He had a career there, restrictions notwithstanding; she had a vacuum to fill.

Marianne's story encapsulates the Pontecorvo enigma. People who knew her before 1950 gave conflicting accounts, some recalling her as vivacious, an extrovert like her husband, while others found her to be withdrawn, morose, secretive. Several media reports in 1950 played up the latter image, implying that she was keeping secrets, aware of the hidden life of her husband, the spy. Marianne was even portrayed as a Mata Hari character, the leader of a communist cell. No source for this allegation was ever given, and there is nothing in the MI5 record to support it. One close relative insisted that she was never a political person and chose to follow Bruno, whom she loved. Gil confirmed this opinion: "I never thought of her as politically minded. I've heard claims that she was, but I doubt it."[1]

Whatever her politics, Marianne missed Sweden terribly. She had left her homeland in 1939 at the start of the war, marrying Bruno just before they began their odyssey to North America as part of the tide of refugees fleeing the Nazis. It is now clear that she experienced frequent emotional crises, suffering highs and lows during their time in Canada and the UK. In the USSR these episodes developed into more serious problems.

Although the three boys were happy and in school, Marianne was beginning to suffer from a profound depression. A catastrophic collapse seems to have begun within her first five years in the USSR, around the time Gil was starting college.[2] As we saw in Chapter 15, Marianne became almost pathologically averse to any form of social life, reading for hours while she lay on the bed and stared out the windows at the surrounding forest. Whether this was because the Pontecorvos had no social interaction with colleagues during this period and Marianne felt like an outcast, or because Bruno chose not to bring those colleagues home due to Marianne's withdrawal, is impossible to know. Whatever the cause, it is certain that what began as shyness, withdrawal, and mild depression developed into a long-lasting and debilitating mental disorder.

At first she left home for short periods to undergo treatment in a clinic. At the end of the 1950s, however, Marianne had "a more serious crisis than previously" and began to spend considerable time in a psychiatric hospital.[3] Bruno was now home alone with the boys, the head of an all-male household.

Even when Marianne was released from the clinic to spend time with her family, her condition was transparent. Friends and family who had known her before 1950, and who managed in later years to see the Pontecorvos in the USSR, noticed her decline. In the middle of the 1960s, Bruno's niece Laura Schwarz visited Moscow with the Italian-Soviet Society. She recalled that Bruno had "the same smile" as always, but Marianne "seemed sick to me, sleepy, almost absent."[4] Laura Schwarz's memory echoes those of Western colleagues, who had known the Pontecorvos in Canada. At Chalk River, as we saw, Bruno had been the "heartthrob of all the single women," and Marianne was his "beautiful vivacious wife."[5] At a conference in Kiev in 1970, the Pontecorvos' Canadian friends were shocked at how "downtrodden and discouraged" Marianne appeared, not at all like the vibrant woman of yore.[6] A photograph from that time shows Bruno, confident and handsome in his shorts and sports shirt, while Marianne wears dark glasses at his side, her mouth pouting, her demeanour passive. Although Bruno was animated, reminiscing about happy times in Canada, Marianne didn't say a word. As the physicist J. David Jackson put it, "Bruno had his physics to sustain him amidst the difficulties of Soviet life. Marianne evidently had nothing."[7] This period, which followed the death of her mother in 1967, must have been especially hard.

IMAGE 18.1. From left to right: Bruno
Pontecorvo, Marianne Pontecorvo, and Maureen
Jackson (daughter of J. David Jackson), 1970.
(J. DAVID JACKSON, COURTESY AIP EMILIO SEGRÈ VISUAL
ARCHIVES, JACKSON COLLECTION.)

RODAM AMIREDZHIBI

One summer, at the end of the 1950s, Bruno found himself alone: Marianne was away in a psychiatric hospital, Gil had gone on holiday in the north of the USSR with friends from the university, while Tito and Antonio were at a youth camp in Crimea. Bruno decided to go on holiday himself, armed with underwater fishing gear, to pursue one of his favourite athletic pursuits at Koktebel, a resort on the Black Sea known as the "Soviet Capri." A physicist colleague, Arkady Migdal, joined him.

During the day they swam and fished. For Bruno this reawakened memories of the summer of 1950, when he and his brother Gillo had done much the same in the waters south of Rome. The privileges of being an Academician gave Bruno access to exclusive hotels. In Koktebel, Bruno and Migdal took advantage of this, staying in a lavish place of lodging frequented by artists and writers. In this elite company, the physicists relaxed over dinner in the evening, and socialized late into the night.

Among the guests was Rodam Amiredzhibi, wife of Mikhail Svetlov, the poet. She was Svetlov's third wife, and this marriage was declining like his previous ones, due to his drinking and womanizing. Rodam and Mikhail were already leading separate lives, and she was in Koktebel with a group of friends. Tall, with dark hair, she looked "like a dark Anita Ekberg."[8] Bruno was smitten, and she, in turn, was attracted to the "sweet, intelligent, sensitive man."

So began an intense and intimate relationship that lasted for the rest of Bruno's life, waxing and waning depending upon whether Marianne was in the sanatorium. A few months later, Marianne was released from the hospital and Bruno resumed his role as husband in Dubna. Then, in Rodam's memory, "Marianne was hospitalized again, and Bruno was back looking for me."[9]

Anecdotal tales of Bruno's keen eye for women were as common during his time in Russia as they had been in Canada. However, the intense relationship with Rodam was more than a passing fancy. A former student recalled that "on Saturdays and Sundays, Bruno came regularly to Moscow," where he and Rodam would spend time together.[10] Some have claimed, erroneously, that she became Bruno's second wife.[11] Their relationship became more public as Marianne's condition declined, with many regarding them as partners.[12] However, a Russian colleague, in halting English, told me enigmatically that she was not his wife "in the precise definition."[13] Even after Rodam's husband died in 1964, Bruno remained married to Marianne.[14] Toward the end of Bruno's life, Miriam Mafai interviewed Rodam. Mafai summarized the situation as follows: "Bruno is able to lead a satisfying life, spending time with Rodam, while Marianne is in the psychiatric clinic, in a world of her own."[15]

BRUNO AND THE STATE

When the Pontecorvos arrived in Russia in 1950, Stalin was still in power, terror was widespread, and the possibility of falling out of favour and being transported to the Gulag was very real. Pontecorvo himself was prohibited from travelling outside the Soviet Union, even to Eastern Europe, let alone visiting family in Italy. Following prolonged lobbying by the director of Dubna, in 1959 Bruno was finally allowed to travel to Eastern Europe, China, and Mongolia. Homesick for Italy, Bruno became depressed that it remained off limits.

Yet, although he suffered from these restrictions, Bruno remained more devoted to the communist cause than many, with one of his colleagues describing him as "convinced in the inspiring force of communism, like a person believing a religious credo."[16] Some Soviet intellectuals denounced their nation's invasion of Hungary in 1956, for instance, but Bruno rationalized the invasion as necessary to "save" the Hungarians. His brother Gillo, by contrast, along with many intellectual communists in the West, left the party in protest. In the USSR, Andrei Sakharov, who played a key role in the development of the Soviet hydrogen bomb, protested the action and was persecuted for years. Although Bruno Pontecorvo remained a party member at this juncture, he refused to sign a letter denouncing Sakharov, even though several other scientists did.

Two days before Bruno's fifty-fifth birthday, in August 1968, Soviet tanks entered Prague. For Bruno, this was a tipping point. He had a soft spot for Alexander Dubczek, a reformist leader who preferred ice hockey games to political meetings. Bruno kept up with events in Czechoslovakia by reading *l'Unità*, which contained serious analysis of the developing crisis and its background. *Pravda*, however, was full of attacks on various Czech leaders, its tone ever more belligerent. Whereas previously Bruno had always toed the party line, on the principle that the leaders had more information than he did, he now felt able to judge the situation for himself. It seemed to him that Dubczek was attempting to combine socialism and democracy in a courageous manner that was supported by the Czech population.

Meanwhile, Soviet radio stations broadcast interminable testimonials from soldiers who had taken part in the liberation of Czechoslovakia in 1945 and now supported the overthrow of the counterrevolution. On Bruno's birthday he reviewed his life, against the backdrop of the news. He later recalled, "The flood of words that spilled from the radio, the emphatic declarations, the solemn speeches, interspersed with classical music, made [me] feel more and more alone."[17] Without explanation, his subscription to *l'Unità* was cancelled and Soviet media became his only source of news. For the first time he began to question whether the party was always correct.

As Bruno approached his sixties, disillusion set in rapidly. He had lived through a period of great change. As a young man in Italy he had seen the ugly spectre of fascism, which threatened to replace culture with barbarism. He had joined a group of intellectuals who saw communism as a way to combat the evil. In these circles, there was an almost irrational belief in a city of the future, built on communist ideals. Bruno Pontecorvo subscribed to that religion, equating capitalism with military adventurism. All such traditional ideologies would have to be overthrown to reach the new world. By Bruno's final years, however, the Soviet Union had fallen apart, as had many of the ideas that had attracted him there. In 1991 he remarked, "For many years I thought communism a science, but now I see it is not a science but a religion."[18]

If facts disagree with a scientific theory, you change the theory; when they disagree with a religious doctrine, you reinterpret the facts. There in a nutshell is the difference. Gillo described himself and his brother, in their youthful certainty, as being "like the early Christians, who believed in something beautiful, which did not exist." He hinted that the ideals that had driven the two brothers in those days were lost, for himself and for Bruno too: "We bet on something which turned out to be false." After glasnost, when contact with the West was at last restored, Bruno bluntly evaluated his previous convictions for a British reporter: "I was a cretin."[19]

ONE RUSSIAN COLLEAGUE, SEMEN GERSHTEIN, REMARKED, "TO EVERYBODY who knew Bruno it was obvious that he could have achieved much more working in the west and could have realized his ideas himself." This led Gershtein to ask the obvious question: "Why did he come to the USSR?"

Even in and around Dubna, no one seemed to know. Some in the 1950s believed that it was because he held the "naïve belief of many foreigners, faithful to the ideals of communism, that the USSR, the country of victorious socialism, was building a communist society." Only later, when it became "more or less safe" to openly discuss these issues did some people call such ideas "stupidity."

Gershtein admitted that a small number of people, rivals of Bruno who envied or disliked him, "adhered to the version [believed] in America— that he was a Soviet spy who fled when the danger arose of his being unmasked." Gershtein, whom I know as a polite and sensitive man, was a friend of Pontecorvo for many years, but never asked him about this issue because he "understood that it might be quite painful." Instead, Gershtein formed his own opinion as to what had happened. One particular event stuck in his mind—a day in the mid-1960s when he sat next to Pontecorvo during a scientific presentation by former atomic spy Klaus Fuchs, who had moved to East Germany after being released from British prison in 1959.

Fuchs was giving what Gershtein regarded as a rather tedious presentation, but Pontecorvo was "very excited." Bruno whispered, "You know— Fermi was very severe in estimating scientists, but he considered Fuchs a star of the first order." Gershtein expected that Pontecorvo would talk to Fuchs after the session ended, but he didn't. As they left the auditorium together, Gershtein recalled that Pontecorvo was "agitated," as if remembering the years just before he came—"or one can say, fled"—to the USSR.

Bruno told Gershtein, "I would be very interested in reading Fuchs' memoirs if he wrote any," and then explained that when Fuchs was arrested, "we were all sure that it was a police provocation against the communists, since we learned that Fuchs was a communist." Pontecorvo claimed that he'd "had no idea that Fuchs was a spy and thought it was a provocation in the spirit of McCarthyism, which had overflowed America and extended to England."

Gershtein felt that this had the ring of truth: "From these words of Bruno it becomes quite clear what he, a communist since 1936, could be afraid of in England after the arrest of Fuchs, and why he decided to change his life so drastically."[20]

For Gershtein, who had lived in the USSR all his life, this was a natural reaction. However, in reality, McCarthyism never reached the United Kingdom. Indeed, Britain often served as a refuge for communist sympathizers from the United States, including the prominent American physicist David Bohm, who became a British citizen. Gershtein's assessment that colleagues who openly regarded Pontecorvo as a Soviet spy were driven by personal motives has some credibility; the opinions of Bruno's friends tend to have been more nuanced. Vladimir Gribov, with whom Bruno wrote his seminal paper on neutrino oscillations, is one example. The two knew one another well, but Gribov claimed, "I never talked about history; intuitively I knew not to."[21] Gribov's view that this was a no-go area was shared by other friends and relatives. This was not because they knew that Bruno had a secret history, for he never admitted this to them;

rather, it was because they suspected intuitively that he might have such a history, and they did not want to damage their relationship with Pontecorvo by confirming this.

Gershtein's bland interpretation of the incident with Fuchs seems to be an example of this. Bruno was usually gregarious with former colleagues, a "hail fellow well met." And he had worked with Klaus Fuchs for two years. Indeed, Fuchs had invited Bruno to the nuclear physics conference in Edinburgh in 1949, and the two men clearly had much in common. Yet Pontecorvo made no effort to talk to him, as if by choice. However, it is now known that the KGB had placed restrictions on contact with Fuchs, and issued clear instructions to that effect. Even the head of the East German Stasi was forbidden to contact Fuchs of his own volition; it is safe to conclude that Pontecorvo too was briefed in advance.

TITO FIGHTS BACK

As we have just remarked, it was as the Soviet Union began to fall that Bruno finally accepted that he had made a bad choice. While the changing world was forcing all Soviets to evaluate their beliefs and establish their place in the new society, for Bruno, who remained a communist, there was also a strong family pressure. The effect of his decision to come to the USSR on Marianne was already clear. As for his sons—who "grew up lonely," in the opinion of Bruno's companion, Rodam Amiredzhibi—they displayed a range of attitudes toward their new home.[22]

Gil, the eldest, followed in his father's footsteps, in terms of both his interest in science and his belief in communist ideals. He explained how in the postwar period it was quite natural for communists to join the party and to appreciate the Soviet Union as a nation that had made great contributions to the war against Nazism. The goal of equality and social justice—the "city in the sky"—had developed during the revolution of 1917 and the USSR was its torchbearer. In Gil's view, that dream still survives, but the means of achieving it has evolved: "It's like an experiment in physics. You set out to find the truth and discover features that lead you to refine the approach. Your opinion on how to complete the experiment may evolve but the goal remains."[23]

Gil has remained in Dubna to this day, having watched over Marianne and working as a nuclear physicist. He has never married. Bruno's youngest son, Antonio, lives in Moscow. Married, with a family, he works as a computer scientist. The middle child, Tito, was "independent, something of a rebel," and quit Russia after Bruno's death to live in the United States.[24] It is ironic that the two sons with traditional Italian names have remained in the former Soviet Union, whereas the one named for a communist hero, Josip Tito, has rebelled and returned to North America.

As it turns out, the seeds of Tito's rebellion go back to the days of the Soviet Union, when he witnessed firsthand his father's strained relations with its oppressive system.

In the mid-1970s, under the leadership of Leonid Brezhnev, the USSR expanded its contacts with the outside world. Bruno began to hope that it might be possible for him to visit Italy, either as a member of a scientific delegation or as a guest of the Italian Communist Party (PCI). However, there were tensions between the USSR and the PCI, many of whose members had resigned in protest over the Soviet invasion of Czechoslovakia. Even though Bruno too was upset by the Czechoslovakian adventure, he did not rock the boat. Nevertheless, his name was never included on the official lists of delegates selected to attend scientific conferences outside the USSR and its satellites, nor did he ever receive an exit visa to travel outside these boundaries. Finally he was forced into action, following a confrontation with Tito.

Tito, by then thirty years old, had completed his university studies in Moscow and begun conducting research in oceanography. On the occasions when his team was working in the Arctic or Pacific, outside the territorial waters of the USSR, Tito was not invited to join them. The first few times this happened, Tito thought it was just a coincidence, but when it became a regular occurrence he realized that his exclusion was a deliberate act. The reason, he felt sure, was that although he was a Soviet citizen, he was the son of Bruno Pontecorvo, had been born in the West, and as such was regarded as a security risk. In the eyes of the Soviets, once in a foreign port, Tito might "escape" and return to Italy or Canada. He had been blacklisted. The obtuse and perverse apparatchiks of the Soviet state were interfering with his career by cutting him off from essential research.

When he raised a protest, he was dismissed with sly disdain: "Is there not enough ocean in the USSR for you?" He confronted Bruno, calling him "an idiot for having left the West to come to the Soviet Union." He accused his father of being blind to what was going on, of being seduced by the money and privilege that went along with his status as an Academician, and ruining Tito's opportunity for a career in science.[25] Years later, Bruno admitted, Tito's accusation still rang in his ears.

Bruno, meanwhile, was stirred into action.

Until then he had meekly accepted the state's right to forbid him from travelling outside the borders of the communist world. Following Tito's outburst, Bruno began to feel that he was being treated unfairly. He knew that no one in Dubna had the authority to intervene, so he decided to see someone in the Politburo, the central governing body of the Soviet Communist Party.[26]

In 1975 there were twenty-three members of the Politburo.[27] It is intriguing that, out of all the possible members, Bruno chose to consult Yuri Andropov, the hawk who had convinced a reluctant Nikita Khrushchev to invade Hungary in 1956 and who had also been a central player in the 1968 invasion of Prague. In 1975, when Bruno Pontecorvo decided to visit, Andropov was head of the KGB.

Announcing himself at KGB headquarters as "Academician Bruno Maximovitch Pontecorvo," Bruno asked to see Andropov in person. After some time, he was informed that Andropov was in a meeting but that his deputy was available. Bruno insisted that he had to see Andropov himself. This was not possible, and Bruno left.

Not long afterward, Bruno received a visitor at Dubna: the KGB officer who had been assigned to him a quarter of a century earlier, when the Pontecorvos had first arrived in Moscow. Bruno never revealed the details of their conversation, beyond the fact that he explained to the KGB officer how Tito was unable to pursue his career, and how Bruno himself was never included in the groups that were allowed to travel to the West. The KGB contact promised to arrange for Bruno to meet someone who could make a change.

More time passed, after which Bruno was summoned to Moscow to meet a "very important person" whose identity remains unknown.[28] Bruno told this person that he was now over sixty years old and wanted to see Italy while there was still time, specifically denying any possibility that he would try to defect. It was no use. A few weeks later, Bruno received a call saying that he could not have an exit visa. Tito's situation also remained the same, which drove him to quit oceanography to rear horses in Dubna.[29]

AMALDI IS SEVENTY

By 1978 Bruno had given up all hope of visiting Italy. One day, however, a message arrived at the Soviet Academy of Sciences, encouraging the group to send a delegate to Rome to take part in a celebration of Edoardo Amaldi's seventieth birthday. Amaldi, with whom Bruno had first shared the discovery that led to the slow-neutron method; Amaldi, the father of CERN, who had indirectly inspired the Warsaw Pact's own high-energy physics programme at Dubna; Amaldi, one of Europe's most influential living physicists. Bruno Pontecorvo was the natural candidate to attend this event.

However, no one in the Academy nominated him, and Bruno himself said nothing. Another delegate was chosen instead.

That, at any rate, is how Bruno described things to Miriam Mafai.[30] However, it is hard to believe that this is the whole story. The episode is reminiscent of Pontecorvo's behaviour when Fuchs visited Dubna, his silence a response to orders from above. Then fortune smiled. At the last minute the chosen delegate fell ill and could not go. Someone (Bruno says he never knew who) suggested that the Academy send "Bruno Maximovitch Pontecorvo" as a replacement. It is not known why the powers that be had a change of heart. There seems to have been a division of opinion at the highest level. Bruno recalled that after he received permission to go, the local KGB tried to dissuade him. They said that it would be safer for him to go to Switzerland or France because in Italy he might be surrounded by police and harassed by the media. He responded unequivocally: "I don't want to go to France; I don't want to go to Switzerland. I want to go to Italy."[31]

Which he did. On September 6, 1978, almost twenty-eight years to the day since he had left Rome for the USSR, Bruno Maximovitch Pontecorvo returned.

WHEN ITALY HEARD THAT THE PRODIGAL SON WAS ABOUT TO RETURN, Bruno got a foretaste of the excitement. On the eve of his departure a journalist telephoned him from Rome. Bruno was polite but firm, explaining that the purpose of his visit was to celebrate Edoardo Amaldi's seventieth birthday, and that he would not give any interviews. The caller pointed out that the return of Bruno Pontecorvo would be such a big story that journalists would besiege him when he arrived. Bruno repeated his statement, and hung up. Perhaps there was some truth to the KGB's warning. The KGB duly provided Bruno with two minders to accompany him to Italy, masquerading as physics colleagues.

On the morning of September 6, Bruno arrived at Rome's Leonardo da Vinci Airport. Press photographers vied for the best shots as he entered the terminal, and the flash bulbs of their cameras nearly blinded him. Journalists fired questions as they thrust microphones under his nose. A newsreel camera whirred into action.

Bruno had prepared a statement, which he repeated in response to the reporters' questions. It confirmed the essential message of his 1955 press conference, the only other occasion since 1950 that he'd spoken to the general public: "I have never worked on the atomic bomb, or the H-bomb, not in the West nor in the USSR."[32] He praised Edoardo Amaldi, Italian physics in general, and Enrico Fermi, who had helped him so much in his career. He said that he had gone to the USSR of his own accord, and had complete freedom of research there. He refused to be drawn into making further comments.

Gillo Pontecorvo, who had come to the airport to see his brother, fought his way through the throng to reach Bruno. There were tears as the two embraced on Italian soil for the first time in nearly thirty years. Bruno finally escaped the crowds and spent the afternoon at his sister Giuliana's home in Cetona, a village near Siena. His KGB limpets came too.

In 1978 Giuliana's son Eugenio Tabet was a young professor of physics, and he suspected that Bruno's bodyguards were not physics colleagues. So he decided to "set them a test." He explained: "At the side of my mother's house was a wood-burning stove. It was a good approximation to what physicists call a blackbody radiator. So I asked the guard what wavelength he thought the radiation would be. I wasn't looking for a precise answer. I just wanted [to get] a feel of whether the question made any sense to him. I can't recall exactly how he responded. But it was total nonsense!"[33]

For many of the physicists at the conference, it was their first meeting with the legendary Bruno Pontecorvo. Many old friends were seeing him for the first time in decades. There is a famous photo from this time, showing Bruno with Edoardo Amaldi and Emilio Segrè. Edoardo's son Ugo, a leading physicist himself, described the reality behind the image: "The photo shows them all smiling. That was just for the camera. I was there the first time Emilio and Bruno met. It was *very* cool. The handshake was formality, no more. Emilio was [still] very unhappy about the patent."[34] Little seemed to have changed since their meeting in Kiev in 1959.

REUNIFICATION

When Bruno returned to Dubna, he shared his experiences with Tania Blokhintseva, daughter of Dmitri Blokhintsev, the first director of JINR at Dubna: "Bruno was absolutely astonished. [It was so long since] he had seen Italy. He said to me, 'You know—the carabinieri—they were very polite.' The amount of traffic made it difficult to cross the road. He was very happy there."[35]

Reading the media accounts of Bruno's return to Italy in 1978, one could be forgiven for thinking that this was the first time a Pontecorvo had emerged from the Iron Curtain to visit the West. However, Bruno's eldest son, Gil, had actually broken the ice three years earlier, with a trip to Italy in 1975.

When Gil mentioned this to me, almost en passant, I was astonished.

He explained that he had travelled to Turin to work with a team of physicists, who have since served as his collaborators in CERN experiments for several decades. Gil confirmed that he had been allowed to leave Soviet territory even though his brother Tito had not: "He was the head of [an oceanographic] team but was not allowed to leave the boat. . . . There is no logic to [Soviet] bureaucracy!"

I imagined Bruno in 1975, still trapped behind the Iron Curtain while his eldest son was in Italy. "When you returned to Dubna, what did Bruno say?" I asked.

"I don't remember it in detail, but you can guess," Gil replied, giving a gentle laugh. He did remember the trip itself, however: "I had a big meeting with the Italian family."

I interjected: "You had seen some in Moscow of course."

"Yes, but this meeting was with the whole clan. There were people there who hadn't seen me since the old times in England." He laughed again, as he recalled what sounded like an inquisition: "It was like a sort of interview. It went on for four hours! Questions like: how can you explain this, how do you explain that? I was sweating at the end. It was all in Italian."[36]

Perhaps Gil's successful exit, and safe return to the USSR, helped loosen the straitjacket that entrapped his father. Three years later, the Soviet authorities must have been satisfied with Bruno's performance too, as his 1978 trip to Italy became the start of a regular pilgrimage. The following year he was back in Rome as a guest of the Italian-Soviet Society. He had been invited to give a physics lecture, and those who knew him noticed that a change was taking place: with his trembling movements, Bruno was showing the first signs of Parkinson's disease.

During this visit his itinerary gave him a chance to travel around some of Italy. After two-thirds of a lifetime, he returned to his hometown of Pisa. The villa where he had grown up was still there, but had been converted into a hotel. He also visited Tuscany, whose countryside he loved but had not seen for decades; he had feared that he would never set eyes on it again.

Now that the Soviets were more relaxed about him, Bruno returned to Italy each year. In 1982, he sought to visit France as well, to attend a symposium on the history of fundamental particle physics. Unfortunately, this was not to be. France refused him.[37]

The scientific community, meanwhile, held him in high esteem. Although he sent a paper to the symposium that was read in his absence, many participants were upset that he could not be there in person. They sent him a card, signed *"vos amis vous saluent"* (greetings from your friends). The signers included old friends such as Edoardo Amaldi and Pierre Auger, Nobel laureates C. N. Yang and Julian Schwinger, as well as future laureates (and fellow neutrino chasers) Leon Lederman and Jack Steinberger.

Most physicists regarded Bruno as an international leader in the field of neutrinos, caring little about his past or insisting that he could never have been a spy. Nonetheless, the governments of France and several other Western nations remained suspicious. These attitudes persisted even though no evidence was ever presented against him.

Peter Minkowski, a particle theorist, recalled one example of the West's hostility toward Pontecorvo. In the early 1980s Minkowski was working at Caltech in Pasadena. He had written a paper that anticipated work by Bruno and Samoil Bilenky, and was eager to meet Bruno to discuss physics. During this time, a senior official in charge of nuclear physics at the

US Energy Department visited Caltech. Minkowski suggested to him that Pontecorvo be invited to the United States. In Minkowski's memory, "He went pale. Then he said we cannot do that; there's a warrant for his arrest."

It is hard to know how reliable this story is. Many years have passed and, if such a warrant ever existed, the proof is either lost or overlaid with black ink in the US security files. In any case, no invitation was issued, so the crisis was avoided. At the end of our conversation, Minkowski was in a reflective mood. He repeated to me, quietly: "He went pale."[38]

In 1984, celebrations were held in Paris to mark the fiftieth anniversary of the discovery of induced radioactivity. Once more Bruno was unable to attend. He sent an article instead, which reviewed his early work in Paris on nuclear isomers. There was no restriction that prevented French citizens from visiting Dubna, however, and in June 1984 Bruno met with Hélène Langevin, the daughter of Frédéric and Irène Joliot-Curie, whom Bruno had taught to ski in 1939. Colleagues recalled that both had tears in their eyes.[39]

The French authorities eventually relented and lifted their ban. In 1989 Bruno returned to Paris for a week. He visited his old haunts at the Curie Institute (formerly the Radium Institute), and on September 11 he led a seminar at the Collège de France. During his talk, he recalled his encounters with Fermi, Joliot-Curie, and other great scientists of the prewar period. Unfortunately, Parkinson's disease limited his range of movement, and whereas in the past he used to walk up and down the stage during lectures, energizing his audience, he now stood at the podium, speaking elegantly and vivaciously but occasionally having to take a rest as the trembling took over.

Ugo Amaldi recalled Bruno's visit to CERN on September 14, 1989. At the time, Ugo was the leader of an experiment at CERN's Large Electron-Positron Collider (LEP), one of whose achievements was to demonstrate that there are indeed three, and only three, varieties of standard neutrinos. Bruno was thrilled as he and Amaldi visited the collider, located 100 metres underground, at the CERN complex near Geneva. The splendour of the apparatus, which consisted of several layers of sophisticated electronics, wrapped in cylinders the size of a cathedral, was breathtaking. One of the features of Parkinson's disease is that its symptoms become more manifest when the sufferer becomes excited, and Bruno began to shake.

Later that day he was scheduled to give a talk in the CERN auditorium, in front of several hundred scientists, who were very eager to hear him. Ugo recalled that, immediately before the talk, "Bruno was really shaking." Nonetheless, Bruno composed himself, and after an introduction by Jack Steinberger he began to speak. His once-vibrant tones were now faint, yet still commanded authority. Suddenly the shaking began, and he paused. "Don't pay any attention to my tremors," he told the audience. "This Parkinson's disease looks much worse than it is." Then, to settle them, he deployed his everlasting sense of humour: "Don't worry, I'm not going to die."[40]

THE LONG CHILL

After the Berlin Wall fell, and the Soviet Union itself began to crumble, Bruno reevaluated his life and the circumstances that had brought him to Russia. In 1990, at the age of seventy-seven, Bruno gave an extensive series of interviews to Miriam Mafai, a communist Italian journalist. These formed the basis for her study of his time in the Soviet Union. Titled *Il lungo freddo* (The long chill), the book paints a picture of Bruno Pontecorvo as he would have wished to be seen. It has also provided invaluable source material for the present work.[41]

Bruno, his eyes perhaps having been opened by his visits to the West, realized what he had lost during his years in the USSR. It was at this stage of his life, when barriers were relaxed thanks to the Soviet policy of détente, that he made his startling admission to a journalist from the London *Independent*: "I was a cretin."[42] He seemed to regret his former unwavering support of the Soviet agenda, and the experience had left him sad, as his extended interviews with Mafai also showed.

He had fled to the USSR in 1950, enthused by the hopes symbolized by the red star atop the Kremlin. He was forbidden from contacting the West for five years, and then allowed only to send letters for twenty. Except for the siblings who came to see him in the USSR, he had no contact with his family during this time. Only in 1978 was Bruno freed to travel beyond the Iron Curtain and make contact with relatives, friends, and colleagues in the West. The most poignant meeting was his reunion with his sister Anna. They had not seen each other since August 28, 1950, the day she had returned to England from their fateful holiday. At long last, in the 1980s, brother and sister were reunited.

If anyone deserved an explanation of Bruno's precipitous disappearance, Anna would be near the top of the list. In 2011, I met her at Cambridge's Churchill College, whose archives contain letters and documents from Bruno's Abingdon home. After I had been waiting for some time, a stooping, white-haired lady came slowly across the quad. Although we had never met, I immediately recognized her: she had the strong nose and long face of her brother.

Anna was now Bruno's only surviving sibling. She smiled wistfully as she recalled seeing Bruno in Rome, where he defied Parkinson's disease by playing the fool and dodging the cars in the road. So what explanation had Bruno offered her for his flight? After so many years apart—nearly three decades—surely she wanted to understand his behaviour? But Anna had not asked. She felt that he had been let down by the USSR and considered himself a fool. "He was so broken by the whole experience," she recalled, "that I did not want to turn another screw."[43]

During Bruno's final years, his friends and relatives seem to have universally treated such questions as a no-go area. They cite Bruno's manifest anguish at his Soviet misadventures as their reason for steering clear. Even Gil, who was twelve at the time of the family's flight to the USSR and old enough to realize that "something was up," seems to never have discussed the subject during the ensuing four decades. Within the circle of Bruno's loved ones, there was almost an omertà, a vow of silence. Colleagues outside the family have certainly expressed the opinion, to me and others, that there is a tacit agreement that the subject of Bruno's departure is taboo. They do not ask why Bruno left so suddenly. In their hearts they suspect the answer, but, by not saying it out loud, it is possible to live within a dream.

LAST MEMORIES

In his seventies, Bruno made several trips to Italy to receive treatment for his Parkinson's. His eightieth birthday came on August 22, 1993. Knowing he would be visiting France and Italy during this time, his colleagues at the Dubna laboratory planned an autumn celebration of the occasion.

In July, he visited France for the last time, and attended an international particle physics conference in Marseille.

By now he often needed assistance, and in Marseilles his nephew Ludo helped him to move around and eat. Lev Okun, a Russian colleague, joined Bruno for lunch at a restaurant, alfresco. The tables were spread across a large patio, at a busy intersection. Bruno chose a table situated so that all the passersby walked between them and the restaurant. Suddenly he turned to Okun as if struck by a sudden insight: "Lev Borisovitch! Have you noticed how the women in Marseilles are not as beautiful as those in Paris?" Okun admitted that he had not, and hinted that it was an opinion for which there was no true evidence. Bruno advised him, "Just count the number of plain women that pass before an attractive one appears."[44]

That was Okun's final memory of his old friend. Nearly eighty, debilitated by Parkinson's, with just weeks to live, Bruno was still the charmer with an eye for a pretty girl, and a scientific approach to all aspects of life.

From Marseilles, Bruno went to Italy. He was expected to remain there for several weeks before returning to Dubna. Irina Pokrovskaya, his personal assistant, recalled the surprise among the workers at Dubna when suddenly, without warning, he returned from Italy ahead of schedule. His arrival was so unexpected that there was hardly time for the laboratory director to arrange for a car to meet him at the airport and bring him home. Tania Blokhintseva, the daughter of a former director of Dubna, described his return "as a deliberate act," as if Bruno was aware that he did not have long to live.[45] An informal gathering took place on his birthday, August 22, even though many of his colleagues were still away for the summer.

Bruno continued to come to the laboratory, but in the first days of September Irina Pokrovskaya noticed "an irreversible shadow" on his face. According to her, Bruno was the first to notice the change. During his visits to the laboratory, he was "serene and calm, not wanting to disturb anybody."[46]

A month after his birthday, he came to the laboratory for the last time. It was evening and he remarked on the beauty of the "yellow birch trees" that were visible through the window of Irina's office. Afterward, she escorted him to the laboratory gate, where his son Tito was waiting. Her last memory of him was the fact that he was still joking as they parted.

Two days later, he died. His famed charisma, modesty, and politeness were with him to the end. As he slipped in and out of consciousness, Bruno's last words to the doctors scurrying around his bed were: "Thank you."[47]

His funeral was held on September 29. In the morning, the chandelier in the funeral parlour of the Palace of Culture at Dubna was draped in black, and as she helped prepare the flowers Irina's "fingers were all thumbs." Then Bruno Maximovitch's casket was brought in. It was a grand occasion, as Dubna celebrated the life of their much-loved Academician. Mourners took turns standing beside the open casket to honour the body.[48] The music of Mozart filled the room. Outside, the weather resonated with *il lungo freddo*, the Russian half life of Bruno Maximovitch. "It was snowing wet snow," Irina recalled.[49] The long chill of the Russian winter had begun.

THE FINAL IRONY

Bruno had arguably lost his chance for a Nobel Prize when Steinberger, Schwartz, and Lederman confirmed his idea that there are distinct varieties of neutrinos. Bruno's death now cost him a more certain share of the 2002 prize, which went to Ray Davis. By official policy, the Nobel is not awarded posthumously. Davis's award was the climax of his forty years of experiments. His endeavours encompassed the full oeuvre of Bruno Pontecorvo, helping to convert the hypothetical neutrino into a precision tool for both physics and astronomy.

Bruno's idea had finally borne fruit in the quest for solar neutrinos, because the sun indeed produces neutrinos, rather than antineutrinos. Although Bruno's original 1945 paper had dismissed the search for solar neutrinos as impractical, his paper from the following year nevertheless provided inspiration for Ray Davis. Once the existence of the particle was confirmed, Davis became the first person to look for neutrinos that had travelled from the sun. Unfortunately for Bruno, it took nearly forty years for the scientific community to accept Davis's experiment as reliable and his results as correct. As we've seen, the reason it took so long was that electron-neutrinos oscillate, changing form en route from the sun— another idea proposed by Bruno Pontecorvo.

What irony. Had there been no such thing as neutrino oscillations, Bruno would have been right once (with his advocacy of chlorine as a detector) and Davis would have observed solar neutrinos at the expected rate. The world would have applauded immediately. The irony is that Pontecorvo was right more than once. Neutrino oscillations diluted Davis's signal to the point that people doubted his results for decades. Bruno Pontecorvo's insight that there is more than one variety of neutrino, and his subsequent suggestion that neutrino oscillations were responsible for the shortfall in Davis's experiments, were largely ignored. It was not until 2001 that the matter was finally settled, with the completion of the SNO experiment in Canada, so near to where Bruno's neutrino quest had begun.[50]

The most far-reaching of Bruno Pontecorvo's ideas is surely his insight that muon-neutrinos and electron-neutrinos are different. This led to the modern Standard Model of particle physics, as well as the prediction that different varieties of neutrinos can swap identities, as long as they have some mass. This prediction, which Bruno developed over several years, reached its mature form in 1968, a full year before Davis discovered the solar neutrino anomaly.

Perhaps it is the neutrino saga that best encapsulates the triumph and the tragedy of Pontecorvo's scientific career. It was because of neutrino oscillations that the sun's neutrinos were diluted before arriving in the chlorine tank. Neutrino oscillations were thus a curse. Ray Davis spent decades trying to figure out why he was seeing so few neutrinos. The explanation wasn't confirmed until the turn of the century, leading to Davis's Nobel Prize, which he received in 2002, at the age of eighty-seven. Bruno Pontecorvo, of course, died in 1993, unaware of the great truths he had expounded.

He didn't live to see the phenomenon of oscillating neutrinos established as a scientific fact. Today, this phenomenon is used to measure the subtlest properties of these ghostly particles. For some theorists, the results suggest that neutrinos may hold the answers to many of the current mysteries of the cosmos, such as why there is more matter than antimatter in the universe at large. The new science of neutrino astronomy, in which neutrinos are recorded by vast detectors under the ice of Antarctica, promises to make quantitative measurements of phenomena whose origins lie far away in the cosmos and in time.

On two occasions, a Nobel laureate has credited Bruno Pontecorvo with the inspiration for his award. We heard Melvin Schwartz say, "His overall contribution to the field of neutrino physics was certainly major." Ray Davis, in turn, provided this epitaph: "[Bruno Pontecorvo] opened everyone's eyes with his original insights."[51] Like so much else in the story of Bruno Pontecorvo's remarkable life, we can only wonder how much further these insights might have gone if he had not fled through the Iron Curtain in 1950.

IMAGE 18.2. Bruno Pontecorvo memorial stone, Rome. (AUTHOR.)

AFTERLIFE

"Midway on our life's journey, I found myself
in dark woods, **the right road lost**."

—*Dante's Inferno*

NINETEEN

THE RIGHT ROAD LOST

MI5's RONNIE REED HAD COMPLETED HIS INITIAL INQUIRY BY December 1950, three months after Bruno Pontecorvo's defection. At the time, the British were oblivious to Philby's duplicity, so no one suspected that the FBI's interest in Pontecorvo was known in Moscow. Although the British government feared that Pontecorvo had fled because he had previously passed classified information to the Soviets, MI5 had no sure evidence, and its investigations led to no certain conclusions. However, in one assessment, at least, Reed was spot-on. In his report he concluded that Bruno was "likely to be seriously disappointed" if he had gone to the USSR in the hope of settling there peacefully.[1]

After spending five years effectively confined to Dubna, and twenty more to the Eastern Bloc, Bruno finally obtained liberty only as the object of his desire—the USSR—began to fragment. Shortly before he died, Bruno admitted, "The Promised Land no longer exists; not here [in the USSR] not anywhere else."[2] He also described a recurring dream: "Sometimes during the night, I imagine there is someone in front of me saying all my scientific work is wrong. Some number, some operation was wrong at the beginning, and I worked all my life at that calculation based on that wrong data. Because of that error, all my works have been useless." He then added, "This did not happen in physics—at least."[3] If "dreams reveal the reality which conception lags behind," this perhaps reflects Bruno's awareness that, in realms unrelated to science, he lost his way.[4]

BRUNO PONTECORVO: SCIENTIST

As a scientist, Bruno Pontecorvo's name will forever be associated with neutrinos. His work on the phantom particles climaxed during the latter half of his life, and scientists today regard it as his legacy. He recognized that the weak interaction is a universal force of nature, he identified the muon as a heavy version of the electron that has its own "flavour," and from this he deduced that there are distinct varieties of neutrinos, an idea that culminated in his theory of neutrino oscillations. Of course, some of Bruno's 1946 neutrino report was based on the ideas of Pryce, Guéron, and Frisch, but history has credited him alone. More recently, the idea that a supernova can be observed via neutrinos has given birth to a new science—neutrino astronomy.

Bruno was unusual in having made contributions both to theory and to experimental physics. Ugo Amaldi once asked, "How many have made great contributions in both theory and experiment? Fermi, Rutherford, Pontecorvo—not many at Nobel level."[5] Nobel laureate Jack Steinberger placed Bruno Pontecorvo high among the great physicists of the twentieth century, although he also regarded Bruno as a fantasist who in his later years sometimes claimed credit for more than his due.[6]

Compare this with the opinion of Semen Gershtein, who rated Bruno as a great physicist whose career was thwarted by his move to the USSR.[7] Certainly Bruno missed out on getting credit for his independent development of the idea of associated production. However, there is no evidence that he was on the right track to detecting the *anti*neutrinos coming from reactors.

Bruno's early work in nuclear physics in the 1930s touches most directly on the political implications of his defection. This was the period when he made his chance observation in Rome, when he worked on isomers, when he was involved in the birth of the heavy-water fission experiments, and when he designed precision instruments for measuring radiation. The extent of his expertise and innovation in that field gives the lie to the British government's public statements of 1950, which tried to downplay his significance by claiming he had not worked on the atomic bomb, and had no recent knowledge of nuclear secrets.[8]

Contrary to this political spin, Bruno Pontecorvo had been at the centre of research that was greatly relevant to the USSR's needs. Enrico

Fermi's remark was nearer the truth when he said that Pontecorvo's expertise would be of great general value in the USSR if he were freely admitted to the atomic project.[9] The Dubna logbooks confirm Fermi's prediction. They reveal the Soviet interest in fission and "H4 particles," and stand as a record of the advice Pontecorvo gave upon arriving at the laboratory in November 1950. Pomeranchuk made use of Pontecorvo's expertise in heavy-water nuclear reactors, and Bruno himself confirmed that he had given advice in this field.

A former member of the US Atomic Energy Commission, who liaised with British intelligence in Washington around 1950, remarked, "Surely [Pontecorvo] must have revealed all after his defection but I have never seen a single piece of evidence about what he might have divulged before [1950]."[10]

BRUNO MAXIMOVITCH PONTECORVO: ENIGMA

It was once believed that the earth was at the centre of the solar system. To explain the planetary orbits required a large number of "epicycles," special refinements added to the model as better data arrived from astronomers. The theory soon became unwieldy. With the single assumption that the planets orbit the sun, however, everything suddenly fits. I take a similar view of the case of Bruno Maximovitch Pontecorvo and his possible role as a spy. One may conclude, based on the absence of evidence against him, that he had no dealings with the Soviets when he was in the West, in which case several independent theses are required to explain the various unresolved questions. On the other hand, if one accepts the hypothesis that Pontecorvo passed secrets before 1950, the kaleidoscope of facts settles into place.

What follows is a list of kaleidoscopic facts, along with possible interpretations made on the basis of this hypothesis:

By 1950 blueprints of the Canadian reactor were in the USSR.

The source was someone other than Nunn May. It is possible that the Soviets convinced Pontecorvo to hand over the blueprints before his defection, to aid their goal of building a nuclear reactor for the social and economic welfare of their citizens. It would have appeared churlish if Bruno refused such a request to help an ally.

A second sample of uranium made its way from Canada to the USSR, and Lona Cohen was its courier.

Whereas Nunn May's source of uranium was the American reactor, the second sample almost certainly came from Canada itself, which was only possible after the main reactor began operating in 1947. The previous year, Bruno had turned down job offers from various prestigious US universities in order to go to Harwell, and then dithered, changed the starting date, and suddenly decided to remain in Canada to work on the NRX reactor. This behaviour could of course reflect genuine indecision, but it also fits rather conveniently with a portrait of a man required to keep the Soviets abreast of developments in the reactor field.[11]

Lona Cohen made visits to the US-Canadian border on various occasions between 1944 and 1948, in order to exchange information with someone based in Canada. Bruno Pontecorvo likewise travelled from Montreal to the US border regularly, ostensibly to keep his application for US citizenship active.[12]

If Bruno was meeting with Lona Cohen, he would have had to take precautions to protect his identity if they were seen together or, worse, if Cohen were ever arrested. We have seen evidence of Bruno's shifty behaviour during this period, turning away when forced to be in photographs at Chalk River, as if afraid to let his face become widely recognized. As the best images on a roll are generally the ones that get published, we must conclude that Bruno refused to face the camera throughout this photographic session. Bruno may simply have been in one of his childish modes, but the images at Chalk River are bizarre.

Geoffrey Patterson sent his letter from Washington, which Philby intercepted, in July 1950, a few days before the family left England prior to Pontecorvo's flight.

The fact that Bruno made a precipitate decision to flee, rather than planning a more orderly move to the Soviet Union, suggests that he was reacting to a major crisis, rather than moving for personal reasons as a matter of principle. The exfiltration of the Cohens to the USSR in July 1950, only weeks before Pontecorvo's arrival, is another intriguing coincidence.

The Soviet reaction to his arrival, which included interrogation and five years under guard, are hardly an appropriate welcome for a hero of socialism.

The fact that Pontecorvo chose Andropov, head of the KGB, as his contact in the Politburo, and refused to deal with anyone else, is another incident that could be quite innocent but nonetheless fits with a pattern. The Kremlin's response is also remarkable: instead of ignoring Pontecorvo, or rejecting his request, they dispatched the same KGB officer that had guarded and helped debrief him immediately after his arrival—the person most conversant with the Pontecorvo affair. This KGB contact then had a lengthy conversation with Pontecorvo, whose details remain unknown.

A lawyer defending Pontecorvo could argue that, as the KGB was all powerful and had helped bring him into the USSR, for whatever reasons, it was natural for Bruno to appeal to the agency for assistance. However, going to Moscow to see Andropov without an appointment would be an extremely naive move, unless Pontecorvo knew that he had considerable leverage. (Would you expect a senior member of your government to see you under such circumstances?)

The Soviets' treatment of Pontecorvo upon his arrival in the USSR suggested to a former head of MI5 that they didn't trust him.[13] This is perhaps understandable, whatever the reasons for his defection. However, this supposed mistrust fits uneasily with the commitment the Soviets had invested in getting him there, and Pontecorvo's precipitate agreement to go along with the plan. The total picture fits more naturally with the idea that the Soviets were punishing Pontecorvo—that Pontecorvo had been an agent who was "trying to disengage" or become independent. He certainly wasn't treated like a hero who had voluntarily chosen to come to the USSR in protest against Western ways.[14]

ULTIMATELY, THE QUESTION OF WHETHER BRUNO PONTECORVO PASSED classified information to the Soviets before 1950 is secondary; once he was there, his know-how proved seminal for the Soviets. For all the hype about Klaus Fuchs and other atomic spies, their information soon became obsolete and was of transitory value at best. Bruno Pontecorvo, by contrast, brought with him a unique expertise, which for five years was exploited ruthlessly by the Soviet authorities to further their nuclear ambitions. Even if he passed no secrets before 1950, Bruno Pontecorvo's later presence in the Soviet Union was potentially as significant as anything that the proven "atom spies" ever did.

Alan Nunn May was convicted of spying in 1946, and spent seven years in British prison before being released. In 1950, Klaus Fuchs was also jailed, and was released after nine years.

After regaining their freedom, both took up science again. Nunn May married, worked for a scientific-instrument company in England, and then became the dean of science at the University of Ghana. He retired in 1978 and returned to Cambridge, where he died in 2003, at the age of ninety-two. Klaus Fuchs returned to East Germany, where he married a friend from his student days. He had considerable success as a scientist. He won the National Prize of East Germany, and became the deputy director of the national nuclear research centre in Dresden. Fuchs died in 1988, at the age of seventy-seven.

By contrast, Bruno Pontecorvo spent forty-three years in Russia, where his scientific career was frustrated, his family was traumatized, and his ideals were slowly crushed in the face of Soviet repression. If Bruno Pontecorvo was a spy, he was punished more than the others.

OCTOBER 1992

In October 1992, a Russian historian was doing research for a documentary about the Cold War, on behalf of the American television network ABC. A guide from the KGB press office took the researcher to a KGB hospital in Moscow, where the elderly Lona Cohen was a patient.[15] She talked about the "young physicists" she had met during her time as a spy, but would not name any names. She confirmed that she went to Niagara Falls around 1945, and again in 1948 or later, "on a sightseeing pretext," to make contact with a source from Canada.

Next, the researcher was taken to another wing of the hospital, where they met with a man named Anatoly Yatskov. Yatskov and Pontecorvo were born just three months apart, and would die within months of each other in 1993. Yatskov confirmed that he met Pontecorvo for the first time while Pontecorvo was "on his way to the USSR." After Bruno arrived, Yatskov served for some time as Bruno's interpreter as well as his aide in entering Soviet life. Given their ages, and Yatskov's experience in atomic affairs, this story is plausible and fits with the descriptions of an unnamed aide that Bruno gave to Miriam Mafai.[16] A fuller biography of Yatskov would record that, in the 1940s, he had been based in New York as the controller of the Soviet atomic spy network, which included Klaus Fuchs, Ted Hall, and Lona Cohen.[17] Yatskov returned to the USSR in 1947. He died of cancer in March 1993, in the same hospital where he met the researcher.

During a conversation between the researcher and the KGB guide, the possibility of interviewing Bruno Pontecorvo came up. The KGB host duly asked Pontecorvo for an interview "just for the record," but Pontecorvo robustly declined. The researcher did not speak to Pontecorvo personally, but his response, as related by the KGB contact, "rung in my ears":

"*Ya khochu umeret' kak velikii fizik, a ne kak vash jebanyi shpion.*"[18]

"I want to die as a great scientist, not as your fucking spy."

Afterword

WHEN I BEGAN THIS INVESTIGATION, I WAS SKEPTICAL ABOUT CLAIMS that Pontecorvo was a spy.

The received wisdom has always been that he had committed some indiscretion, and that the KGB called him back to Moscow "just in time," as the "British/Italian/Canadian [take your pick] police were about to arrest him." However, even a cursory examination of the evidence shows such claims to have no sound foundation.

Statements that Pontecorvo was a spy, made in 1951 by the US Congress's Joint Committee on Atomic Energy, and propagated in the media at the time, were driven by McCarthyism, or based on theses that were manifestly inaccurate—such as the claim that Pontecorvo was named by Gouzenko (he was not). Often, these charges were simply asserted by fiat without any backup. It is now clear that Reed's report was exploited by MI5, which, through selected journalists, put out versions designed to present the security services in a positive light. There are clues to MI5's influence in many of these stories; for example, Alan Nunn May's name is sometimes misspelled as "Allan" in the early literature, as it was in Bruno's personal address book, which had been made available to what in effect was an MI5 ghostwriter.

More than sixty years later, the secrecy persists: FBI papers on Pontecorvo remain classified, with whole pages blacked out, and several documents in the UK archives have their sources redacted. Others appear to have been mislaid, or misfiled. One example is the pivotal letter from

Washington of July 1950, which led Philby to give Moscow the tip-off. This had been described as "lost" until December 2012, when, following persistent inquiries, I learned via Peter Hennessy that it had become "available for inspection."

All available sources confirm that there was no evidence in 1950 that would enable the UK authorities to "purge" Pontecorvo. The MI5 files, along with the personal diaries of senior intelligence officers, reveal the hit-or-miss reality of their trade. MI5 often had strategic successes when it infiltrated known left-wing or right-wing extremist organizations, but identifying lone spies—such as scientists who periodically passed information to the USSR for reasons of principle—was like searching for a needle in a haystack. In Pontecorvo's case MI5 was reduced to what was euphemistically called "amicable elimination."

Guy Liddell encapsulated the problem in his diary. On June 2, 1950, he reviewed the case of Klaus Fuchs:

> The FUCHS case showed that another man of his kind might well be recruited for a secret project. Once the decision to hire [such a spy has been] made, there are two [opportunities] for detection: where he gets the information and when he passes it on. As the former was in [the scientist's] brain, and he had it legally, it was impossible [to catch him in the act]. As to the latter, in FUCHS' case there was a period of a year when he didn't operate at all and thereafter only made contact once every three months. Unless you were on his tail for three months without detection—which was very difficult—there was very little chance of a result.
>
> When you considered that there were literally hundreds of cases of a prima facie kind where the evidence was far stronger than the case of FUCHS, it would be realised we were up against a formidable problem.[1]

Based on Liddell's diary and MI5 documents, it is clear that if Fuchs had not made a confession at the start of 1950, he would not have been prosecuted but instead would have been "amicably eliminated"—transferred from Harwell to a university outside the ring of security. By the spring of that year, the UK authorities were preparing to take that same action in the case of Bruno Pontecorvo; the above remarks could equally well have been made about him. MI5 was essentially impotent if faced with a lone spy who kept his head down. In the case of Pontecorvo, MI5 had suspicions but no proof.

The most notable subsequent evidence came from Oleg Gordievsky. Gordievsky himself, who joined the KGB in 1963, had no dealings with Pontecorvo, but he claims to have known KGB officers who did. These individuals allegedly told Gordievsky that Pontecorvo had been a hugely valued agent during and after the war, and this anecdotal evidence is now preserved in print. However, when Gordievsky was smuggled out of the USSR into Finland in the trunk of a British diplomatic car (a mirror image of Pontecorvo's trip thirty-five years earlier) he brought no documents bearing Pontecorvo's name. At best, Gordievsky's assertions appear to be based on KGB files he once read, or memories of earlier conversations. No details of the information that Pontecorvo passed to the USSR have been given, and the possibility of genuine error, due to misheard or misremembered accounts, is clearly present.

To this day, no one who claims that Pontecorvo was a spy has ever produced verifiable evidence, nor even identified what he is supposed to have done. In *Half Life* I have identified two pieces of classified material that passed from Chalk River to the USSR—a sample of uranium transferred later than Nunn May's, and blueprints of the nuclear reactor. Although there is no proof that Pontecorvo was the source of these materials, he had both the motive and the opportunity, and on balance can be identified as the prime suspect.

If Bruno was not approached by the Soviets during his time in Canada, then they missed an open goal. Kurchatov had followed Bruno's career in nuclear physics closely from 1935 onward, and his first action upon taking charge of the Soviet atomic bomb project had been to set up a list of potential contacts in North America.

Considering how efficiently the KGB responded to Kurchatov's demand to find sources within the Western atomic project, it is unlikely that they overlooked Bruno Pontecorvo. When Kurchatov learned the names of the scientists at Chalk River, he would have recognized Bruno immediately from their joint interest in isomers. Bruno was close with Frédéric and Irène Joliot-Curie—whose political sympathies were no secret—and it would also be on record that Bruno had attracted the attention of Communist Party members in Paris, and joined the party himself in 1939.

If he was approached, what can we assume about Bruno's reaction?

Later in his life, Bruno provided some clues when he commented on the case of Klaus Fuchs: "I would not have condemned Fuchs. He did what he thought was right and at that time the USSR were allies and not enemies of America."[2] Bruno also gave hints of his own allegiance to the communist ideal when his brother Gillo left the party following the invasion of Hungary. When Miriam Mafai interviewed Bruno thirty years later, she reported his reactions as follows: "He can not understand how Gillo can challenge the analysis of events as taken by Moscow and the leadership of the USSR. Gillo seems to have forgotten that loyalty to the USSR is one of the fundamental principles of the communist conscience."[3]

The last sentence summarizes the dogma of any truly committed communist in that era: if Moscow calls, you obey. It is the closest thing to a confession that Bruno Pontecorvo ever made.

The KGB maintained a vice-like grip on anyone who gave even a morsel of classified information to Soviet agents. Even a casual gesture on Bruno's part, such as providing blueprints to help his Soviet allies build a nuclear power plant for the benefit of their citizens, would be enough to put the squeeze on him later. Bruno's statement about the importance of obedience to the USSR suggests that the refusal of any such request would have been inconceivable. This is the strongest evidence that exists of Bruno's espionage, unless we take at face value Oleg Gordievsky's claims that Bruno Pontecorvo was a long-term spy of major importance to the USSR who voluntarily proffered secret documents to the Soviet embassy "probably in Ottawa."[4]

Six decades after the flight from Abingdon, Gil is remarkably relaxed about the matter, as if nothing unusual took place. He also has a wry sense of humour. When I told a group of Pontecorvo relatives about Bruno's difficulties in getting visas to exit France and reach the United States in 1940, Gil laughed and said that this probably explained why Bruno was so relaxed in the USSR. A cousin then joked that Bruno's migration to the USSR "had been handled by a travel agency."[5]

So why did Bruno Pontecorvo leave for the Soviet Union, and who were the "travel agents"?

In my judgment, something sudden and unexpected prompted Pontecorvo's flight. Gil told me that he too felt that the decision was made on the spur of the moment. Bruno's favourite brother, Gillo, had expected to join him on an underwater fishing trip, but Bruno "had cancelled at the end of August."[6] Based on the available evidence, the most likely catalyst was the Patterson letter. We can assume that Philby warned Moscow about it, and that Moscow warned Pontecorvo, possibly through the intermediary of Sereni.

In his 1951 report, when considering who might have orchestrated Pontecorvo's defection, Ronnie Reed identified Sereni as his prime suspect. Bruno's brother Guido had told Reed, "Giuliana and her husband Tabet and Laura might have influenced Bruno but could not have organised anything, whereas Emilio Sereni was powerful enough to do so and quite possibly may have done."[7]

Although Bruno's sister Anna never asked him directly for an explanation of his behaviour, she thought deeply about the saga in the course of her long life. She told me, "I thought he had been kidnapped." As to who might have kidnapped him, she had no idea, and as to how he was spirited to the USSR, she offered nothing—until after we had said goodbye. Just as her bus arrived, she unexpectedly remarked, "I never did trust Emilio Sereni."[8]

The full details of Sereni's role in Pontecorvo's defection may never be known. However, the possibility that Philby was its midwife now seems most probable.

As we've seen, Philby discovered that the FBI was interested in "PONTECORVO's" communist activities. It's also been stated that he "passed on every secret" to Moscow.[9] Indeed, he had tipped off Moscow about Nunn May, Fuchs, and another Los Alamos spy, Arthur Adams, so we can assume the same to be true in Pontecorvo's case. Thus, by the end of July 1950, Moscow would be aware that the FBI was pursuing Pontecorvo.

By that stage, Pontecorvo was on a camping trip and would not be in contact with the communist wing of his family until the end of August. It must have been when he made one of his visits to Rome by car that Moscow alerted him. The rest of the scenario is easy to imagine: aware of the FBI interest in him, Pontecorvo suddenly defects.

It is hard to sustain a case that he would have acted so precipitously if he were totally innocent. He later claimed that a rising tide of anti-communist hysteria in the United Kingdom made him fear that innocence wouldn't protect him, but the claim does not survive scrutiny. Being exposed as communist at that time in the United States was a serious matter, but not in England. If Bruno had been based in the US, this explanation would make sense, but given that he was living in the UK, and was already in the process of being shifted from Harwell to the safety of Liverpool, the explanation seems implausible. On balance, I feel that it would take a more certain threat to make Bruno uproot his wife and children, and his own life and work, so completely. When Philby alerted Moscow, the Soviets would have delivered a stark message to Bruno: "You're about to be arrested. We've seen the proof."

What can we conclude about the enigmatic half life of Bruno Pontecorvo? What further evidence might come to light?

I have met many of the surviving actors from the 1950s, but there are loose ends nonetheless. Some individuals did not reply to my requests; others I was unsuccessful in reaching. I hope that any who read this account and have something significant to add will make contact. If the KGB files were to be opened, many questions might be answered. Meanwhile, the FBI files on Bruno Pontecorvo remain blacked out. Reading between the lines of the MI5 files, which reference collaboration with the FBI, one can infer that at least some of these redactions were driven by a desire to obscure failings by the bureau, as well as cover-ups undertaken with their British counterpart.

The papers in London that assess the implications of Pontecorvo's defection are notable for their failure to mention the hydrogen bomb. Given that Pontecorvo was an expert on tritium and heavy-water reactors, this is either an oversight (as the concept of the hydrogen bomb was already public knowledge), or a result of the fact that the relevant papers remain classified due to continued sensitivity about the weapon. It was this latter possibility that led me to ask whether other papers on the Pontecorvo affair existed, which in turn led to Peter Hennessy's query, and the discovery of the "lost" file revealing Philby's role.

Marianne's story remains to be told. I am grateful to the Pontecorvo family for granting me access to Bruno's half of the correspondence between the pair in 1938–1939. Marianne's letters, however, remain private in Sweden, and only her diary was made available to me. Reading Bruno's half of the discourse is like overhearing one end of an intriguing phone call, and making guesses about the content of the other half. There has been speculation that Marianne was a committed communist too, and that she played an active and willing role in their decision to flee. Others, including members of the family, doubt this. A more complete understanding of Bruno's induction into the Communist Party, and of Marianne's attitude toward politics, could be buried in their correspondence from this time.

Emilio Sereni's role has been explored by Simone Turchetti. Sereni's diaries, however, include encoded material, so there may be further opportunities there to determine what part Sereni played in organizing the defection.

When I spoke to people who had some knowledge of the affair, but not of Pontecorvo himself, they almost invariably thought that the espionage question had been settled long ago: popular opinion had condemned Bruno Pontecorvo, notwithstanding the lack of evidence against him. This shows the power of the ex cathedra assertions of guilt made long ago, which have gained an aura of established truth through repetition. However, when I asked Bruno's colleagues and friends whether he could ever have been a spy, they almost universally insisted, "Not Bruno! Impossible." One colleague who had worked with Pontecorvo every day at Harwell was "very vehement" and "would stake his life that Bruno was not a spy."[10] All who knew him mentioned his openness, his childlike naïveté, and his indomitable charisma, which all contributed to their conviction that he could not possibly have had a secret life. Some claimed that Bruno was actually on record denying that he had ever been a spy.

However, I have not seen any such explicit denial. At the 1955 press conference, for example, such questions were not allowed, and on other occasions they were avoided. In his 1992 article on Pontecorvo for the *Independent*, Charles Richards asked, "Had he spied for Moscow? He still does not talk about it."[11]

In any case, the character testimonials from friends and colleagues cannot be taken at face value. Far from being naive, Bruno successfully kept secrets from his closest colleagues for years. In particular, he hid his Communist Party membership from almost everyone, including Henry Arnold, the security officer at Harwell. Later, Bruno always insisted that he was against atomic weapons, yet during his time in England, when the scientific community started an active protest against militarization in nuclear physics, Bruno kept his thoughts to himself. Charisma and duplicity can coexist within the same person, as Kim Philby ably demonstrated.

In his rapport with his colleagues, Bruno is actually similar to three established atomic spies: Alan Nunn May, Klaus Fuchs, and Ted Hall. Each of them hid their clandestine work from their fellow scientists, who reacted to the subsequent exposures with incredulity.

If the Soviet defector Igor Gouzenko had not exposed Alan Nunn May, the security authorities might never have been aware of Nunn May's existence. His colleagues, including Bruno Pontecorvo, were astonished at the news of his treachery. When MI5 informed Wallace Akers, the director of Tube Alloys, about Nunn May, he too was "deeply shocked." If Akers had been asked to rank the scientists employed in Canada on the basis of their integrity, he confirmed that he "would have placed May at the top."[12]

Fuchs too only surfaced due to the actions of outside parties. In his case, decrypted Soviet messages mentioned specific details that allowed him to be identified as CHARLZ. He had fooled everyone, not least his fellow spy at Los Alamos, Ted Hall, who had thought himself to be "the only one." As Hall's widow explained to me, "He didn't know about Fuchs."[13]

This takes us back to the beginning of my interest in Bruno Pontecorvo, and my conversation with Rudolf Peierls. Rudi and Genia Peierls had taken Klaus Fuchs into their home, viewing him with sympathy as a fellow émigré from fascism. They treated him almost as a member of the family, only to discover that he had fooled them. I still recall Rudi Peierls's sadness when I asked him whether Pontecorvo too had been a spy. He replied, "You never can tell."

Acknowledgments

I MUST ACKNOWLEDGE AT THE OUTSET MY SPECIAL GRATITUDE TO THE members of the extended Pontecorvo family, in particular two first-hand witnesses: Gil Pontecorvo and Anna Newton. Gil, Bruno's eldest son, was twelve years old when the family fled from Abingdon-on-Thames (my current hometown) to the Soviet Union. Gil has first-hand memories of what really happened, in contrast to the many myths that were propagated at the time. A physicist himself, based in Russia, he also helped with translation and access to research notes and papers from Bruno's early years in Dubna. Bruno's sister, Anna Newton, was also invaluable, as she was with Bruno and his family in the days immediately before his defection, and her testimony helped clarify some inconsistencies in the MI5 accounts.

In addition, I am indebted to many individuals and organizations who provided advice during my research for *Half Life*, and to those who have read the manuscript in part or in its entirety. The list of helpful parties includes many scientists, both in the West and the former USSR, along with the families of former spies, members of the intelligence community, residents of Abingdon who knew the Pontecorvo family in 1950, and a host of others whom I would never have had the pleasure of meeting were it not for the surprising receipt of that query about MI5, which came from Peter Hennessy, Baron Hennessy of Nympsfield.

So thanks to:

Joseph Albright
Ugo Amaldi
Christopher Andrew
Lorna Arnold

Marlene Baldauf
Alessandro Bettini
Samoil Bilenky
Benny Birnberg
Tania Blokhintseva

Mark Bretscher
Paul Broda
Franco Buccella
Frances Cairncross
Duncan Campbell
Rino Castaldi
Diana Cobban May
Chris Collins
Gordon Corera
Tam Dalyell
Thelma Druett
Sven-Olof Ekman
Graham Farmelo
Giuseppe Fidecaro
Maria Fidecaro
Anthony Gardner
Paul Gardner
Michael Goodman
Oleg Gordievsky
Jeremy Grange
Vladimir Gribov
Tom Griffin
Joan Hall
David Hanna
Roger Hanna
Joe Hatton
Peter Hennessy
Peter Higgs
Gregory Hutchinson
Boris Ioffe
David Jackson
George Kalmus
Valery Khoze
Jasper and Rita Kirkby
Peter Knight
Anneke Lawrence-Jones
David Lees
Luigi di Lella

Lev Lipatov
Harry Lipkin
Fedele Lizzi
John Maddicott
Miriam Mafai
Luciano Maiani
Neil Maroni
Victor Matveev
Art McDonald
Hamish McRae
Matt Melvin
Charles Miller
Peter Minkowski
Giorgio Panini
Rudolf and Genia Peierls
Don Perkins
Michel Pinault
Chapman Pincher
Anna Pontecorvo (Newton), Antonio Pontecorvo, Barbara Pontecorvo, David Pontecorvo, Gil Pontecorvo, Gregory Pontecorvo, Ludo Pontecorvo, Simone Pontecorvo
Kate Pyne
Nicholas Reed
Richard Rhodes
Stella Rimington
John Rowlinson
Yves Sacquin
John Sandalls
David Saxon
Gino Segrè
Brian Smith
Godfrey Stafford
Jack Steinberger
Kellogg Stelle
Eugenio Tabet
Stan Taylor

I. Todorov

Simone Turchetti

William Tyrer

David Wark

Forbes Wastie

Peter Watson

Nigel West

Christine Wootton

Nino Zichichi

Sarah Wearne, Michael Triff, and the archives at Abingdon School

Allen Packwood and the staff at the Churchill College archives, Cambridge

Carlo Dionisi and the organizers of the Bruno Pontecorvo centenary conference, Rome

Claire Daniel, Martin Hendry, Lesley Richmond, and David Saxon for research at the University of Glasgow archives

Jens Vigen and Tullio Basaglia at the CERN library and their liaison with Dubna

Giuseppe Mussardo and Luisa Bonolis for video interviews in Russia, and the research and evaluation of material from historical archives in France, Italy, and Russia

Three anonymous sources in the former Soviet Union, and two in the United Kingdom

Finally, *Half Life* owes much to editorial advice from T. J. Kelleher, John Searcy, and Sandra Beris at Basic, and Robin Dennis and Sam Carter at Oneworld, as well as Michael Marten, who read my first draft in full and encouraged me to rewrite it, and my agent, Patrick Walsh, who helped this project come to life.

Acronyms

AIP: American Institute of Physics oral history archive (online)
BPSSW: *Bruno Pontecorvo: Selected Scientific Works*
CAC: Churchill Archives Centre, Churchill College, Cambridge University
TNA: The National Archives, London

Notes

PROLOGUE

1. Guy Liddell diary, 1950, TNA KV 4/472.

2. Recounted in Chapman Pincher, *Treachery*, chap. 59.

3. John Rowlinson interview, November 6, 2012.

4. L. Fermi, *Atoms in the Family*, p. 257.

5. This is my memory of Alexei's remarks from four decades past. The quotes are his own words, taken from a memorial for Bruno. Sadly, Alexei Norairovich Sissakian himself died in 2010 and so was unable to develop his memories with me further. The quote as written is excerpted from the memorial. It is consistent with my memory of his remarks to me.

6. Thelma Druett interview, February 2012.

7. Alan Moorehead's book *The Traitors* was first published in 1952 and cast Fuchs, Nunn May, and Pontecorvo as the three "traitors." It is now clear that Moorehead was given free rein by MI5 to include Pontecorvo as an atomic spy without any hard evidence. Nonetheless, his book established the perception that Pontecorvo escaped to the USSR as the net was about to close around him. In any event, none of the trio was a traitor to the United Kingdom as the Soviet Union was its ally at the time when Fuchs and Nunn May passed information. As neither Moorehead nor MI5 ever specified what Pontecorvo was supposed to have done, or when, there is no basis for classifying him as a traitor either. Furthermore, he was not a British citizen until 1948.

8. N. Reed, *My Father, the Man Who Never Was*. On page 117 Ronnie Reed recalls being "put in charge of counter-espionage against the Russians in 1951." This seems to be a false memory; TNA archives show he was in place in 1950, and Reed himself recalled that this meant he was in "charge of cases like Klaus Fuchs," which straddled the end of 1949 and beginning of 1950. My understanding of Reed's role is also based on e-mails and phone conversations with Nicholas Reed on December 17, 2012, March 12, 2013, and March 20, 2013.

CHAPTER 1

1. Anna Pontecorvo interview, November 11, 2011.

2. Turchetti, *The Pontecorvo Affair*, p. 16

3. Bruno Pontecorvo, Autobiographical notes; Anna Pontecorvo interview, November 11, 2011.

4. Anna Pontecorvo interview, November 11, 2011.

5. Anna Pontecorvo interview, November 11, 2011.

6. Bruno Pontecorvo, Autobiographical notes.

7. Ludo Pontecorvo interview, September 12, 2013. The accuracy of this memory is unclear. Bruno's younger brother Gillo was indeed an international competitor, who took part in the tournament at London's Queen's Club in 1939. Bruno could play a good game, by all accounts, and was the winner of several tournaments at the club level, but never reached the same stratospheric heights as his brother.

8. Gil Pontecorvo interview, January 25, 2014. Maria's family belonged to the Chiesa Evangelica Valdese, an evangelical Protestant group that had broken from the Catholic Church.

9. Anna Pontecorvo interview, November 11, 2011.

10. Neil Maroni interview, December 23, 2013. A remark by an unnamed family member, as quoted by Neil Maroni, son of Giovanni Pontecorvo.

11. Another cousin, Eugenio Coloni, joined the antifascist group Justice and Liberty. He was later arrested and eventually killed by fascist militants in 1944. His significance for Bruno's life was limited.

12. These were Lev and Xenia Zilberberg. See Chaliand and Blin, *The History of Terrorism*, or Smadar Sinai, "Manya Shochat and Xenia Pampilova's Experience of Immigration to Eretz Yisrael: A New Identity or a New Garment?," accessible at www.aisisraelstudies.org/papers/AIS2007_SinaiSmadar.pdf, which states, "In 1926, [Xenia's] daughter Xenichka married Emilio Sereni . . . who became one of the leaders of the communist party in Italy."

13. Guido Pontecorvo, as quoted in TNA KV 2/1888.

14. L. Fermi, *Atoms in the Family*, p. 97.

15. Bruno Pontecorvo, Autobiographical notes.

16. The discovery of the neutron in 1932 was key. See Rhodes, *The Making of the Atomic Bomb* for a description of the discovery and its implications.

17. This is essentially the atomic model proposed in 1803 by English chemist John Dalton. His model explained the laws of chemical combination and inspired quantitative experiments. This led scientists to measure the relative weights of the atoms of different elements, which range from hydrogen, the lightest element, to uranium, the heaviest normally found on earth.

18. Dig uranium from the ground and, in 993 out of every 1,000 atoms, the nucleus will have 146 neutrons, making 238 "nucleons" in total—the isotope U-238. The remaining seven atoms will most probably contain only 143 neutrons, making U-235. Even the most common form of uranium, U-238, is radioactive. Half of a sample of the stuff will decay in 4 billion years—we call this its half-life, which

means that about half of the uranium found in the rocks of the newborn Earth, 5 billion years ago, has now decayed. The isotope U-235, by contrast, has a half-life of about 700 million years. This time span, which is very long on a human scale, is nonetheless relatively short compared to the age of the earth and the half-life of its sibling, U-238. Thus, over the eons, much more U-235 than U-238 has disappeared from primeval rocks. This explains the dominance of U-238 in natural ores today. This U-238 also acts as a blanket that impedes the fission of U-235, making a succession of fissions rare, and the chance of a chain reaction negligible in uranium ores. Uranium first has to be "enriched" by increasing the percentage of U-235 before it can be used as a practical source of energy.

19. The half-life is the time after which half of the atoms of a sample will have decayed.

20. There is an exception, however. When more than 10^{58} neutrons come together, the assembly can be held together by the force of gravity. This is a neutron star.

21. Hans Bethe, quoted in Rhodes, *The Making of the Atomic Bomb*, p. 165.

CHAPTER 2

1. Pauli originally called it a "neutron," but when the constituent of the nucleus was discovered and given that name in 1932, Enrico Fermi proposed the name *neutrino* for the particle emitted in beta decay.

2. Close, *Neutrino*, pp. 22–24.

3. Half a century later, the editors admitted that this was their greatest blunder.

4. L. Fermi, *Atoms in the Family*, p. 84.

5. Frédéric Joliot, Nobel Prize address, 1935. See http://www.nobelprize.org /nobel_prizes/chemistry/laureates/1935/joliot-fred-lecture.pdf.

6. Ugo Amaldi e-mail, January 12, 2014. Laboratorio di Sanità Pubblica, headed by Trabacchi, was located in the same building on the Via Panisperna and kept one gram of uranium on hand for medical purposes.

7. Ugo Amaldi e-mail, January 12, 2014. Amaldi recalled that, during this holiday, Fermi explained his theory of beta decay to the "Boys."

8. The idea seems to have originated with Ettore Majorana, a brilliant theorist on Fermi's team. Majorana disappeared in 1938 and was presumed dead. See Turchetti, *The Pontecorvo Affair*, p. 24.

9. The positron is the antimatter sibling of the negatively charged electron. It has the same mass as an electron but is positively charged.

10. Bruno Pontecorvo, Autobiographical notes.

11. Edoardo Amaldi as told to Ugo Amaldi; Ugo Amaldi interview, September 12, 2013.

12. Fermi's memory of events as told to Subrahmanyan Chandrasekhar. See Fermi, *Note e memorie*, p. 927; quoted in Robotti and Guerra, "Bruno Pontecorvo in Italy."

13. Not October 22, contrary to popular wisdom. See later comments about this.

14. Ugo Amaldi e-mail, January 12, 2014. This seems to be a leitmotif for Fermi, as eight years later in Chicago, when his nuclear reactor was about to reach criticality, he said, "And now we go for lunch."

15. Ugo Amaldi e-mail, January 12, 2014. Ugo recalls his father, Edoardo, saying that Fermi reached an understanding of the phenomenon during lunch, and that on the way back to the laboratory he explained the idea of slow neutrons to them.

16. Edoardo Amaldi interview, AIP, http://www.aip.org/history/ohilist/4485_1 .html. Amaldi describes this episode and also considers whether Fermi's insight was inspired by his earlier analysis of electrons bouncing from atoms, which he had made following the measurements of atomic spectra by Amaldi, Segrè, and Pontecorvo.

17. L. Fermi, *Atoms in the Family*, chap. 11. See also note 20 below, which casts doubt on her version.

18. Ugo Amaldi e-mail, January 12, 2014. Ugo Amaldi confirms this. These papers were drafted in the Amaldi apartment when Ugo was asleep, at the age of two months. On more than one occasion his mother later told him that there was so much excitement she was afraid he would be woken.

19. L. Fermi, *Atoms in the Family*, p. 101.

20. Historian Alberto De Gregorio studied Enrico Fermi's logbooks, which give October 20 as the date. See his "Chance and Necessity in Fermi's Discovery of the Properties of the Slow Neutrons" (http://arxiv.org/ftp/physics/papers /0201/0201028.pdf) for a detailed history and analysis of different memories of these events. I am grateful to Ugo Amaldi for bringing this document to light. Ugo also doubts the goldfish pond episode: his father, Edoardo, never mentioned it, and Emilio Segrè claimed to Ugo that it never happened.

21. Ugo Amaldi e-mail, January 12, 2014. The history of the water bucket is confused. The variable results of Amaldi and Pontecorvo, which started the saga, were at one point blamed upon a bucket of water, which the cleaner would leave "now under one table and another time under another."

22. In November he repeated some of the tests and found the same results as "those of October 21 [sic]."

23. Ugo Amaldi interview, September 11, 2013; Ugo Amaldi e-mail, January 12, 2014.

24. Ugo Amaldi interview, September 11, 2013.

25. Robotti, "The Beginning of a Great Adventure."

26. Bruno Pontecorvo, "On the Properties of Slow Neutrons" [in Italian], *Il nuovo cimento* 12, no. 4 (April 1935): 211–222.

27. Graphite is carbon, which is light enough to slow neutrons efficiently, and also cheap. Heavy water, however, is expensive and not readily available.

28. Turchetti, *The Pontecorvo Affair*, chap. 1, note 54.

29. Bruno Pontecorvo, Autobiographical notes.

30. Frisch, *What Little I Remember*, quoted in Rhodes, *The Making of the Atomic Bomb*, p. 227.

31. In 1932 Dmitri Ivanenko, a theoretician in Moscow, invented a model of the atomic nucleus consisting of neutrons and protons, independently of Bohr,

but no one seems to have been aware of this. Ivanenko's lack of recognition made him bitter, and in 1949 his morbid obsession led him to attack Soviet physicists who were acknowledging Western ideas.

32. Maurice Goldhaber, interview by Gloria Lubkin and Charles Weiner, January 10, 1966, transcript, Niels Bohr Library, American Institute of Physics, New York, p. 27.

33. Holloway, *Stalin and the Bomb*, p. 39.

CHAPTER 3

1. Quote from FBI memorandum, November 9, 1949, TNA KV 2/1888.

2. Date stamp in Italian passport 467675, TNA KV 2/1888.

3. Irène served in Blum's government from June to September 1936, when she gave up the position due to ill health, as she continued to suffer from tuberculosis.

4. Mafai, *Il lungo freddo*, p. 78. As Malraux had left his wife, and at that time lived with Josette Clotis, with whom he later had two children, this would indeed seem to be a classic example of meeting one's mistress in the hour between leaving work and going home.

5. Bruno Pontecorvo, Autobiographical notes.

6. Robotti and Guerra, "Bruno Pontecorvo in Italy"; Luisa Bonolis e-mail, January 19, 2014.

7. Mafai, *Il lungo freddo*, p. 85.

8. Mafai, *Il lungo freddo*, p. 84.

9. Ronnie Reed report paper 206A, dated January 16, 1951, TNA KV 2/1890: "Gilberto told Guido that Bruno was a very ardent communist and had in fact been responsible for converting both Gilberto and Giuliana to communism in 1940."

10. Bruno Pontecorvo, Autobiographical notes.

11. Ernest Hemingway, quoted by his biographer A. E. Hotchner, http://en .wikipedia.org/wiki/A_Moveable_Feast.

12. Bruno Pontecorvo papers, CAC; Gil Pontecorvo interview, September 22, 2011: "She worked for a family of rich Swedes, babysitting."

13. Marianne Nordblom passport, CAC.

14. Diary in Bruno Pontecorvo papers, CAC.

15. Until 1984 Paris had a mail service consisting of a network of tubes powered by compressed air. See John Vincour, "Paris Pneumatique Is Now a Dead Letter," *New York Times*, March 31, 1984, http://www.nytimes.com/1984/03/31 /style/paris-pneumatique-is-now-a-dead-letter.html.

16. The phrase "pile ou face" (heads or tails) in Marianne's diary appears to refer to a game of tossing coins.

17. Mafai, *Il lungo freddo*, p. 93.

18. Emling, *Marie Curie and Her Daughters*, p. 34.

19. Quoted in Emling, *Marie Curie and Her Daughters*, p. 93.

20. Lew Kowarski interview, AIP, http://www.aip.org/history/ohilist/4717_1 .html.

21. Luisa Bonolis interview, January 19, 2014; see also Robotti and Guerra, "Bruno Pontecorvo in Italy."

22. Isomerism had been observed long before, but it was Kurchatov who first convincingly demonstrated the phenomenon in the new era of induced radioactivity and inspired a new field of inquiry.

23. Close, *The Infinity Puzzle*, p. 21.

24. A photon in the green part of the visible spectrum has an energy of about two electron volts. An electron volt (eV) is the energy that an electron gains when accelerated by a potential of one volt. Thousands and millions of eV are denoted keV and MeV, respectively.

25. Emling, *Marie Curie and Her Daughters*, p. 151.

26. His family had fled after the revolution. He joined Joliot-Curie's group in 1934, and became a French citizen on November 16, 1939.

27. Lew Kowarski interview, AIP, http://www.aip.org/history/ohilist/4717_2 .html. The *von* moniker was much loved by Hans von Halban, but we shall use the basic surname *Halban* from here on.

28. Lew Kowarski interview, AIP, http://www.aip.org/history/ohilist/4717_2 .html. Another colleague, quoted in Broda, *Scientist Spies*, p. 115, described Halban as "a playboy, rich from his father and from marrying a banker's daughter."

29. CAC. The *certificat de domicile* issued on February 8, 1940, shows that she lived there from January 4 to September 10, 1938, when her French visa was about to expire, and returned on August 28, 1939.

30. This is an anecdotal memory from a local source who wished to remain anonymous. If there is any written record, it has not been made public.

31. Mafai, *Il lungo freddo*, p. 93.

32. Sven-Olof Ekman interview, November 28, 2013.

33. Letter from Bruno Pontecorvo to Marianne Nordblom, October 12, 1938, Bruno Pontecorvo files CAC.

34. Letter from Bruno Pontecorvo to Marianne Nordblom, October 13, 1938, CAC.

35. Letter from Bruno Pontecorvo to Marianne Nordblom, October 15, 1938, CAC.

36. Quoted in obituary, *The Guardian*, October 14, 2006.

37. Stable relative to beta decay. The paper is "On the Possible Existence of Beta-Stable Isomers" [in French], presented at Actes du Congrès International du Palais de la Dècouverte, Paris, 1937 (listed in Bruno Pontecorvo bibliography at http://pontecorvo.jinr.ru/bibliography.html). At these same proceedings, George Placzek suggested that nuclear levels may be grouped into individual classes that do not intercombine—see note in Bruno Pontecorvo, "Isomeric Forms of Radio Rhodium," *Nature* 141, no. 3574 (April 30, 1938): 785.

38. Bruno Pontecorvo and M. Dode, "On a Radioelement Produced in Cadmium under the Action of Fast Neutrons" [in French], *Comptes Rendus de l'Académie des Sciences* 207, no. 4 (1938): 287–293.

39. The distribution of these photons in space is related to the difference in angular momentum of the original isomer and the state into which it decays. Technically, this is called their angular distribution, or multipolarity.

40. Bruno Pontecorvo, "Isomeric Forms of Radio Rhodium," *Nature* 141, no. 3574 (April 30, 1938): 785.

41. A minibiography of the Joliot-Curies, which mentions the work at Ivry, can be found at http://www.analytik.ethz.ch/praktika/radiochemie/unterlagen /biografien/Joliot-Curie.pdf.

42. Bruno Pontecorvo and A. Lazard, "Isometric Nuclei Produced by Continuous X-ray Spectra" [in French], *Comptes Rendus de l'Académie des Sciences* 208, no. 2 (1939): 99–101.

43. Bruno Pontecorvo, Autobiographical notes.

44. The isomers actually have slightly different masses, but the difference is too small to be measured directly, hence the sobriquet *isomer*, for what appeared to be "equal masses." Each rung on the energy ladder has a slightly different energy, and by the mass-energy equivalence, this implies a different mass. However, the difference in their masses is equivalent to one part in several thousand. It is this nugatory difference that enables photons to be emitted as the isomers tumble down the energy ladder.

45. Lew Kowarski interview, AIP, http://www.aip.org/history/ohilist/4717_1 .html.

46. Holloway, *Stalin and the Bomb*, p. 50, note 5. Rhodes, *Dark Sun*, p. 27 contains the letter sent to Abram Ioffe, with whom Kurchatov began experiments.

The KGB double agent Oleg Gordievsky claimed to me on April 30, 2013, that the KGB used Pontecorvo as an agent during this period in Paris. This seems unlikely. Pontecorvo was peripheral to the fission research, unlike Joliot-Curie, who had already told Kurchatov about the phenomenon. None of this work was secret at first.

47. He repeated the test with rings of various sizes and verified that the intensity of radiation was less for large rings than for small ones, and that with the largest ring there was no radioactivity at all. This confirmed that the radiation was indeed the result of the uranium and not some other source.

48. Pinault, "*Frédéric Joliot, la science et la société,*" p. 152; Michel Pinault e-mail, January 21, 2014.

49. Letter from Bruno Pontecorvo to Marianne Nordblom, January 27, 1939, CAC. The mention of Ivry is confusing. The fission demonstration was made at the Collège de France; it seems unlikely that Bruno would choose to refer to his own work in such splendid terms at that very moment. Later, on March 27, he unambiguously refers to the fission work in another letter to Marianne.

50. Rhodes, *The Making of the Atomic Bomb*, p. 290.

51. The idea of a chain reaction originated with Hungarian physicist Leo Szilard in 1934. He had not foreseen that uranium would be key, and mistakenly thought that a light element such as beryllium would do. He shared his insight with Rutherford. The concept was thus in the open. The story is told in many books, e.g., Rhodes, *The Making of the Atomic Bomb*, p. 28 and Farmelo, *Churchill's Bomb*, p. 73 et seq.

52. Holloway, *Stalin and the Bomb*, p. 57.

53. Manham, *Snake Dance*, p. 189, which appears to be based on René Brion and Jean-Louis Moreau, *De la mine à Mars: la genèse d'Umicore* (Brussels: Lannoo, 2004).

54. Letter from Bruno Pontecorvo to Marianne Nordblom, February 1939, CAC.

55. Letter from Bruno Pontecorvo to Marianne Nordblom, June 26, 1939, CAC.

56. Legal document signed by Bruno Pontecorvo, CAC.

57. Ludo Pontecorvo interview, April 18, 2013.

58. Guy Liddell diaries, TNA KV 4/472.

59. Sven-Olof Ekman e-mail, May 15, 2013, and copy of FBI file on Bruno Pontecorvo.

60. Letter from Bruno Pontecorvo to Marianne Nordblom, CAC.

61. Guido Pontecorvo to MI5, January 12, 1951, TNA KV 2/1888.

62. Mafai, *Il lungo freddo*, p. 186.

63. Weart, *Scientists in Power*, p. 114.

64. Lew Kowarski interview, AIP, http://www.aip.org/history/ohilist/4717_1 .html.

65. Lew Kowarski interview, AIP, http://www.aip.org/history/ohilist/4717_1 .html.

66. The fact that we are not in the presence of naturally occurring atomic explosives is the result of a delicate balance. In Russia, Yakov Zel'dovitch and Yuli Khariton calculated that if the amount of U-235 were slightly higher (about twenty rather than seven in every thousand atoms), a chain reaction could happen as long as the neutrons had been slowed (moderated) with water. Their calculations were correct, except for one thing: they assumed water to be a much more efficient moderator than it is in practice.

67. Lew Kowarski interview, AIP, http://www.aip.org/history/ohilist/4717_1. html.

68. Quote from vetting report dated July 6, 1940, TNA KV 2/1888, memo 117B. This was also noted by Ronnie Reed in his report of 1951. See also Turchetti, *The Pontecorvo Affair*, p. 71.

CHAPTER 4

1. Weart, *Scientists in Power*, p. 153.

2. Mafai, *Il lungo freddo*, p. 105.

3. Turchetti, *The Pontecorvo Affair*, p. 44.

4. Segrè, *A Mind Always in Motion*, p. 160.

5. Eugenio Tabet interview, September 12, 2013.

6. Bruno Pontecorvo files, CAC.

7. Bruno Pontecorvo files, CAC.

8. See Némirovsky, *Suite Française*; confirmed by Météo France.

9. Mafai, *Il lungo freddo*, p. 106.

10. Lew Kowarski interview, AIP, http://www.aip.org/history/ohilist/4717_1.

11. See Némirovsky, *Suite Française*, p. 4.

12. Mafai, *Il lungo freddo*, p. 105.

13. Sonia Tomara, *New York Herald Tribune*, June 14, 1940, http://spartacus -educational.com/2WWtomara.htm.

14. Mafai has this as *Muftar*, which doesn't exist and appears to be Bruno's phonetic memory of *Mouffetard* fifty years later.

15. Mafai, *Il lungo freddo*, p. 106.

16. Mafai, *Il lungo freddo*, p. 107.

17. Sonia Tomara, *New York Herald Tribune*, June 14, 1940, http://spartacus-educational.com/2WWtomara.htm.

18. The card in the Churchill College archives was written in French; the translation is mine. Bruno and Gillo later told Mafai that they left Paris early on the morning of June 13, which is as I have written here. The card was written on June 14. The distances travelled are consistent with two days of travelling by bicycle, and their memory of a June 13 departure.

19. Details of Luria's Nobel Prize are at http://www.nobelprize.org/nobel_prizes/medicine/laureates/1969/luria-bio.html.

20. Lew Kowarski interview, AIP, http://www.aip.org/history/ohilist/4717_1.

21. Bruno Pontecorvo passport visa, CAC.

22. Bruno Pontecorvo passport visas, CAC. Marianne's visa number is 1209087 and Gil's is 1209090. Marianne is noted as Swedish, and Gil as Italian.

23. Mafai, *Il lungo freddo*, p. 110.

24. Bruno Pontecorvo passport, CAC.

25. Bruno Pontecorvo passport, CAC.

26. Mafai, *Il lungo freddo*, p. 111.

27. The bridge that spans the Narrows is known as the Verrazano-Narrows Bridge. However, the waterway itself is simply called the Narrows, not the Verrazano Narrows, contrary to a common misconception.

28. FBI/INS records of arrival number 65-5650 list the date as August 19, and the intended destination as the home of Paolo Pontecorvo at 503 W. 121 St. Mafai, *Il lungo freddo*, p. 112 has August 20. TNA KV 2/1888, memo 109A claims that Bruno entered the US on August 18, 1950. I have written the date as August 19, using the actual US entry forms as the primary documentary source. The reference to humidity obscuring the view originates with Bruno Pontecorvo as told to Mafai on the page referenced above. This is consistent with the weather data recorded at http://weatherspark.com/history/31081/1950/New-York-United-States.

29. Gillo Pontecorvo obituary, *The Guardian*, October 14, 2006, http://www.theguardian.com/news/2006/oct/14/guardianobituaries.obituaries.

30. Gillo Pontecorvo remarks in Pontecorvo family film, viewed September 12, 2013.

CHAPTER 5

1. Frisch and Peierls memorandum, quoted in Rhodes, *The Making of the Atomic Bomb*, p. 324.

2. Bureaucracy, politics, and the sheer scale of the challenge delayed action until late in 1941. See Farmelo, *Churchill's Bomb*.

3. The story of these experiments is found in Rhodes, *The Making of the Atomic Bomb*, pp. 348–351.

4. Sir John Cockcroft, "The Early Days of Canadian and British Atomic Energy Projects," www.iaea.org/Publications/Magazines/Bulletin/Bull040su/04004701820su.pdf.

5. Holloway, *Stalin and the Bomb*, p. 70.

6. Holloway, *Stalin and the Bomb*, p. 54.

7. Kurchatov left his name off the paper, although he had inspired and overseen the experiment, so that the young duo could get the full credit. In 1960, Flerov became the director of the laboratory of nuclear reactions at JINR in Dubna. See also Holloway, *Stalin and the Bomb*, p. 55.

8. Holloway, *Stalin and the Bomb*, p. 387, note 77.

9. This independent discovery remained undisclosed for forty years. See Holloway, *Stalin and the Bomb*, pp. 66–67 and note 98.

10. H. York, quoted in Rhodes, *The Making of the Atomic Bomb*, p. 327.

11. L. Fermi, *Atoms in the Family*, p. 254.

12. It's not clear how sure Bruno Pontecorvo was that this job was guaranteed when he left Paris. Segrè had said at the time that he could "get you a job." Segrè wrote to Pontecorvo again in August, by which time the latter was in the US, and urged him to "get there quickly." It seems that there may have been some confusion, as another physicist, Sergio Benedetti, thought that he had also been offered the job. Comments at Bruno Pontecorvo centenary meeting, Rome, 2013.

13. Mafai, *Il lungo freddo*, p. 115.

14. On the order of an electron volt or less.

15. E. Amaldi and E. Fermi, "On the Absorption and the Diffusion of Slow Neutrons," *Physical Review* 50 (November 1936): 899.

16. R. Fearon, W. Russell, and B. Pontecorvo, "Preliminary Field Experiment in Scattered Neutron Well Logging; 25 June 1941," S. Scherbatskoy papers, box 1, folder 6, Smithsonian archives. Referenced in Turchetti, *The Pontecorvo Affair*, p. 45.

17. Bruno Pontecorvo, "Neutron Well Logging," *Oil and Gas Journal* 40 (1941): 32–33.

18. Bruno Pontecorvo, "Radioactivity Analysis of Oil Well Samples," *Geophysics* 7, no. 1 (1942): 90–92.

19. Turchetti, *The Pontecorvo Affair*, p. 45.

20. Mafai, *Il lungo freddo*, p. 115.

21. Mafai, *Il lungo freddo*, p. 117.

22. Mafai, *Il lungo freddo*, p. 115.

23. L. Fermi, *Atoms in the Family*, p. 163.

CHAPTER 6

1. Holloway, *Stalin and the Bomb*, p. 77.

2. The separation was accomplished by gaseous diffusion. Fuchs's Russian contact was with the GRU. After late 1943, when Fuchs moved to the United States, he operated with the NKGB (Andrew and Gordievsky, *KGB*, p. 314). See also note 3 in the "Interlude" chapter of the present book and p. ix of Andrew and Gordievsky for some guidance to the labyrinthine histories of the various Soviet security services

3. See Holloway, *Stalin and the Bomb*, p. 79 for more on Flerov's efforts and the slow response of various authorities.

4. This was in December 1941.

5. James Chadwick interview, AIP, http://www.aip.org/history/ohilist/3974_4 .html. Farmelo, *Churchill's Bomb*, p. 179.

6. Via Fuchs and also possibly John Cairncross, secretary to the Minister without Portfolio, Lord Hankey, who received Chadwick's report.

7. See comments by Alan Nunn May in Broda, *Scientist Spies*, p. 111 et seq. The ICI patents are discussed in Weart, *Scientists in Power*, pp. 171–181.

8. Broda, *Scientist Spies*, p. 122.

9. Turchetti, *The Pontecorvo Affair*, p. 47, note 28.

10. L. Fermi, *Atoms in the Family*, p. 168.

11. S. Scherbatskoy papers, box 1, folder 6, Smithsonian archives; see Turchetti, *The Pontecorvo Affair*, p. 47, note 27.

12. Harry Lipkin e-mail, February 18, 2011.

13. The name was changed to Eldorado Mining and Refining Limited.

14. For more information about Boris Pregel, see Turchetti, *The Pontecorvo Affair*, p. 57 or http://en.wikipedia.org/wiki/Boris_Pregel.

15. Letter from Boris Pregel to Bruno Pontecorvo, November 24, 1942, CAC.

16. Bertrand Goldschmidt, quoted in Weart, *Scientists in Power*, p. 196.

17. Bruno Pontecorvo, Autobiographical notes.

18. TNA KV 2/1888.

19. Mafai, *Il lungo freddo*, p. 130.

CHAPTER 7

1. Turchetti, *The Pontecorvo Affair*, p. 52.

2. TNA KV 2/1887.

3. Gowing, *Britain and Atomic Energy*, p. 277.

4. TNA AB 1/361.

5. TNA KV 4/242, memo 23a, dated October 23, 1950.

6. TNA KV 2/1888.

7. TNA KV 4/243.

8. The report was "lost" for seven years, and only reappeared in 1950. This visit by the FBI seemed inconsequential at the time, but would later prove central to Bruno's sudden decision to flee to the USSR in 1950—see Chapter 13.

9. Burke, *The Spy Who Came In from the Co-op*, p. 132.

10. Burke, *The Spy Who Came In from the Co-op*. The claim that "Chalk River was penetrated by Soviet agents" is also asserted by Andrew and Mitrokhin, *The Mitrokhin Archive*, p. 174.

11. Weather history for January 1943 can be found at http://www.weather -warehouse.com/WeatherHistory/PastWeatherData_TulsaIntlArpt_Tulsa_OK _January.html.

12. Wallace, "Atomic Energy in Canada," p. 126. (I am indebted to J. D. Jackson for this source.)

13. Wallace, "Atomic Energy in Canada," p. 127.

14. These arguments occurred between Sir John Anderson of the British War Cabinet, and Vannevar Bush, head of the US Office of Research and Development. The USORD oversaw the Manhattan Project, of which General Groves was the director.

15. See Farmelo, *Churchill's Bomb*, especially pp. 211–223. Also, John Angus's first name and status per email from Archive Sources, University of Glasgow, September 25, 2014.

16. Broda, *Scientist Spies*, chap. 12.

17. This has been well documented in descriptions of life at Los Alamos, so there is no reason to doubt that it was the case in Chicago also. At Los Alamos, the research director, J. Robert Oppenheimer, lobbied against the policy. His colleagues supported him. He organized weekly colloquiums where technical staff from the entire project discussed their ideas freely. Overall, the personnel at Los Alamos shared information in ways that horrified General Groves, the director of the Manhattan Project.

18. Goldschmidt, *Les rivalités atomiques*, p. 44, quoted in Weart, *Scientists in Power*, p. 198. See also Gowing, *Britain and Atomic Energy*, p. 160; Broda, *Scientist Spies*, p. 125.

19. Broda, *Scientist Spies*, p. 108.

20. Gibbs, "British and American Counter-Intelligence."

21. Quoted in Broda, *Scientist Spies*, p. 115.

22. Broda, *Scientist Spies*, p. 126.

23. Broda, *Scientist Spies*, p. 126.

24. Broda, *Scientist Spies*, p. 113.

25. Broda, *Scientist Spies*, p. 124.

26. Presumably, the Germans were under so much pressure that the threat of a German invasion of the UK—for which radar was the intended defense—was now greatly reduced. Also, with Hitler now on the defensive, there was fear that he might develop his own atomic bomb and hold the Allies hostage in a last stand.

27. Paul Broda interview, October 11, 2013.

28. Broda, *Scientist Spies*, p. 131.

29. Broda, *Scientist Spies*, p. 144.

30. Sudoplatov et al., *Special Tasks*.

31. Sudoplatov did not suggest that these major scientists were Soviet agents, only that they had leaked information to those who were. The possibility that Fermi "may have spoken indiscreetly on some occasion to his old pupil and colleague Bruno Pontecorvo is far from implausible" (anonymous analysis of Sudoplatov claims, June 16, 2003, http://www.freerepublic.com/focus/fr/929408/replies ?c=191). This remark has some support from our Chapter 6, where Pontecorvo admits that data he had received from Fermi "had not been published, and cannot be published for a long time to come, because of their confidential character." Sudoplatov misidentified the Soviet agent MLAD as Pontecorvo, whereas this was in fact Ted Hall. Hall was not publicized as an atomic spy until the appearance of *Bombshell* in 1997. Sudoplatov's claims were made before 1994, and he may have included disinformation to protect Hall. An exchange of opinions between Jerrold and Leona Schecter (Sudoplatov's coauthors on *Special Tasks*) and Thomas

Powers on Sudoplatov's more extreme claims appeared in the *New York Review of Books* in September 1994 (http://www.nybooks.com/articles/archives/1994/sep/22/were-the-atom-scientists-spies-an-exchange/). This exchange references a denial by John Cairncross that he transmitted news of the Tube Alloys project to the Soviets; Cairncross attributes this leak to Donald Maclean.

32. Remarks by Gillo Pontecorvo as reported by Guido Pontecorvo to MI5, January 1951, TNA KV 2/1888.

33. Mafai, *Il lungo freddo*, p. 130. Niels Bohr was Danish; Marianne Pontecorvo was Swedish. *Nils* is the Swedish variant of *Niels*.

34. Before the project started, it was known that light elements, such as boron and nitrogen, absorb slow neutrons. Today this property is exploited in control rods, made of elements such as boron, which can prevent a nuclear fission reaction from getting out of control.

35. This was on January 6–7. Bruno Pontecorvo was also in the US on January 12 and in New York on January 20–24. TNA KV 2/1888.

36. Broda, *Scientist Spies*, p. 130.

37. TNA AB 2/643, AB 2/645, AB 2/647.

38. TNA KV 2/1888.

39. This is in part what Fermi had in mind, following Pontecorvo's defection, when he said that Pontecorvo's presence in the Soviet Union was far more important than any information that he might have passed before 1950.

40. Broda, *Scientist Spies*, p. 128.

41. Broda, *Scientist Spies*, p. 134.

42. One idea was to look for helium, as the presence of this gas is a sign of alpha particles—the nuclei of helium atoms.

43. The issue was not finally resolved until later that year, after the war had ended. This investigation by B. Pontecorvo and D. West is reported in TNA AB 2/318.

44. TNA AB 2/653.

45. Broda, *Scientist Spies*, p. 133 and TNA AB 2/653.

46. Lew Kowarski interview, AIP, http://www.aip.org/history/ohilist/4717_4.html.

47. Lew Kowarski interview, AIP, http://www.aip.org/history/ohilist/4717_4.html.

48. Kowarski recalls that the party began shortly before 4:00 p.m.

49. If Alan Nunn May had passed information to an enemy, he could have faced the death penalty for treason. However, the Soviet Union was an ally, and he was convicted of breaking the Official Secrets Act.

50. Letter from Geoff Hanna to Giuseppe Fidecaro, October 24, 1996. I am indebted to David Hanna for access to this letter and other correspondence linked to Chalk River.

CHAPTER 8

1. B. Pontecorvo, "Inverse Beta Process," Report PD-205, National Research Council of Canada, Division of Atomic Energy, Chalk River, November 13, 1946.

2. B. Pontecorvo, "On a Method for Detecting Free Neutrinos," Report PD-141, National Research Council of Canada, Division of Atomic Energy, Chalk River, May 21, 1945.

3. This discussion must have occurred when Otto Frisch visited Chalk River from Los Alamos; papers of Otto Frisch, TNA CSAC 87.5.82/A.64.

4. When it was declassified in 1964, it was lodged among papers at the National Archives, unnoticed. Most physicists, myself included, were either unaware of its existence, or assumed that it contained nothing more than the public 1946 paper. The received wisdom on the history of Bruno Pontecorvo and solar neutrinos has thus been based on the 1946 paper.

5. Pryce's calculation dealt with neutrinos that are produced by the fusion of protons in the sun. The energy of these neutrinos is too low to convert chlorine into argon. There are neutrinos with higher energies, produced by other processes in the sun, for which the chlorine method works. Their quantity, however, is relatively trifling. Ray Davis succeeded in finding these neutrinos, but only after several decades of refinements in his experiment. That story has been told elsewhere, and American theorist John Bahcall plays a leading role in it; see for example Close, *Neutrino*.

6. H. R. Crane, "The Energy and Momentum in Beta-Decay and the Search for the Neutrino," *Reviews of Modern Physics* 20 (1948): 278.

7. In 1951, Frederick Reines, who had worked on atomic explosions in the aftermath of the war, expressed interest in looking for neutrinos produced by an atomic blast, but in 1952 he decided to use a reactor instead. Fermi pointed out that a reactor had an advantage over an atomic blast—one could repeat the experiment. Reines and Cowan's first attempt to detect neutrinos at the Hanford nuclear reactor was unsuccessful, but in 1955 they began their successful attempt at the Savannah River reactor in South Carolina, around the same time that Davis was pursuing his own quest. Reines later won the Nobel Prize in Physics in 1995. Clyde Cowan, who died in 1974, missed out, as Nobel Prizes are not awarded posthumously.

8. Marcello Conversi, Ettore Pancini, and Oreste Piccioni announced their discovery in *Physical Review* 71 (1947): 209.

9. A single neutrino paired with an electron would be unable to balance the rotary angular momentum. A pair of photons would also be unlikely compared to a single photon.

10. B. Pontecorvo, "Nuclear Capture of Mesons and the Meson Decay," *Physical Review* 72, no. 246 (1947).

11. E. P. Hincks and B. Pontecorvo, "The Absorption of Charged Particles from the 2.2-μsec. Meson Decay," *Physical Review* 74 (1948): 697.

12. J. Steinberger, "On the Range of the Electrons in Meson Decay," *Physical Review* 74 (1948): 500.

13. The history of muon decay is as follows: Hincks and Pontecorvo had shown that muon decays don't make an electron and a heavy particle, but they didn't eliminate the possibility that it decays into an electron and a photon. Steinberger measured a spectrum that was consistent with decay into three particles.

By January 1949 Steinberger stated that he had "some evidence" that the decay produces three light particles (*Physical Review* 75: 1136). In March, R. Leighton, C. Anderson, and A. Seriff gave "strong evidence [that the muon] decayed to an electron and two neutrinos" (*Physical Review* 75: 1432). They showed that the spin of the muon can be the same as the electron. (Technically, they showed that it is "half-integer," like the electron.) They also measured the mass of the muon. Not until later that year did Hincks and Pontecorvo explicitly mention the decay into an "electron and two neutrinos" (*Physical Review* 77: 102).

14. This is widely credited to G. Puppi, *Nuovo Cimento* 5 (1948): 587. However, Puppi was not the first to come up with this hypothesis. Bruno Pontecorvo had elucidated such ideas in a letter on May 8, 1947, to Gian Carlo Wick (copy in Fondo Wick, Archivio Scuola Superiore Normale di Pisa). This includes the remarks *"se ne deduce una similarita tra processi beta e processi di assorbimento ed emissione di mesoni, che, assumendo non si tratti di una coincidenza, sembra di carattere fondamentale"* (A similarity can be deduced between beta processes and the absorption or emission of mesons [i.e. muons], which, assuming that it is not a coincidence, seems to be of fundamental character.") In *Physical Review* 72 (1947): 246, Pontecorvo proposes a "fundamental analogy between beta processes and the process of emission or absorption of charged mesons."

15. http://www.universitystory.gla.ac.uk

16. S. C. Curran, J. Angus, and A. L. Cockcroft, "Beta Decay of Tritium," *Nature* 162 (1948): 302–303. D. H. W. Kirkwood, B. Pontecorvo, and G. C. Hanna. "Fluctuations of Ionisation and Low Energy Beta Spectra," *Physical Review* 74 (1948): 497–498.

17. http://www.universitystory.gla.ac.uk. Curran, Pontecorvo, and their respective teams had developed the concept in order to measure the energy of beta particles. A proportional counter for alpha particles was invented in 1943, by John Simpson in Chicago. See https://www.orau.org/ptp/collection/proportional%20counters/simpson.htm

18. Giuseppe Fidecaro, "Bruno Pontecorvo: From Rome to Dubna," in BPSSW, http://www.df.unipi.it/~rossi/PONTE_5.pdf.

19. TNA KV 2/1887.

20. Henry Arnold memo of January 5, 1948, TNA KV 2/1887 8a.

21. TNA KV 2/1889.

22. Wallace, "Atomic Energy in Canada," p. 131.

CHAPTER 9

1. Unless otherwise stated, all correspondence is from TNA KV 2/1888 or TNA KV 2/1889, or Churchill College papers.

2. Turchetti, *The Pontecorvo Affair,* chap 2, note 47; Bruno Pontecorvo letter to John Cockcroft, February 2, 1946, CAC.

3. On December 29, 1945, Bruno told George Uhlenbeck at the University of Michigan that he would be writing to Segrè for advice, but felt it to be unlikely that he would come to Michigan, although "I am not yet decided." On January

5, 1946, Michigan expressed regret that he wouldn't be coming, and upped their offer to that of a full professorship at $5,500 a year. In response, on January 17, Bruno said that he had "not yet made a decision" but would do so immediately after a physics meeting in New York, due to take place at the end of the month.

4. Formal letter of employment sent June 5, 1946; Pontecorvo accepted on July 8, 1946. TNA KV 2/1892.

5. Confirmed in Cabinet Office telegram to British Joint Services Mission, Washington, October 21, 1950, and report to prime minister, TNA KV 2/1888, memo 25a.

6. Godfrey Stafford interview, December 20, 2012.

7. L. Fermi, *Atoms in the Family*, p. 254.

8. Bruno Pontecorvo, Autobiographical notes.

9. According to Laura Fermi, this was the "end of 1948" (*Atoms in the Family*, p. 255). The APS meeting in Chicago was held November 26–27 and Bruno was present. TNA KV 2/1888.

10. L. Fermi, *Atoms in the Family*, p. 255.

11. J. David Jackson e-mail, August 28, 2012.

12. TNA KV 2/1889.

13. Peter Watson e-mail, January 2, 2014.

14. Anonymous source, 2013.

15. Mafai, *Il lungo freddo*, p. 127.

16. The letters exchanged between Bruno and Marianne in 1938 refer more than once to some unspecified illness, and hint that Marianne changed travel plans at the last minute (see Chapter 3). I am grateful to Sven-Olof Ekman, a Swedish journalist from Marianne's hometown, for help with my research.

17. F. W. Marten memo, TNA KV 2/1888.

INTERLUDE

1. Holloway, *Stalin and the Bomb*, p. 90.

2. Kurchatov as quoted in Holloway, *Stalin and the Bomb*, p. 90.

3. Holloway, *Stalin and the Bomb*, p. 94. KGB stands for *Komityet Gosudarstvennoy Bezopasnosti* (Committee for State Security), and the agency of that name was established in 1954. The KGB succeeded the MGB, the Soviet Ministry of State Security, which was itself formed in 1946. The MGB was immediately preceded by the NKGB, in existence from 1943 to 1946, and before that by the NKVD, or *Narodnyi Kommissariat Vnutrennikh Del* (People's Ministry of Internal Affairs), which operated from 1934 to 1943. For a guide through this labyrinth, see Andrew and Gordievsky, *KGB*, p. 9. For ease of comprehension, I shall refer to all these entities in the main text as the KGB, to distinguish them from the second, independent intelligence agency of the Soviet Union—the GRU, or *Glavnoye Razvedyvatelnoye Upravleniye*. The GRU was founded in 1926 as the Chief Intelligence Directorate of the Red Army.

4. The security corps of the Manhattan Project suspected that several scientists were passing information to the Soviet embassy in San Francisco. Holloway, *Stalin and the Bomb*, p. 103.

5. Oleg Gordievsky and Christopher Andrew in various works; Oleg Gordievsky in communications, April 27 and April 30, 2012; Sudoplatov et al., *Special Tasks*.

6. Joint Committee on Atomic Energy, "Soviet Atomic Espionage," April 1951, http://archive.org/stream/sovietatomicespi1951unit/sovietatomicespi1951unit _djvu.txt.

7. See Sudoplatov et al., *Special Tasks*, p.182, and discussion in footnote on p. 81 of Rhodes, *Dark Sun*.

8. Andrew and Gordievsky, *KGB*, p. 318; Oleg Gordievsky e-mail, April 30, 2012.

9. Oleg Gordievsky e-mail, April 30, 2012.

10. References are either circular—Gordievsky being cited as the source of the claim in his own book—or anecdotal. There appears to be no clear answer as to how Gordievsky knows all this. See discussion of Gordievsky in Afterword for further commentary.

11. Holloway, *Stalin and the Bomb*, p. 99.

12. Holloway, *Stalin and the Bomb*, p. 104.

13. Fuchs's confession to Michael Perrin on March 2, 1950, notes that before August 1944 he had "no real knowledge of the pile process or of the significance of plutonium." This is reprinted in Appendix B of Williams, *Klaus Fuchs*.

14. Andrew and Mitrokhin, *The Mitrokhin Archive*, p. 174.

15. Boris Ioffe e-mails and Skype interviews, February 16, March 9, and July 19, 2011; Boris Ioffe interview with Giuseppe Mussardo, 2012. The source seems to be someone other than Nunn May, in that the blueprints were apparently transmitted after he had left Canada. There is also the evidence of Nunn May's deathbed confession. His stepson, Paul Broda, has described this in detail in *Scientist Spies*. In Broda's judgment, Nunn May's statements at that singular time were "very honest"; if Nunn May did not mention some significant action at that point, then "he didn't do it" (Paul Broda interview, August 12, 2012).

16. Andrew and Gordievsky, *KGB*, p. 318.

17. Comment from Skinner to Ronnie Reed, TNA KV 2/1888. Bruno Pontecorvo's stated reason for these periodic trips was that he needed to deal with his ongoing desire to retain US residency status. The dates in his passport confirm these visits occurred at intervals of roughly six months.

18. Gibbs, "British and American Counter-Intelligence."

19. As mentioned earlier, technically this was the Soviet NKVD ("People's Ministry of Internal Affairs"), which later transformed into the KGB ("Committee for State Security").

20. Pincher, *Treachery*, chap. 8. Hall's courier, Lona Cohen, worked with the KGB, controlled via the Soviet consulate in New York. Fuchs had worked with the GRU during his time in the UK, but in the US his control was transferred to the KGB. His courier during his time at Los Alamos was Harry Gold. Fuchs and Hall were completely independent. Hall had no knowledge that Fuchs was a spy—Ted's wife, Joan, confirmed that "Ted thought he was the only one" (Joan Hall interview, May 1, 2013). There was no direct testimony in Fuchs's trial, but all the evidence suggests that he was also ignorant of Hall's role.

21. Report of the Canadian Royal Commission on Espionage (the Kellock-Taschereau Commission). The full title of the report is "The report of the Royal Commission appointed under Order in Council PC 411 of February 5, 1946, to investigate the facts relating to and the circumstances surrounding the Communication by Public Officials and Other Persons in Positions of Trust, of Secret and Confidential Information to Agents of a Foreign Power." Access to the document may be traced through http://catalog.loc.gov/vwebv /holdingsInfo?&bibId=8343655&searchId=6517&recPointer=1&recCount=100.

22. A Nunn May confession, reported in Broda, *Scientist Spies*, p. 140; Paul Broda interview, August 12, 2012, and October 10, 2013.

23. Broda, *Scientist Spies*, p. 142.

24. During the previous months, Nunn May had become worried about the quality of the people he was dealing with. They offered him money, which he declined. He began to wonder if his information was actually being delivered to qualified scientists in Moscow. So he decided to assess the Soviet reaction by handing over samples of uranium. This would give Moscow proof that he was actually working at an atomic project, and the samples would have to be sent to proper experts to be evaluated. This is why he handed over the minute samples of U-233 and U-235, which he had originally intended to retain as trophies of his time on the project.

25. Joint Committee on Atomic Energy, "Soviet Atomic Espionage," April 1951; reprinted from report of the Canadian Royal Commission, June 27, 1946, which states, "These samples were considered so important by the Russians that upon their receipt, Motinov flew to Moscow with them" (pp. 447–457). The quoted texts in these paragraphs are accessible online at http://www.ebooksread.com /authors-eng/united-states-congress-joint-committee-on-atomic/soviet-atomic -espionage-tin.shtml on pages 9 and 10.

26. Pincher, *Treachery*, Chapter 2, Kindle edition location 421.

27. Andrew and Gordievsky, *KGB*, p. 317.

28. However, there is no public record to support this claim. It is clear that although MI5 later expressed strong suspicion of Pontecorvo's motives and freely encouraged authors to accuse him of having spied, it had no evidence that Pontecorvo ever transmitted any classified information to Soviet contacts. This is clear from the released files along with the frank diaries of Guy Liddell (TNA KV 2/1887–1891 and KV 4/472). Gordievsky's claims are doubly anecdotal, in that they are his memory of remarks that others had made to him.

29. Japan surrendered on August 15, 1945. The formal documents were signed on September 2.

30. Philby, *My Silent War*, Introduction: "I regarded my SIS [Secret Intelligence Service] appointments purely cover jobs, to be carried out sufficiently well to ensure my attaining positions in which my service to the Soviet Union would be most effective. My connection with SIS must be seen against my prior total commitment to the Soviet Union."

31. Gibbs, "British and American Counter-Intelligence," note 253 discusses at length Philby's role in the Nunn May case.

32. In the summer of 1949, Meredith Gardner, an American linguist, cracked the Soviet diplomatic codes. The resulting decrypts were known by the code name VENONA.

33. KGB is a modern name, which, as remarked in note 3, I shall use for convenience. Before 1943 its analogue was known as NKVD. KGB stands for *Komityet Gosudarstvennoy Bezopasnosti* (Committee for State Security). The second, independent intelligence agency of the Soviet Union was GRU, or *Glavnoye Razvedyvatelnoye Upravleniye*, founded in 1926 as the Chief Intelligence Directorate of the Red Army.

34. Broda, *Scientist Spies*, p. 146.

35. Philby's tip-off about Nunn May is discussed in Andrew, *Defence of the Realm*, p. 344.

36. Letter from Bruno Pontecorvo to GEC, February 1946, CAC.

37. Albright and Kunstel, *Bombshell*, p. 100.

38. The first mention of MLAD was picked up in codes in 1946. The Fuchs case took up much of the code-breakers' attention, and it was probably in the spring of 1950 that MLAD was identified as Hall. See Albright and Kunstel, *Bombshell*, chap. 22.

39. Andrew and Mitrokhin, *The Mitrokhin Archive*, p. 173.

40. Morris Cohen had become a Soviet agent during the Spanish Civil War, while serving in the International Brigades.

41. The story of these adventures has been told in Albright and Kunstel, *Bombshell*. Hall's information was at least as important as that of Fuchs, and possibly more so. There is also speculation that Stalin was wary of Fuchs, who was German, and hence confirmation via Hall proved important.

42. Albright and Kunstel, *Bombshell*, p. 332, note 135; "first chilly months" p. 133.

43. Albright and Kunstel, *Bombshell*, pp. 179–181.

44. Joseph Albright e-mail, September 17, 2012; Albright and Kunstel, *Bombshell*, p. 134.

45. Joan Hall interview, January 25, 2013; Albright and Kunstel, *Bombshell*, p. 191.

46. There is speculation that Hall may have been involved in some activity from 1948 to 1950 (Albright and Kunstel, *Bombshell*, pp. 189–192). The possibility that Hall met Lona Cohen again, in February 1950, is discussed in *Bombshell*, p. 221. The 1949 meeting between the Halls and the Cohens took place in New York's Central Park, according to Morris Cohen in *Bombshell*, p. 200. The possibility that there were two meetings between Ted Hall and one or more of the Cohens in 1949 is also mentioned there. This could be consistent with Joan Hall's recollection that the four of them met in a park in New York, but "not Central Park" (Joan Hall interviews, January 25, 2013, and May 1, 2013).

47. Interview reported in Albright and Kunstel, *Bombshell*, p. 134; Joseph Albright e-mails, September–December 2012.

48. Krasnikov family archive, 1993, quoted in Albright and Kunstel, *Bombshell*, p. 134; I confirmed this with Joseph Albright via e-mail communications, from

September to December 2012. There is some uncertainty about when Cohen's uranium odyssey occurred. One account (that of V. N. Karpov, as related in *Bombshell*, p. 134) places it in "late 1944 or 1945." If this is correct, then it is possible that Lona could have overlapped with Bruno Pontecorvo before the Gouzenko debacle. This scenario would imply that he too had obtained uranium from Chicago, as none was being produced in Canada until much later. However, the dates do not fit with other accounts, which state that Cohen was not reactivated until mid-January 1945. On balance it seems more likely that the uranium originated from Canada in 1948.

49. T. M. Samolis, ed., *Veterany vneshnei razvyedki Rossii: Kratkiye biograficheskiye spravochniki* [Veterans of Russian Foreign Intelligence: Short Biographical Summaries] (Moscow: Russian Foreign Intelligence Service, 1995), as reported in Albright and Kunstel, *Bombshell*, p. 134.

CHAPTER 10

1. The British security files on Pontecorvo show that the authorities were always nervous about his alien background, notwithstanding the fact that he had adopted British nationality. Klaus Fuchs was already under surveillance when, on October 19, 1949, Max Born of the University of Edinburgh wrote to him, encouraging him to send "one of your men" (Pontecorvo) to attend a physics conference in the city in November. Born wanted Bruno to speak about his research on cosmic rays. This was open research with no significance for national security, but MI5 nonetheless made a note in Bruno's security file because the invitation had come via Fuchs (19b in TNA KV 2/1887). It is ironic therefore that after Bruno and his family disappeared behind the Iron Curtain in 1950, the version of events put out for public consumption was that Pontecorvo had no relationship with Fuchs at all. *Additional note to readers:* All quotes in this chapter come from file KV 2/1887 in the National Archives unless stated otherwise.

2. TNA KV 4/282.

3. TNA KV 2/1888.

4. Segrè was not in Paris during Pontecorvo's residency, nor was he on the ship *Quanza*, as he was based at Berkeley in 1940. He was based in Palermo in 1937, and visited Copenhagen and Germany, but there is no record that he visited Paris or met Pontecorvo anywhere. He immigrated to the United States in June 1938. See Segrè, *A Mind Always in Motion.*

5. TNA KV 2/1888.

6. AN FBI report recorded by MI5 in TNA KV 2/1888, memo 97a: "Bruno Pontecorvo's sister Giuliana lived at 1839 Wallace Ave Bronx. At 1845 Giuseppe Berti lived—alleged comintern agent and Italian Communist leader."

7. Klaus Fuchs files, TNA KV 2/1248–1250.

8. Copy of invitation in Bruno Pontecorvo files, CAC.

9. TNA KV 2/1889 and AB 6/635; Turchetti, *The Pontecorvo Affair*, p. 108.

10. This news may be what stimulated Harwell to send all employees a questionnaire about their families. The advertised reason was the arrest of Fuchs. In

turn, this enabled Arnold to talk with Pontecorvo without raising any suspicion. Michael Goodman interview, October 7, 2013.

11. Godfrey Stafford interview, December 20, 2012.

12. Lorna Arnold interview, January 4, 2013.

13. TNA KV 2/1887, memo 20 and 20a.

14. TNA KV 2/1887, memo 21a.

15. Other documents point to the source as being an MI6 representative in the British embassy in Stockholm.

16. TNA KV 2/1887, memo 26a.

17. This suggests either some misunderstanding by Arnold, relating to the events of 1946, when Bruno turned down offers from the US in favour of Harwell, or a clever piece of disinformation by Pontecorvo, as there is no record among the extensive Pontecorvo papers of any such opportunity arising in 1950. It seems unlikely that this was a misunderstanding on Arnold's part, first because of his experience as an interrogator, and second because he has carefully said that Bruno "already toyed" with Rome, and is "at present" considering America. It is unclear what advantage Bruno saw in making such a claim, unless it was to emphasize that he was much sought after.

18. TNA KV 2/1887.

19. TNA KV 2/1887, minute sheet note 36 (see Image 10.1). In other words, they suspected that he had passed information, but had insufficient proof to arrest him.

20. Michael Goodman interview, October 7, 2013.

21. Weart, *Scientists in Power*, p. 259. The CIA claim comes from "National Intelligence Survey: France," CIA archives, 1952, pp. 73–73, 73–79; US Department of State, *Foreign Relations of the United States: 1949*, Volume I, pp. 466, 488, 626, cited in Weart, *Scientists in Power*, p. 328, note 4. Joliot-Curie's speech was reported by Jacques Fauvet in *Le Monde*, April 4–5, 1950, and *L'Humanité*, April 6, cited in Weart, *Scientists in Power*, p. 328, note 6.

22. Weart, *Scientists in Power*, p. 261.

23. TNA KV 2/1887.

24. TNA KV 2/1887.

25. Chapman Pincher interview, November 13, 2013. Some—e.g. Chapman Pincher in *Treachery*—have argued that Hollis was in reality a KGB double agent, or at the very least complacent and inept. Although I offer no opinion on the more extreme version of this claim, the lack of action on the Pontecorvo file during the summer of 1950 seems in accord with such criticisms.

26. TNA KV 2/1887.

CHAPTER 11

1. Laura Arnold interview, March 18, 2013.

2. Gil Pontecorvo interview, June 12, 2011.

3. The Abingdonian archive; Sarah Wearne interview, October 7, 2011.

4. Abingdon record: "After taking his degree in summer 1949 Mr J F H Barker BA stayed up at Oxford for an extra term"

5. Today this grand house is the head teacher's residence.

6. Section A, or A Branch, was involved with bugging, phone tapping, covert entry and specialized secret photography. See http://www.powerbase.info/index.php /MI5_A_Branch.

7. He was made a Companion of the Bath—a British military order of chivalry.

8. One student said that a career in MI5 planting bugs was appropriate for Barker "because he was a nasty man." Another said he was the best teacher in the school.

9. Macintyre, *Agent ZigZag*, p. 70.

10. Interview with anonymous source, August 20, 2013.

11. Pincher, *Treachery*, chap. 54 and Chapman Pincher interview, November 13, 2013. Chapman Pincher commented that in those times MI5 had a secret informant in every Fleet Street office. Another person confirmed that MI5 came to them for background, and so it's possible that "someone could think that I was MI5." Having someone in place around Harwell would be consistent practice.

12. Guy Liddell diary, July 10, 1946, TNA KV 4/467.

13. The memories of former pupils vary. When Barker left the school, one account states that they were told he was "to return to the Navy"; another that he was "to go to intelligence."

14. Anthony Gardner interview, February 25, 2013.

15. Bruno Pontecorvo, Autobiographical notes.

16. Anthony Gardner interview, February 25, 2013; Paul Gardner interview, March 4, 2013.

17. David Lees interview, October 15, 2013.

18. Anthony Gardner interview, February 25, 2013.

19. TNA KV 2/1888, memo 117.

20. TNA KV 2/1888. The fiancée was named Jean Archer. The engagement was later terminated.

21. TNA KV 2/1888, memo 118b.

22. TNA KV 2/1888.

23. TNA KV 2/1888.

24. TNA KV 2/1887, memo 42a.

25. Abingdon School archives. Sarah Wearne interview, October 7, 2011.

26. The fees for the term itself totalled fourteen pounds. The remainder consisted of an outstanding four shillings and six pence for a photograph, and one penny owed to the school snack bar.

27. TNA KV 2/1888.

28. TNA KV 2/1887, note 70d.

29. His passport subsequently showed that August 6–8 was spent in Austria.

30. Letter from parents to Guido; translation by Guido in TNA KV 2/1889. The original Italian version was returned to the parents.

31. TNA KV 2/1888. Bruno's parents arrived in Chamonix on August 24 and started a search when they failed to find him there.

32. TNA KV 2/1889.

33. Transcriptions of telegrams, translated by MI5, TNA KV 2/1888.

34. Caledonian Insurance Company, policy 4/49BM279, for travel in "France Italy and Switzerland for the period 25 July to 24 August." TNA KV 2/1888.

35. Even today, with the Mont Blanc Tunnel, it takes nearly eight hours.

36. Date and time confirmed in TNA KV 2/1888, file 72. These are transcripts, the English translations having been made by Guido for MI5. I have not altered the English. The originals were returned to Bruno's parents in 1951. The MI5 file notes that the letters and telegrams were "Bruno's last communication to them and they would like to retain them for sentimental reasons."

37. TNA KV 2/1889, note 161A.

38. TNA KV 2/1888.

39. TNA KV 2/1888 and Anna Pontecorvo interview, November 11, 2011.

40. Both quotes are as recounted by Guido to MI5, TNA KV 2/1888.

41. Reed's report notes this in TNA KV 2/1888.

42. Years later Bruno claimed to Ugo Amaldi that he feared that a world war was about to erupt, and that he fled to the USSR as he "didn't want to be on the wrong side" (Ugo Amaldi interviews, April 18, 2013, and September 12, 2013). This seems fanciful, given Bruno's sudden decision to flee and the other details described here.

43. Historian Simone Turchetti has made a compelling case that, after months of pressure, the patent dispute was a tipping point for Bruno. (See Turchetti, *The Pontecorvo Affair*.) While it seems certain that this played a role, people who knew Bruno are skeptical that the patent dispute alone could have been responsible. Whereas his colleague Emilio Segrè was famously concerned about the financial value of the patent, Bruno himself regarded it as a bonus but not in any obsessive way. Giuseppe Fidecaro interview, April 17, 2013, and Ugo Amaldi interview, April 18, 2013.

44. There are some uncorroborated claims that Sereni then went to Moscow for about two months, immediately after Bruno's flight. On November 10, Sereni landed at Le Bourget Airport in Paris, having arrived from Prague, but his passport showed that he'd stopped in Prague in transit from Moscow. There is no direct link to Bruno's activities, but MI5's analysis points to suspicion that in mid-August Moscow alerted Sereni to the need for contact with Bruno, and also that later in the year Sereni may have been a contact in the USSR once Bruno arrived there. See TNA KV 2/1889. Sereni's diaries, however, don't support this (Simone Turchetti interview, January 29, 2013). TNA KV 2/1889, memo 289A cites an unnamed source who came over to the West from the USSR in 1952. However, this source also claimed that Pontecorvo had gone to the USSR via Austria. We now know that this is incorrect, so some of the reports about Sereni's movements in TNA may also be suspect.

45. Simone Turchetti interview, January 29, 2013.

46. Evidence for this will appear in Chapter 13.

47. On March 24, 1953, Reed made his final assessment of the evidence and concluded, "I now think it may have been Emilio Sereni who persuaded Pontecorvo to defect." TNA KV 2/1891.

48. The MI5 files leave many questions open. In a letter to his parents Bruno mentioned that Antonio was in Ladispoli, which has been interpreted by MI5

and others to suggest that Antonio remained there alone. However, an innocent explanation of how the children got to Ladispoli is that all three were there with Giuliana's children all along. The accounts in TNA, as told to MI5, do not paint a clear picture. There is also confusion about Gil, who in the MI5 files appears to be with his parents in Circeo, yet when Bruno later writes to his parents he says that Gil is in Ladispoli. Gil informed me that he had no memory of Circeo and, moreover, was sure that he had never been there. This has no great significance for this biography but may have relevance for anyone specifically interested in analyzing Bruno's behaviour and movements during this period.

49. Ronnie Reed summary, TNA KV 2/1888.

50. Anna had no recollection of seeing any car damage, but after so long could not regard this as significant. She said that the children were ferried to and fro between Rome and Ladispoli by coach (Anna Pontecorvo interview, March 21, 2012). Gil also has no memory of ever visiting Circeo. So it is possible that Bruno and Marianne were in Circeo alone, while the children were with Giuliana in Rome or Ladispoli.

51. Anna Pontecorvo interview by MI5, TNA KV 2/1888.

52. TNA KV 2/1888–1891.

53. Anna Pontecorvo interview by MI5, TNA KV 2/1888.

54. Bruno Pontecorvo told Miriam Mafai forty years later that his car went into the garage on August 29 (Mafai, *Il lungo freddo*, p. 11). Bruno would know that the car only went into the garage immediately prior to his departure from Rome, so this date seems reliable. It also agrees with the garage owner's statements made to Italian police, who interviewed him on behalf of MI5. Mafai's e-mail on March 7, 2012, confirmed that Gillo and his sister Giuliana told her "the same story."

55. Anna Pontecorvo interview, November 11, 2011, and phone conversations 2011–2012.

56. Gil Pontecorvo suspected this and later confronted her. Gil Pontecorvo interview, November 22, 2011.

57. Mafai, *Il lungo freddo*, p. 11; Miriam Mafai e-mail March 7, 2012.

58. August 29 cable from SAS in Rome to Munich, TNA KV 2/1888. Thus the date of August 30, quoted in some accounts, based on a *Sunday Express* story published at the time, is wrong.

59. TNA KV 2/1888, memo 105a.

60. TNA KV 2/1888, memo 119c.

61. Marianne had visited her family in 1938, and had not done so again until 1949. Bruno did not accompany her on the latter occasion, and "had difficulty getting her back" (Herbert Skinner quote in TNA KV 2/1888, memo 121c). Letters from Marianne's mother to Marianne can be found in TNA KV 2/1888 et seq. and CAC. Letters from Bruno to Marianne are in CAC. Marianne's letters to Bruno are not publicly available.

62. Letter from Marianne's mother to Marianne, July 1, 1950, TNA KV 2/1889.

63. TNA KV 2/1888.

64. Bruno was by now a naturalized British citizen, as were his sons. Marianne was still Swedish.

65. TNA KV 2/1888.

66. Sent August 31, 1950, at 3:15 p.m., TNA KV 2/1889.

67. SAS confirmed to MI5 that Bruno Pontecorvo's ticket booking was made in one office and that of Marianne and the boys was made in another. There is no record of when Bruno's ticket was issued or of which specific office was involved. The passenger manifest confirms the different number sequences. TNA KV 2/1888.

68. On October 28, the *Daily Herald* claimed that Bruno was afraid to return to the UK after a mystery encounter at Lake Como in August with two unknown men, one Italian and the other Czech. He was supposedly overheard to have said to Marianne, "I dare not go back. I should be sent to prison if I did." There is no source given for this story.

69. Chapman Pincher notes that a KGB source confirmed that this pattern was used in the defection of Burgess and Maclean, in 1951. Two KGB officers met them as soon as they were out of British territory. Chapman Pincher interview, November 13, 2013.

70. At 8:50 p.m., according to MI5 records, TNA KV 2/1888.

71. According to Henrico Attavilla, the Stockholm correspondent of *Il Tempo*, MI5 files, TNA KV 2/1888.

72. Gil Pontecorvo interview, November 22, 2011.

73. MI5 sources discovered that, upon landing in Stockholm, Bruno declared the money in his possession to total 1,369 Italian lira, 685 Swedish kroner, and 199 Danish kroner, the equivalent of about 18 pounds. However, by the time he arrived in Finland, he had 1,360 lira, 100 Swedish kroner, 80 Danish kroner, and in addition had accumulated 436 US dollars (TNA KV 2/1888). Thus he would appear to have spent kroner to pay for food and accommodations, and acquired a gift of both airline tickets and US dollars. Possible explanations emerged later, when MI5 discovered that Emilio Sereni left for the USSR a few days before Bruno's departure and remained there for a month. MI5's source claimed that Sereni went to the USSR to introduce Bruno to Soviet officials (TNA KV 2/1888). MI5 did not obtain this information until three years later, and there are some reasons to doubt its accuracy, but it was the best information available at the time. MI5 commented in 1953, "This report is interesting as it fits the theory that [Ronnie Reed] considered the most likely" when he first reviewed the case in 1951. Reed's judgment was that the good offices of the PCI (Italian Communist Party) or the Soviet embassy had provided the money for Bruno to go to Stockholm. "The further suggestion was that he there met a Russian representative who persuaded him to go to Finland and subsequently to Russia." By 1953, the information in MI5's hands led Reed to conclude, "I now think it may have been Emilio Sereni who persuaded Pontecorvo to defect" (Reed review, TNA KV 2/1888). This also resonates with the suspicions of Anna Pontecorvo (Anna Pontecorvo interview, November 11, 2011).

74. Finnish customs official report to police, TNA KV 2/1888.

75. TNA KV 2/1888.

76. FBI File WF 65-5650.

CHAPTER 12

1. TNA KV 2/1888.

2. TNA KV 2/1888.

3. TNA KV 2/1888. This is not a direct quote of Bruno's mother's letter, which was written in Italian, but a transcription made by MI5 of Guido's English translation.

4. TNA KV 2/1888, memo 118b.

5. TNA KV 2/1889.

6. Anthony Gardner interview, February 25, 2013.

7. John Candy as told to Christine Wootton and Dunmore School around 1985, Christine Wootton e-mail January 13, 2014.

8. Joe Hatton interview, August 30, 2011.

9. Sir John Cockcroft diary, CAC.

10. Caldirola's card was found by MI5 and transcribed: TNA KV 2/1888.

11. Mark Bretscher interview January 24, 2013.

12. The record states: "Gil Pontecorvo—fees paid on *20 August 1949* for the Michaelmas term, which began on 16 September 1949. Pontecorvo—fees paid *4 January 1950* for the Lent term, which began 17 January 1950. Pontecorvo—fees paid *20 April 1950* for the Summer term, which began 28 April 1950." This completed Gil's first year at Abingdon School (Roysse's). I am indebted to Sarah Wearne for this research.

13. Abingdon School (Roysse's) records, Abingdon School archives, c/o Sarah Wearne, viewed October 7, 2011.

14. Received at Harwell on Monday, September 4, 1950.

15. TNA FO 372/84837.

16. TNA KV 2/1887, memo 40a.

17. Edoardo Amaldi as recounted to Donald Perkins. Donald Perkins interview, September 16, 2011.

18. Amaldi's name went on file, however. In 1953 he arrived in the UK, to see Patrick Blackett and Lord Cherwell to encourage the UK's interest in CERN. At London Airport his name was "checked against a list." His bags were searched, and he was asked if he had heard from Pontecorvo. Donald Perkins interview, September 16, 2011.

19. TNA KV 2/1887, memo 42a.

20. TNA KV 2/1887, memo 48a.

21. TNA KV 2/1889, memo 169b.

22. TNA KV 2/1889, memos 188/190.

23. TNA KV 2/1889, memo 181a.

24. Saturday cinema was a feature of the 1950s, and this is the weekend when the story broke.

25. Paul Gardner interview, March 4, 2013.

26. Anthony Gardner interview, February 25, 2013.

27. Anthony Gardner interview, February 25, 2013; Paul Gardner interview, March 4, 2013; David Lees e-mail, October 7, 2013; David Lees interview, October 15, 2013.

28. Newspaper headlines of October 27, 1950, from assorted online sources.

29. Newspaper quotes are from *Sydney Morning Herald*, October 22 and 25, 1950; *Palm Beach Post*, October 24, 1950; Melbourne *Age*, October 27, 1950.

30. *Sydney Morning Herald*, October 25, 1950.

31. Anthony Gardner interview, February 25, 2013.

32. Ludo Pontecorvo interview, September 12, 2013.

33. Ugo Amaldi interview, September 12, 2013.

34. Gil Pontecorvo interview, November 22, 2011.

35. TNA KV 2/1889, memo158a; TNA KV 2/1888, memo 121c.

36. In her letter of October 26, Giuliana tells Guido that this was "about 15 days ago." TNA KV 2/1888, memo 120a.

37. Giuliana's involvement seems assured, for her account raises further questions: How did the Pontecorvos get from her house to the airport if their car was in a garage? The "return to England in stages" and the request to pay the garage expenses are mutually inconsistent, unless Giuliana already knew they had left by plane. Furthermore, the idea that she received this request in a letter, sent from Rome in mid-September, some time after Bruno's departure, is fanciful. On the other hand, there could be an innocent explanation of the confused dates, or at least a partial one. If she consulted a calendar during the interview, she could have identified the wrong week: the date of August 31, given for the car repair, could be a mistake for August 24, and the dates of the Pontecorvos' departure, given as September 5 or 6, could be a mistake for the actual date of September 1.

38. TNA KV 2/1889.

39. TNA KV 2/1888.

40. TNA KV 2/1888.

41. Guy Liddell diary, October 21, 1950, TNA KV 4/472.

42. TNA FO371/84837.

43. Telegram sent October 24, 1950, TNA KV 2/1888, memo 71a.

44. TNA KV 2/1889, memo 171a.

45. TNA FO371/84837.

46. TNA KV 2/1887, memo 62a.

47. Guy Liddell diary, October 23, 1950, TNA KV 4/472. Liddell's brief is given in KV 2/1887, memo 62a.

48. Pincher, *Treachery*, Kindle edition location 6298; Chapman Pincher interview, November 14, 2013.

49. TNA KV 2/1887, memo 29A.

50. TNA KV 2/1888 and quoted in Pincher, *Treachery*, Kindle edition location 6298. Chapman Pincher interview, November 14, 2013.

51. The title of the debate is recorded in TNA reports. The admission that Pontecorvo might have "atomic secrets of value to the enemy" contradicts other official statements about Pontecorvo's significance at the time. Indeed, the British administration, having been given false information by MI5, downplayed Pontecorvo's significance. Their position was that he had not been involved in secret work, at least none that would have much interest to an enemy. After the revelation that Fuchs and Nunn May were actually Soviet agents, the British government was nervous that the revelation of a third spy would harm relations with the

United States. At the time, Britain was desperate to have full access to American atomic know-how.

52. Laura Arnold interview, August 30, 2013. Perrin was at this stage the deputy controller of atomic energy at the Ministry of Supply. It was to Perrin that Fuchs made his initial confession.

53. TNA FO371/84837.

54. TNA KV 2/1889, memo 121B.

55. Statements were made in the House of Commons on October 23 and November 6, 1950. On November 19, 1951, member of parliament Frederick Erroll asked the new Minister of Supply, Duncan Sandys, questions about atomic scientists. Sandys confirmed that Pontecorvo was the only atomic scientist to have disappeared, to which Erroll asked, "Will the Minister make sure no other disappearances take place?" *Hansard* 494, no. 18 (November 26, 1951).

56. TNA FO371/84837.

57. Guy Liddell diary, October 15, 1951, TNA KV 4/473.

58. TNA KV 2/1888, memo 121d.

59. TNA KV 2/1889.

60. TNA KV 2/1890, note 279a.

61. TNA KV 2/1890, note 283a.

62. Barbara Pontecorvo e-mail, August 19, 2011.

63. TNA KV 4/242, note 26a, and KV 2/1887, note 65a. On October 23, 1950, Dick White of MI5, in a phone call to Geoffrey Patterson of the British embassy in Washington, said that he had "learned that morning that Paul [was] employed on research work connected with radar and that this should be reported to the FBI."

64. Jack Steinberger interview, Rome centenary meeting, September 2013.

65. Giannini's lawsuit sought a total claim of $10 million, for both past research uses during the Manhattan Project ($7.5 million) and the anticipated future production of fissile materials, including plutonium ($2.5 million). Simone Turchetti e-mail, August 17, 2014.

66. *L'Unità*, August 23, 1950; quoted in Turchetti, *The Pontecorvo Affair*, p. 112.

67. *Calgary Herald*, October 24, 1950.

68. *News Chronicle*, November 3, 1953, stored at TNA KV 2/1888, memo 303a. In the 1980s a lawsuit in the United States recovered some of the money. However, much of it was taken up in legal fees. Gil Pontecorvo interview, September 11, 2013.

CHAPTER 13

1. Peter Hennessy letter to author, December 4, 2012.

2. It is the information in those documents that formed the narrative in Chapter 10.

3. Chapman Pincher confirmed that he had not seen it during his own research, which spanned several decades. Chapman Pincher interview, November 13, 2013.

4. Chapman Pincher interview, November 13, 2013.

5. Andrew and Mitrokhin, *The Mitrokhin Archive,* p. 203: "For almost a year [after his arrival in late 1949] Philby's sole contact with the Centre was via messages sent to Burgess in London."

6. He came under suspicion and resigned from MI6 in 1951. However, colleagues in MI6 refused to believe that he could have been a double agent, and it was not until 1963 that Philby's duplicity was made obvious to all, following his defection to the USSR. Philby travelled to the USSR in 1963 from Beirut, where for several years he had been a newspaper correspondent. Although his role as a traitor was suspected by MI5 in 1951, at which time he was "amicably eliminated" (removed from the service without prosecution), his colleagues in MI6 continued to believe he was an innocent victim until 1963. This shows that the opinions of colleagues are not always reliable.

7. Philby, *My Silent War,* Introduction.

8. Gibbs, "British and American Counter-Intelligence," note 253.

9. MLAD was Ted Hall, and QUANTUM (or KVANT) was identified in 2009 as Boris Podolsky, a Russian-born American physicist. See Haynes et al., *The Rise and Fall of the KGB in America.* Some authors have misidentified one or the other of these as Bruno Pontecorvo. QUANTUM was an appropriate name for Podolsky, who is famous in quantum physics for his part in the EPR (Einstein-Podolsky-Rosen) paradox.

10. Andrew and Mitrokhin, *The Mitrokhin Archive,* p. 204.

11. Andrew and Mitrokhin, *The Mitrokhin Archive,* p. 194.

12. Bruno Pontecorvo, Autobiographical notes; remarks made by Bruno Pontecorvo to Edoardo Amaldi, recounted in Ugo Amaldi interview, September 12, 2013.

13. Patterson's identification as security liaison officer, is taken from Andrew, *Defence of the Realm,* p. 377.

14. M. Marten, *Tim Marten,* p. 129; Michael Marten interview and access to audiotape of Tim Marten, July 3, 2013.

15. TNA KV 2/1887, memo 39a.

16. Chapman Pincher has alleged that Philby removed critical evidence showing that Pontecorvo was a spy, and then "told his Soviet controller [which] may well have led the Soviets to warn Pontecorvo of his predicament" (Pincher, *Too Secret Too Long,* p. 152; discussed in Turchetti, *The Pontecorvo Affair,* p. 200). Pincher provided no sources, evidence, or even clear identification of exactly when Philby took these actions. He told me that the account represented his judgment based on what he had unearthed. When he wrote that book, he had not seen the letter of July 1950, which I showed him on November 13, 2013. He described it as a "very plausible" factor in Pontecorvo's sudden flight.

In order to evaluate Philby's role, which is probably pivotal in separating Pontecorvo's life into its two halves, a brief review of these events is worthwhile (see also Turchetti, *The Pontecorvo Affair,* p. 200 et seq.). First, in line with Turchetti, we can dismiss the idea that Philby was involved in this mini-saga of 1943, when news of the FBI's discovery of communist literature in Pontecorvo's Tulsa home was first transmitted to the British authorities in New York. At that time, Philby was employed at the Special Operations Executive in Beaulieu, England, and was in no position to remove letters in North America. Turchetti also dismisses the possibility that Philby "sat on" the 1949 FBI report that recorded Segrè and Thornton's information about Pontecorvo; as we have seen, this report was in the hands of MI5 by December 1949. However, now that the letter of July 13, 1950, has come to light, a new interpretation of Philby's role becomes possible, especially if one imagines things from Philby's perspective.

The allegations that Bruno "engaged" in communist activities are sufficiently general that they could cover anything from reading the *Daily Worker* to full-blown spying, including the murky middle ground of being a member of the Communist Party. When he read the letter, Philby had no means of knowing the reason for the FBI's interest, or the level of Pontecorvo's activities. His subsequent actions can therefore be seen as either ironic or fortuitous, depending on whether Bruno was innocent or guilty of spying. In any event, the result was that Pontecorvo fled to the USSR, which was without doubt to the Soviet Union's great advantage.

17. Today there are copies of two letters from the FBI in the British archives, dated February 2 and February 19, though whether these represent the full extent of the correspondence is unclear. Also murky is whether there is indeed a further letter from the FBI, or whether the letter of February 10 is a reply sent by the British. The public evidence is rather bland, at least to modern eyes: in 1943 the FBI reported finding communist literature in Pontecorvo's house, something that would hardly merit such a rapid-fire exchange of letters at the time.

If that was the sum total of the FBI's evidence, Bruno Pontecorvo had little to fear, although in the McCarthyite frenzy of 1950 there would have been some turbulence. Conspiracy theorists might argue that Philby found these letters in July 1950, destroyed them, informed Patterson that he couldn't find any trace, and then "alerted Soviet agents," in line with Pincher's aforementioned claim.

This seems unlikely for the following reason: For Philby to have taken such action, the content of the letters would have to have been exceedingly serious, whereas copies of letters from the FBI to BSC are, as we have seen, rather bland. Philby had no wish to draw unnecessary attention to himself. So what content could have led Philby to destroy the letters? It is hard to sustain a plausible case. If the FBI had evidence that Pontecorvo was a spy, it is unlikely that this would have arisen in February 1943, when he had yet to begin serious work in Canada; it is even less likely that such damning evidence would have remained dormant until 1950, and then merely inspire a vague request to the British regarding possible "Communist activities."

Although Philby has been credited with almost magical powers of duplicity and stealth, this sequence seems far-fetched. The thesis that Philby would put himself at such serious risk is barely credible. He would have realized that the FBI was hardly likely to let such evidence fade away, and that they would doubtless come back to him with severe questions. Due to VENONA, UK intelligence was already aware that a network of double agents existed within its own organizations, including the Washington embassy itself. Fully aware of this, Philby had no wish to risk demonstrating that he himself held one of the starring roles in the VENONA decrypts.

18. In TNA KV 2/1888, file 16a, G. R. Mitchell notes that Cimpermann, the American FBI liaison in London, delivered archival copies of these letters to MI5 on October 23, 1950. File 103a records their receipt by the agency on October 28, as well as Ronnie Reed's confirmation that he personally received them on October 30. Memo 23a states that these letters are "the only material on PONTECORVO in his [FBI] files," which supports the thesis that, in 1950, there was no evidence in either the UK or US that Pontecorvo had broken the Official Secrets Act.

19. TNA KV 2/1888.

20. Because the only evidence against Hall was from VENONA, and the authorities were reluctant to reveal their success in breaking the Soviet code, the only way they could prosecute Hall was if he confessed. Hall declined to do so, was never prosecuted, and remained unknown until the final years of his life, when the VENONA decrypts were made public. Fuchs, as we have seen, confessed when confronted by the security authorities in England, and went to jail for fourteen years. Some have suggested that QUANTUM was Bruno Pontecorvo, but this is fanciful: in 1943, QUANTUM passed information about the gaseous diffusion process, under development at Oak Ridge, Tennessee. Pontecorvo was never at Oak Ridge, and had no expertise in gaseous diffusion. His specialties were neutrons and nuclear reactors. The information on QUANTUM's activities is consistent with the biography of Boris Podolsky. See also Andrew, *Defence of the Realm*, p. 375, note 58.

21. TNA KV 4/472.

22. Anthony Gardner interview, February 25, 2013.

23. The MI5 suspicion that these could be the mysterious Messrs Wittka and Allegrini, who accompanied Bruno from Rome to Stockholm, might have substance, however.

24. Gil Pontecorvo interview, September 4, 2013.

25. Boris Ioffe described this interview in the Kremlin about physics as an "interrogation" (Boris Ioffe e-mail, August 5, 2011, and video interview of Ioffe by Giuseppe Mussardo, 2012. I am grateful to Giuseppe Mussardo for a copy of this interview). Another former Soviet scientist, who wished to remain anonymous, told me that soon after 1955, when Pontecorvo's presence became known, a well-connected source (who in the opinion of my informant was extremely reliable) mentioned that Pontecorvo had also been interviewed by the KGB. This is consistent with the claims of former KGB agent F. D. Popov, who goes further and states that Pontecorvo "was regarded as so important that he was interviewed by Beria" (quoted in Pincher, *Treachery*, chap. 48; confirmed in Chapman Pincher interview, November 14, 2013).

26. F. D. Popov, quoted in Pincher, *Treachery*, chap. 48.

27. Guy Liddell diary, September 12, 1950, TNA KV 4/472.

28. Lorna Arnold interview, 2013.

CHAPTER 14

1. Gil Pontecorvo interviews, February 24, 2011 and September 4, 2013; Bruno Pontecorvo in Mafai, *Il lungo freddo*, p. 159.

2. Gil Pontecorvo interviews, February 24, 2011, and October 12, 2013.

3. I have debriefed Gil Pontecorvo about this on several occasions. He is sure the route was not via Porkkala, as the border was unremarkable: "There were no major military installations for example" of the kind that one would have expected at the Porkkala base. At most, he recalls passing an unremarkable sentry post in the middle of the woods, after which Bruno emerged from the trunk. A direct journey to the USSR would have taken the family east, whereas Porkkala is to the west. Gil again: "I did not check the direction of the sun!" Although Porkkala cannot be ruled out, it seems most likely that the Pontecorvos took a direct eastward road journey to the USSR. Gil confirmed that stories about ships or submarines are nonsense: "I have never been on a submarine!"

4. Mafai, *Il lungo freddo*, p. 159.

5. Background in this section is drawn from Mafai, *Il lungo freddo*; interviews with Miriam Mafai in 2012; and interviews with Gil Pontecorvo throughout the project.

6. Gil Pontecorvo interview, February 24, 2011.

7. Bruno Pontecorvo, Autobiographical notes.

8. Mafai, *Il lungo freddo*, p. 163.

9. It was a rare privilege for a Soviet citizen to meet a foreigner at that time. Even two decades later, in my experience, it was a natural reaction for sophisticated Soviets in that situation to be interested in how events were perceived in the West.

10. Gil Pontecorvo interview, September 4, 2013.

11. Mafai, *Il lungo freddo*, p. 164.

12. Gil Pontecorvo interview, January 20, 2012.

13. This is the date as Gil recalled it, sixty years later. Actually September 16 is the date he started school in 1949. In 1950 the term started on September 19. This illustrates both the detailed level of Gil's memories, as well as the subtle tricks the mind plays over time.

14. Gil Pontecorvo interview, August 25, 2011.

15. Sereni's diary shows that he went back and forth between the USSR and the West during this period. This was because he was a major party member, whom the Soviets saw as key to spreading communist influence in the world. It is thus possible that Bruno did not fully realize that he would be cutting himself off so thoroughly; for him, the move to the USSR could have been merely another stop on his life's journey. Simone Turchetti interview, October 5, 2011.

16. The early years of Laboratory Number Three are described in Boris Ioffe, "The First Dozen Years of the History of ITEP Theoretical Physics Laboratory," *European Physical Journal H* 38, no. 1 (January 2013): 83–135.

17. Ioffe, "A Top Secret Assignment," p. 31; Boris Ioffe e-mail, August 5, 2011; Giuseppe Mussardo interview, December 12, 2012.

18. The eventual success of the Soviet H-bomb occurred because Andrei Sakharov found a way (known as the "layer cake") to make a bomb using a minimal amount of tritium, and Vitaly Ginzburg had the insight that tritium could be made within the bomb itself by bombarding lithium deuteride with neutrons.

19. Ioffe, "A Top Secret Assignment," p. 31; Boris Ioffe e-mail, August 5, 2011; Giuseppe Mussardo interview, December 12, 2012.

20. Laboratory Number Three was officially established in December 1945, but during 1946 "no real work" occurred. Serious work happened only from 1947 onward (Boris Ioffe e-mail, August 5, 2011). Nunn May had been exposed in 1945. There is no mention of any blueprints in his deathbed statement (Broda, *Scientist Spies*, and Paul Broda e-mail, October 10, 2013). Only the final stage of the blueprints' journey is known. Yakov Terletsky, a physics professor at Moscow State University, was a part-time employee of the KGB. His role was to filter all material on atomic projects coming from abroad, and ensure that the information reached the relevant teams in the USSR.

21. Ioffe, "A Top Secret Assignment," p. 31; Boris Ioffe e-mail, August 5, 2011; Giuseppe Mussardo interview, December 12, 2012.

22. This comes via a trusted scientific colleague from the former Soviet Union, who was involved in the nuclear physics programme in the 1950s and heard this from "the grapevine" decades ago. While I am confident of the reliability of my colleague, who also trusts his source, now dead, I cannot assess the accuracy of his source or his memory after so long.

23. According to Popov, Pontecorvo "was regarded as so important that he was interviewed by Beria" (quoted in Pincher, *Treachery*, chap. 48 and Chapman Pincher interview, November 14, 2014). Given Pontecorvo's significance as a nuclear scientist, the fact that the interview allegedly took place in the Kremlin, and Beria's role in the Soviet atomic programme, this is plausible. However, my informant was unable to confirm Beria's presence. "Grey and mustard" is artistic license based on my personal experience of Soviet government decor from the 1970s. Henry Ford's cars were famously available in any colour as long as it was black. A similar uniformity seems to have applied to walls in the USSR. I have assumed that the same was true on this occasion.

24. In Joan Hall's opinion, if Ted had been forced to choose between fleeing to the USSR or facing ten years in jail "or worse, the electric chair," his decision would have been obvious. Joan Hall interview, May 1, 2013.

25. Bruno's version of his arrival in the USSR was given decades later. Whether he was genuinely a willing participant, or if this was an example of revisionist history, only he knew. Clearly, if Bruno had the intention to move to the USSR before the Pontecorvos left England, they could have taken more suitable belongings. Anna had asked if she could hitch a ride with them en route to Italy. Her presence made space tight. She had to repack and leave a case behind in England. If Bruno were already planning to leave England forever, it would have been easy for him to politely tell Anna that there was no space for her, and then pack for his own needs more appropriately. Clothing and financial support could be provided by the Soviets, but personal memorabilia and family documents could not. Ronnie Reed of MI5 concluded as much in 1951.

26. Remark by anonymous relative, confirmed by Anna Pontecorvo, March 21, 2012.

27. This is the account Bruno gave to Miriam Mafai years later. The idea that Gil, at age twelve, had such a sophisticated understanding of socialism as to "scold" his mother seems like propaganda. If true, however, it would suggest that the family had discussed communist ideology in the home for some time.

28. Gil Pontecorvo interview, February 24, 2011.

29. Mafai, *Il lungo freddo.*

30. Gil Pontecorvo interview, August 25, 2011.

CHAPTER 15

1. Pollock, *Stalin and the Soviet Science Wars*, p. 91, as quoted in Turchetti, *The Pontecorvo Affair*, p. 184.

2. Ioffe, "A Top Secret Assignment"; Boris Ioffe e-mail, August 5, 2011; Giuseppe Mussardo interview, December 12, 2012.

3. Dubna could accelerate deuterons to a kinetic energy of 280 MeV, alpha particles to 560 MeV, and protons to 680 MeV.

4. Venedict Dzhelepov, in BPSSW. There is also some unsubstantiated gossip that Dzhelepov, who later became director of the laboratory, might have played a role in recruiting Bruno Pontecorvo to Dubna. Ugo Amaldi and Giuseppe Fidecaro interviews, November 9, 2013.

5. TNA KV 2/1888.

6. Joint Committee on Atomic Energy, "Soviet Atomic Espionage," April 1951, http://archive.org/stream/sovietatomicespi1951unit/sovietatomicespi1951unit_djvu.txt.

7. For example: "Prof Pontecorvo is said to have been seen [at Kamenice]" where "fresh uranium deposits have turned up" ("Two New Czech Atom Plants," *Daily Telegraph*, March 21, 1951).

8. Boris Ioffe video interview by Giuseppe Mussardo, 2012 Boris Ioffe e-mail, August 5, 2011; Ioffe's interview by Giuseppe Mussardo, viewed by author December 12, 2012.

9. Samoil Bilenky interview, October 12, 2013.

10. Boris Ioffe video interview by Giuseppe Mussardo, 2012, viewed by author December 12, 2012.

11. This was very similar to the "classical superbomb" developed by Edward Teller in the United States. The idea came to the USSR via Klaus Fuchs. See also Rhodes, *Dark Sun*, p. 256; In "A Top Secret Assignment," p. 25, Ioffe argues that the idea was developed by Soviet physicists but that its true origin was known only to a handful.

12. Ioffe, "A Top Secret Assignment," p. 28.

13. For the history of Pomeranchuk's book as well as his career, see http://www.kipt.kharkov.ua/itp/akhiezer/en/recollections/pomeranchuk/.

14. Burke, *The Spy who Came In from the Co-op*, pp. 13 and 122.

15. Lorna Arnold interview, March 18, 2013.

16. Bruno Pontecorvo press conference, 1955; quote is translation of Mafai, *Il lungo freddo*, p. 193.

17. And there the collection of logbooks stayed for sixty years. In 2013, in preparation for celebrations of Bruno Pontecorvo's hundredth birthday, some documents became available. It was at the celebration, held in Rome in September, that I met Gil and saw their content for the first time.

18. The hydrogen in heavy water consists primarily of deuterium, whose atomic nuclei consist of a proton linked to a neutron. As there are two constituents in this hydrogen nucleus, it is sometimes referred to as H2. Tritium, which consists of a proton and two neutrons, is thus H3. "Quadium," or H4, is a proton accompanied by three neutrons.

19. Dubna could generate collisions between alpha particles and solid targets. Alpha particles consist of two protons and two neutrons, which offered the hope that H4, which consists of one proton and three neutrons, might be present in the debris.

20. The material regarding H4 has no relevance other than as an interesting dead end in the Soviets' thermonuclear strategy. Having spent four decades as a nuclear and particle physicist, I can assert that H4 would be low on anyone's list of interests, if it even appeared at all. If one were to ask any competent students of nuclear physics about H4, they would immediately recognize that it would probably be so unstable as to be in effect nonexistent. Basic quantum theory tells us that at most two neutrons can coexist in a relatively stable state in hydrogen isotopes. H3 (tritium) is already radioactive and unstable. A third neutron, as in H4, would have to be on a higher rung of the energy ladder. The most probable way for H4 to decay would be to release a neutron and leave tritium, which it does in less than a thousandth of a billionth of a billionth of a second.

Bruno Pontecorvo, in 1950, was well versed in this field. It would be remarkable if this did not occur to him. Although the theory of the nuclear "shell model" is standard fare today, it was formulated only in 1948–49 by Maria Goeppert-Mayer and Hans Jensen, for which they later won the Nobel Prize. It is possible that this work was either not known or not accepted in the USSR in 1950. On balance it would seem likely that the Soviets were interested in the strategic possibility of producing "super-tritium" (H4), and the question was how to detect it. Bruno responded to this as an experimental challenge, ignoring the fact that well-established theory placed the existence of H4 in great doubt.

21. It was during this period that stories linking Pontecorvo with uranium appeared in the Western media and in intelligence reports delivered by Western embassies. For examples, see the *Daily Telegraph*, March 21, 1951, and the secret report sent from Trieste to MI5 on April 13, 1951: "PONTECORVO is alleged to be directing extraction of uranium pitchblende in KAMENICE about 100 km SE of PRAGUE" (TNA KV 2/1889, memo 234).

22. H4, in particular, is a topic raised at Dubna, in which Bruno had no prior interest. There is no record that anyone at Harwell, Bruno included, had shown any interest in this unusual isotope. The fusion reaction rate of deuterium with H4, were it stable, would exceed that of deuterium and tritium.

23. The alternative theories have various problems. One possibility is that he was ill and unable to work, but there is no evidence for this. Marianne did fall ill, but there is no evidence that this unduly hindered Bruno's physics career, and certainly not during these first months. Overall, it is clear that from at least April 1951, Bruno Pontecorvo was focused totally on particle physics. Claims in the Western media that he was involved in other activities can thus be eliminated, except during these first months. That his expertise would be called upon

seems certain; there is no reason to delay, so the gaps in the early record may be indicative, though not proof. It is intriguing, as in note 7 and note 21 above, that claimed sightings of Pontecorvo at uranium sites point towards early 1951, with no such reports after April 1951.

24. One further document from this period has come to light: a brief note from the Soviet Academy of Sciences, signed by Igor Kurchatov on July 27, 1951, and marked with the word "Approved." This document refers to a "Consolidated Programme of Research Work for Installation M." It appears to be an outline for a research programme in particle physics, based on his work recorded in his logbook from April 1951, and which later formed the core of the work reported in Bruno's logbooks from September 1951 onward. Source: M. G. Sapozhnikov, "Seminar on B. Pontecorvo's Life and Ideas," JINR, Dubna, undated.

25. Quoted in Close et al., *The Particle Odyssey*. p. 75.

26. Soviet Academy of Sciences, "Report of the Institute for Nuclear Problems" [in Russian], 1951, referenced in *Journal of Experimental and Theoretical Physics* 29 (1955): 265–273, ref. 2; p. 129 in BPSSW.

27. I am grateful to Gil Pontecorvo for accessing and translating this report.

28. For example, see http://www.britannica.com/EBchecked/topic/438712/Abraham-Pais and http://www.webofstories.com/play/murray.gell-mann/68.

29. Strange particles contain one or more strange quarks (or strange antiquarks). Particles without strangeness either have no strange quarks, or the number of strange quarks and antiquarks exactly balance, so that their individual positive and negative amounts of strangeness cancel.

30. Mafai, *Il lungo freddo*, p. 177.

31. Gil Pontecorvo interview, September 4, 2013.

32. Mafai, *Il lungo freddo*, p. 192.

33. Stella Rimington comment to author, June 14, 2013.

34. Pincher, *Treachery*, chap. 55 and Chapman Pincher interview, November 13, 2013.

35. Mafai, *Il lungo freddo*, p. 189; Miriam Mafai e-mail, 2012.

36. Mafai, *Il lungo freddo*, p. 189.

37. Mafai, *Il lungo freddo*, p. 186.

38. Mafai, *Il lungo freddo*, p. 181.

39. Stalin, who had engineered plots throughout his career, imagined conspiracies all around him and used the case of the Jewish doctors as an excuse to launch an anti-Semitic campaign. The obvious question of why these senior medical professionals, some of whom were also university professors, would commit such crimes was answered by a simple, blunt assertion: "They were agents of foreign powers."

40. Mafai, *Il lungo freddo*, p. 181.

41. Gil Pontecorvo interview, September 4, 2013.

42. Mafai, *Il lungo freddo*, p. 187.

CHAPTER 16

1. Mafai, *Il lungo freddo*, p. 192.

2. The two conversational interviews conducted by Arnold hardly merit such a colourful description. Bruno's reaction is more understandable, however, if he

had learned in August 1950 that the FBI interest in him had resurfaced, and that the Americans were pressing MI5 for action.

3. Quotes taken from British media reports of the *Isvestia* article and of the press conference, in TNA KV 2/1888.

4. Guido Pontecorvo to MI5, in TNA KV 2/1888.

5. The scientists' statement ("Atomic Energy Control," *The Times*, January 21, 1947) was mentioned in a speech to the UN Security Council by Andrei Gromyko, the Soviet Union's deputy minister of foreign affairs. Quoted in Laucht, *Elemental Germans*, chap. 6.

6. *Picture Post*, February 18, 1950: "British scientists answer the question all world is asking: what should be done about the hydrogen bomb?"; *Daily Mirror*, March 25, 1950, front page: "Scientists jib at H bomb jobs."

7. Years later, after the Krogers had been arrested in the UK, released, and returned once again to the USSR, Morris Cohen/Peter Kroger happened to run into George Blake, another infamous KGB spy, in Moscow. They arranged to meet. The authorities immediately ordered that no such meetings were to take place.

8. Mafai, *Il lungo freddo*, p. 192.

9. Bruno won the Stalin Prize in 1953. The cash value of the prize in 1950 was about a thousand times a typical monthly income. In other words, the winner received, in effect, a lifetime's salary in advance. Life in the Soviet Union had been hard following the war, and although by the 1950s food was plentiful, there was "no variety of other goods" (Sacha Adriana interview, January 23, 2012). In addition, the award also gave the recipient access to special shops, which under the circumstances would have been invaluable. As a leading Academician and member of the party, Bruno also had access to foreign goods, theatre tickets, special hotels, and other privileges off limits to the majority of citizens.

10. Reports of press conference, for example Mafai, *Il lungo freddo*, p. 191 et seq.

11. Gil still qualifies for a UK passport; it was Bruno that crossed the line and took his family with him. Gil Pontecorvo interview, August 25, 2011.

12. Mafai, *Il lungo freddo*, p. 275.

13. Mafai, *Il lungo freddo*, p. 147.

14. Mafai, *Il lungo freddo*, p. 205.

15. Ioffe, "A Top Secret Assignment"; Bruno Ioffe interview, December 2012.

16. Semen Gershtein in BPSSW.

17. Its proton kinetic energy was 600 MeV or 0.6 GeV. At CERN's synchrotron in 1959, the energy was 28 GeV.

18. Nino Zichichi interview, June 30, 2013.

19. Gian Carlo Wick, as reported in Mafai, *Il lungo freddo*, p. 212; Ugo Amaldi interviews, April 18, 2013, and September 12, 2013.

20. Mafai, *Il lungo freddo*, p. 212; Miriam Mafai interview, March 2012.

CHAPTER 17

1. Jack Steinberger, quoted in U. Dore and L. Zanello, "Bruno Pontecorvo and Neutrino Physics," available at http://arxiv.org/pdf/0910.1657.pdf.

2. Bruno Pontecorvo is one of a select group of scientists who have an equation inscribed on their tomb or memorial stone. Ludwig Boltzmann's headstone in Vienna records the thermodynamic passage from order to decay. Paul Dirac's eponymous equation for the electron is set in stone in London's Westminster Abbey.

3. The Cowan and Reines discovery paper is available at http://www.science mag.org/content/124/3212/103.

4. In the acknowledgments at the end of the paper, Pontecorvo thanks Pomeranchuk for a comment about an effect of relativity. This appears to be one of only two cases where he and Pomeranchuk had any discussion about fundamental physics; the other dealt with strange particles.

5. Venedict Dzhelepov in BPSSW.

6. Given that the K-zero turns into an anti-K-zero through a two-step process, what would the analogous mechanism be for a neutrino? Bruno's ideas on this subject seem to have matured when he considered the fact that an electrically neutral atom can be made from a negatively charged electron and a positively charged muon; he realized that a negatively charged muon accompanied by a positron forms a similar system. In 1947, he had showed that a muon decays into an electron and two neutrinos—actually, he now realized, one neutrino and one antineutrino. This implied that the atom (made of a positive muon and an electron) could convert into a neutrino and an antineutrino. In turn, this latter pair of neutral entities could materialize as a negative muon and a positron. The result would be the conversion of an atom into an anti-atom. This does not happen in reality, for reasons that became clear a few years later. However, this idea was an important step in the development of Bruno's thinking about neutrinos. In theory, the process described above could have allowed a neutrino to convert to an antineutrino, or vice versa.

7. Bruno Pontecorvo in *Journal of Experimental and Theoretical Physics* 36 (1959): 1615. Quotation taken from English-language version, "Universal Fermi Interaction and Astrophysics," reprinted in BPSSW, p. 164.

8. Schwinger combined relativity and quantum theory and applied them to QED (quantum electrodynamics). He won the prize in 1965 for work done in 1947.

9. Not entirely. There are subtle problems that remain. The full solution of "the Infinity Puzzle" is outlined in my book of that name.

10. Half a century later, the limit is about one in a trillion.

11. Published in *Journal of Experimental and Theoretical Physics* 37 (1959): 1751.

12. S. Oneda and J. C. Pati, "V-A Four-Fermion Interaction and the Intermediate Charged Vector Meson," *Physical Review Letters* 2 (February 1, 1959): 125.

13. This is true for pions that carry electric charge. There is an uncharged variety, which decays into two photons.

14. Quotes in this page come from Igor Zheleznykh, "Early Years of High-Energy Neutrino Physics in Cosmic Rays and Neutrino Astronomy (1957–1962)," Proceedings of the ARENA workshop, May 2005.

15. Bruno Pontecorvo, "Electron and Muon Neutrinos," in BPSSW, p. 167.

16. Dubna report P-376. Gil Pontecorvo was one of the interpreters at the conference, where he provided simultaneous translation from Russian to English. Gil Pontecorvo interview, September 22, 2011.

17. The switch from antineutrino to neutrino might appear trivial, but in practice electrons arising from neutrino interactions are more difficult to isolate. This is because of their ubiquitous presence in matter, which increases background noise. We live in a world of matter; this makes antimatter signals easier to identify.

18. Melvin Schwartz, Nobel Address, 1988.

19. M. Schwartz, "Feasibility of Using High-Energy Neutrinos to Study the Weak Interactions," *Physical Review Letters* 4 (1960): 306.

20. Melvin Schwartz, Nobel Address, 1988.

21. Leon Lederman noted that "Bruno Pontecorvo addressed the right question but with a hopeless approach." Namely, Bruno made "an interesting error" by assuming that the low number of neutrinos would be fatal—and therefore assumed that he would need an accelerator of huge intensity.

22. Where did Kurchatov obtain the crucial helium-3? It is notable that Pontecorvo's paper gave all the necessary details except one: it did not reveal how much helium-3 he had used. This fact had been declared a state secret. The reason, as everyone realized, was that the precious helium-3 was the "waste" from the production of tritium—the key ingredient of the hydrogen bomb. Therefore the amount was not revealed, although experimental nuclear physicists could probably have deduced it from the details of the experiment.

23. The inspiration here was an American theorist, John Bahcall, who in 1962 pointed out that these higher-energy neutrinos might be detectable. Davis learned of Bahcall's observation and made use of the chlorine-detector idea. This story is told in detail in Close, *Neutrino*, p. 69.

24. Ray Davis recollection, date unknown.

25. B. Pontecorvo and V. Gribov, *Physics Letters B* 28 (1969): 493 and B. Pontecorvo, *Journal of Experimental and Theoretical Physics* 26 (1968): 984.

26. I first used this analogy in my book *Neutrino*, which tells the story of Ray Davis.

27. 1 followed by 58 zeroes, which is ten billion trillion trillion trillion trillion, or ten octodecillion.

28. Ludo Pontecorvo interview, April 18, 2013.

29. Our galaxy, the Milky Way, has a diameter of roughly 100,000 light-years. The Large Magellanic Cloud is a satellite of the Milky Way, located about 160,000 light-years away. It is visible to the naked eye in the Southern Hemisphere.

30. When the electron-neutrinos convert the neutron in deuterium to an easily visible proton, a direct measurement is possible.

31. Art McDonald interview, August 30, 2011, and e-mail, December 17, 2011.

CHAPTER 18

1. Gil Pontecorvo interview, September 22, 2011.

2. Mafai, *Il lungo freddo*, p. 200.

3. Mafai, *Il lungo freddo*, p. 216.

4. Mafai, *Il lungo freddo*, p. 277.

5. Interview with former colleague from Canada, name withheld, 2013.

6. Jackson, "Snapshots of a Physicist's Life," p. 23.

7. Jackson, "Snapshots of a Physicist's Life," p. 23, and J. David Jackson e-mail, August 28, 2012.

8. Gillo Pontecorvo, quoted in Mafai, *Il lungo freddo*.

9. Rodam Amiredzhibi, quoted in Mafai, *Il lungo freddo*, p. 218.

10. Giuseppe Longo, presentation at Bruno Pontecorvo centenary conference, Rome, September 12, 2013. At this same conference, various former colleagues confirmed the intensity of Bruno's relationship with Rodam.

11. Contrary to a common misconception, they never married. (The Wikipedia biography of Bruno Pontecorvo repeated this misconception, until corrected by me in 2014.)

12. Bruno stayed married to Marianne, even late in life. He eventually became concerned that he had no documentary proof of their marriage—the certificate was yet another item they had failed to bring on their flight. Remark by anonymous colleague.

13. Interview with Russian colleague, name withheld, 2012.

14. Mafai, *Il lungo freddo*, p. 218.

15. Mafai, *Il lungo freddo*, p. 223.

16. Gillo Pontecorvo, as related to the author by an anonymous colleague of Bruno Pontecorvo, 2013. This is consistent with Bruno's own statement to Charles Richards, *The Independent*, August 2, 1992.

17. Mafai, *Il lungo freddo*, p. 256.

18. Semen Gershtein in BPSSW.

19. Interview with Charles Richards, *The Independent*, August 2, 1992.

20. Semen Gershtein in BPSSW.

21. This is memory of a remark Gribov made to me around 1990; it was corroborated by Valery Khoze on December 6, 2012.

22. Mafai, *Il lungo freddo*, p. 193.

23. Gil Pontecorvo interview, August 26, 2013.

24. Anna Pontecorvo interview, November 11, 2011.

25. Tito Pontecorvo, quoted in Mafai, *Il lungo freddo*, p. 280.

26. Mafai, *Il lungo freddo*, p. 281.

27. A list of Politburo members according to Wikipedia can be found at http://en.wikipedia.org/wiki/List_of_members_of_the_Politburo_of_the_Communist_Party_of_the_Soviet_Union_in_the_1970s.

28. Mafai, *Il lungo freddo*, p. 282; Miriam Mafai interview, March 2012.

29. Tito bred Akhal-Teke stallions, and became outstandingly successful at it, as this video shows: http://www.youtube.com/watch?v=aaae5M3NgaI.

30. Mafai, *Il lungo freddo*, p. 271.

31. Tania Blokhintseva interview, August 24, 2012.

32. Mafai, *Il lungo freddo*, p. 284.

33. Eugenio Tabet interview, September 12, 2013; Eugenio Tabet e-mail correspondence, October 2013.

34. Ugo Amaldi interview, September 12, 2013.

35. Tania Blokhintseva interview, August 24, 2012.

36. Gil Pontecorvo interview, August 26, 2013. The visit was during New Year's 1975, and the events described took place in the first week of January.

37. J. Laberrigue-Frolow in BPSSW, p. 466.

38. Peter Minkowski interview, around July 1, 2013. The visitor from the Energy Department was the late Peter Rosen, a physicist who had been the head of the theoretical physics division at Los Alamos.

39. Photo in BPSSW, p. 467.

40. A recording of Pontecorvo's talk at CERN is available at http://cds.cern.ch /record/1002188?ln=it.

41. I was able to talk with Miriam Mafai shortly before she died in 2012 to verify that her narrative came verbatim from Bruno Pontecorvo, and discuss the extent to which it reflected her own vision. In her opinion the comments about life in the Soviet Union reflected Bruno's philosophy, adding enigmatically, "There are things you cannot understand, *sauf si vous étiez communiste*." Miriam Mafai interview and e-mails, March 1, 2012.

42. Interview with Charles Richards, *The Independent*, August 2, 1992.

43. Anna Pontecorvo interview, November 11, 2011.

44. Lev Okun in BPSSW, p. 501.

45. Tania Blokhintseva interview, August 24, 2012.

46. Irina Pokrovskaya in BPSSW, p. 508.

47. Tania Blokhintseva in BPSSW, p. 494.

48. Ludo Pontecorvo interview, April 18, 2013.

49. Irina Pokrovskaya in BPSSW, p. 508. Marianne Pontecorvo died in 1995.

50. The full story involves experiments conducted in Japan, which detected neutrinos produced by cosmic rays, as well as solar neutrinos. This is covered in more detail in Close, *Neutrino*.

51. J. Bahcall and R. Davis, "The Evolution of Neutrino Astronomy," *Essays in Nuclear Astrophysics* (Cambridge: Cambridge University Press, 1982), pp. 243–285.

CHAPTER 19

1. TNA KV 2/1888.

2. Mafai, *Il lungo freddo*, p. 295. Luisa Bonolis, an Italian historian of science, tried to obtain original tapes or transcripts of Mafai's interviews, but was unsuccessful. Bonolis assessed Bruno's responses to Mafai, as reported in the original Italian, and found them to be ambiguous, as if he did not want to face the reasons for his flight. Luisa Bonolis interview, December 13, 2012.

3. Mafai, *Il lungo freddo*, p. 293.

4. Quote originates with Franz Kafka.

5. Ugo Amaldi interview, April 18, 2013.

6. Jack Steinberger interview, September 13, 2011. At the Rome conference in September 2013, following a presentation by Bilenky, Steinberger publicly attacked the claims that Pontecorvo had demonstrated that the muon does not decay into an electron and a photon, calling them "inventions of Mr Pontecorvo" made "later in life as he remembered [events]. Their work was published a few months after mine, which found the electron spectrum to be continuous [i.e., incompatible with decay into an electron and a photon] and there is no statement in Hincks and Pontecorvo that muon does not decay to electron and photon, which I did later with an Italian student." A written version of Steinberger's critique can be found in *CERN Courier*, February 24, 2014, available at http://cern courier.com/cws/article/cern/56229.

7. Semen Gershtein in BPSSW.

8. While the H-bomb was still in the future at the time of Pontecorvo's defection, its awful potential was already recognized. Yet there is no mention in any of the available British documents of any expertise that Pontecorvo had that would be relevant to the weapon. President Truman had publicly announced the decision to develop the H-bomb in January 1950, months before the Pontecorvo affair erupted. By the time of his defection, the United Kingdom was already aware of a number of technical details, including the advantage that heavy-water reactors could have in breeding tritium, which is the key ingredient of the bomb. In 1950, Pontecorvo had been working on such a reactor for six years, and was one of the world's leading experts in the field. When he arrived in the USSR, he had already performed experiments involving tritium, and, we now know, he subsequently advised the Soviets on the details of heavy-water reactors. Yet there is not a whiff of concern about these facts in the documents from 1950 that the British have released. It is hard to believe that informed government scientists would have overlooked such a link.

This peculiar silence contrasts with the very vocal concerns about cosmic rays that were expressed in the media at the time. The discovery of strange forms of matter with weird properties in cosmic rays led to speculation that they might be a route to even-more-devastating weapons. If these unfamiliar particles held such power, and if the Soviets mastered them first, it was possible that the West could be held hostage and subjugated. This very subject was Bruno's latest passion when he defected—which could be seen as quite alarming. Fortunately, cosmic rays are of no use in building weapons of mass destruction, but no one knew that in 1950. This was therefore another aspect of Bruno Pontecorvo's expertise that the British government downplayed at the time.

9. Enrico Fermi as quoted by Herbert Skinner, TNA KV 2/1888.

10. Remark by Arnold Kramish to Michael Goodman, Michael Goodman interview, October 7, 2013.

11. One possible scenario is that during 1945 Bruno had planned to move to Harwell, but that his plans were disrupted by Nunn May's exposure at the end of that year, and arrest in 1946. With Nunn May out of the picture, Moscow no longer had an expert intimately involved within the Canadian project. Around this same time, Klaus Fuchs, who had been actively spying for Moscow at Los Alamos, was transferred to Harwell. Almost at once, Pontecorvo decided

to postpone his own move to the British laboratory. Instead, he announced his desire to remain in Canada. This placed Bruno Pontecorvo at the heart of the Canadian reactor project, during the period when it was producing the novel forms of uranium and other elements essential for atomic weapons and power. During this period, it appears that the Soviets' courier, Lona Cohen, was secretly given uranium for transmission to Moscow. In the meantime Klaus Fuchs continued to pass information to his Soviet contacts from his new post at Harwell. By 1949, Pontecorvo's work in Canada was completed, after which he also moved to Harwell as planned. There, he was confronted with the catastrophe of Fuchs's exposure, in 1950.

12. KGB documents that were declassified during Yeltsin's presidency. Albright and Kunstel, *Bombshell*, pp. 133–134 describes Cohen's Canadian visits. Lona Cohen also described them to a Russian historian in 1992 (confirmed to author in an e-mail from that Russian historian, an anonymous source, March 12, 2014). Bruno's travels are confirmed in TNA KV 2/1888.

13. Comment by Stella Rimington, June 14, 2013.

14. Herbert Skinner remark to MI5, TNA KV 2/1888: "The circumstantial evidence for him having handed over information to an agent while in Canada is extremely strong."

15. Lona Cohen died on December 23, 1992.

16. Yatskov could have been the "man of about forty" who first met the Pontecorvos in Leningrad on or around September 3, 1950. Yatskov's claim that he served as an aide-de-camp to Bruno also fits with Bruno's descriptions of the KGB employee who helped him in 1975, and whose identity Bruno refused to reveal.

17. Yatskov's career is described in Albright and Kunstel, *Bombshell*.

18. William Tyrer e-mails, February 28, 2014; e-mail from anonymous source, March 2, 2014.

AFTERWORD

1. Guy Liddell diary, June 2, 1950, TNA KV 4/472.

2. Mafai, *Il lungo freddo*, p. 153.

3. Mafai, *Il lungo freddo*, p 205.

4. Oleg Gordievsky e-mail correspondence, April 30, 2012.

5. Conversation with Pontecorvo family members, Rome, September 22, 2013.

6. As recalled by Gillo's son, Ludo Pontecorvo, interview April 18, 2013.

7. Guido Pontecorvo remarks to MI5, TNA KV 2/1888.

8. Anna Pontecorvo interview, November 11, 2011.

9. Ben MacIntyre, remark in lecture about his book *A Spy Among Friends: Kim Philby and the Great Betrayal* (Bloomsbury 2014).

10. Unnamed Harwell scientist, quoted by Godfrey Stafford, interview December 20, 2012.

11. Interview with Charles Richards, *The Independent*, August 2, 1992.

12. Quoted in Guy Liddell diary, September 18, 1945, TNA KV 4/466.

13. Joan Hall interview, May 1, 2013.

Bibliography

Albright, Joseph and Marcia Kunstel. *Bombshell: The Secret Story of America's Unknown Atomic Spy Conspiracy.* New York: Times Books, 1997.

Andrew, Christopher. *The Defence of the Realm: The Authorized History of MI5.* London: Allen Lane, 2009.

Andrew, Christopher and Oleg Gordievsky. *KGB: The Inside Story.* London: HarperCollins, 1990.

Andrew, Christopher and Vasili Mitrokhin. *The Mitrokhin Archive: The KGB in Europe and the West.* London: Penguin, 2000.

Arnold, Lorna. *My Short Century: Memoirs of an Accidental Nuclear Historian.* Palo Alto, CA: Cumnor Hill Books, 2012.

Bickel, Lennard. *The Deadly Element: The Men and Women Behind the Story of Uranium.* London: Macmillan, 1979.

Bonolis, Luisa. "Bruno Pontecorvo: From Slow Neutrons to Oscillating Neutrinos." *American Journal of Physics* 73 (2005): 487–499.

Broda, Paul. *Scientist Spies: A Memoir of My Three Parents and the Atom Bomb.* Leicester, UK: Matador, 2011.

Burke, David. *The Spy Who Came In from the Co-op: Melita Norwood and the Ending of Cold War Espionage.* Woodbridge, UK: Boydell Press, 2009.

Chaliand, Gerard and Arnaud Blin. *The History of Terrorism: From Antiquity to Al Qaeda.* Berkeley: University of California Press, 2007.

Close, Frank. *The Infinity Puzzle: Quantum Field Theory and the Hunt for an Orderly Universe.* New York: Basic Books, 2011.

Close, Frank. *Neutrino.* Oxford: Oxford University Press, 2010.

Close, Frank, Michael Marten, and Christine Sutton. *The Particle Odyssey: A Journey to the Heart of Matter.* Oxford: Oxford University Press, 2002.

Emling, Shelley. *Marie Curie and her Daughters: The Private Lives of Science's First Family.* New York: Palgrave Macmillan, 2012.

Farmelo, Graham. *Churchill's Bomb: A Hidden History of Science, War and Politics.* London: Faber and Faber, 2013.

Fermi, Enrico. *Note e Memorie (Collected Papers)*, Vol. 1, *Italy 1921–1938*. Chicago: University of Chicago Press, 1965.

Fermi, Laura. *Atoms in the Family: My Life with Enrico Fermi, Architect of the Atomic Age*. Chicago: University of Chicago Press, 1954.

Frisch, Otto. *What Little I Remember*. Cambridge: Cambridge University Press, 1979.

Gibbs, Timothy. "British and American Counter-Intelligence and the Atom Spies 1941–50." PhD thesis, University of Cambridge, 2007.

Goldschmidt, Bertrand. *Les rivalités atomiques, 1919–1966*. Paris: Fayard, 1967.

Goodman, M. *Spying on the Nuclear Bear: Anglo-American Intelligence and the Soviet Bomb*. Redwood City, CA: Stanford University Press, 2008.

Gowing, Margaret. *Britain and Atomic Energy, 1939–1945*. New York: St. Martin's Press, 1964.

Haynes, John Earl, Harvey Klehr, and Alexander Vassiliev. *Spies: The Rise and Fall of the KGB in America*. New Haven: Yale University Press, 2009.

Holloway, David. *Stalin and the Bomb*. New Haven: Yale University Press, 1996.

Ioffe, Boris. "A Top Secret Assignment." In *At the Frontier of Particle Physics: Handbook of QCD: Boris Ioffe Festschrift*, edited by M. Shifman. Singapore: World Scientific, 2002.

Jackson, J. D. "Snapshots of a Physicist's Life." *Annual Review of Nuclear and Particle Science* 49 (1999): 1–33.

Jeffery, Keith. *MI6: The History of the Secret Intelligence Service, 1909–1949*. London: Bloomsbury, 2010.

Jeffreys-Jones, Rhodri. *In Spies We Trust: The Story of Western Intelligence*. Oxford: Oxford University Press, 2013.

Johnson, Charles W. and Charles O. Jackson. *City Behind a Fence: Oak Ridge, Tennessee 1942–1946*. Knoxville: University of Tennessee Press, 1981.

Lamphere, Robert J. and Tom Shachtman. *The FBI-KGB War: A Special Agent's Story*. London: W. H. Allen, 1987.

Laucht, Christoph. *Elemental Germans: Klaus Fuchs, Rudolf Peierls and the Making of British Nuclear Culture, 1939–59*. Basingstoke, UK: Palgrave Macmillan, 2012.

Lipkin, Harry J. *Andrei Sakharov: Quarks and the Structure of Matter*. Singapore: World Scientific, 2013.

Macintyre, Ben. *Agent Zigzag: The True Wartime Story of Eddie Chapman, Lover, Betrayer, Hero, Spy*. London: Bloomsbury, 2007.

Macintyre, Ben. *A Spy Among Friends: Kim Philby and the Great Betrayal*. London: Bloomsbury, 2014.

Mafai, Miriam. *Il lungo freddo: storia di Bruno Pontecorvo, lo scienziato che scelse l'Urss*. Milan: Arnondo Mondadori Editori, 1992.

Manham, Patrick. *Snake Dance: Journeys Beneath a Nuclear Sky*. London: Chatto and Windus, 2013.

Marten, Michael. *Tim Marten: Memories*. Raleigh, NC: Snipe Books, 2009.

Moorehead, Alan. *The Traitors: The Double Life of Fuchs, Pontecorvo, and Nunn May*. New York: Dell, 1952.

Némirovsky, Irène. *Suite Française*. New York: Vintage, 2007.

Pais, Abraham. *Inward Bound: Of Matter and Forces in the Physical World*. Oxford: Oxford University Press, 1986.

Peierls, Rudolf. *Bird of Passage: Recollections of a Physicist*. Princeton, NJ: Princeton University Press, 1985.

Philby, Kim. *My Silent War: The Autobiography of a Spy*. New York: Modern Library, 2002.

Pinault, Michel. "Frédéric Joliot, la science et la société," thesis, University of Paris, February 1999.

Pincher, Chapman. *Too Secret Too Long*. New York: St. Martin's Press, 1984.

Pincher, Chapman. *Treachery: Betrayals, Blunders, and Cover-ups: Six Decades of Espionage Against America and Great Britain*. New York: Random House, 2009.

Pollock, Ethan. *Stalin and the Soviet Science Wars*. Princeton, NJ: Princeton University Press, 2008.

Pontecorvo, Bruno. Autobiographical notes (in Russian), 1988. Italian version: "Una nota autobiografica." In *Enciclopedia della Scienza e della Tecnica*. Milan: Arnoldo Mondadori, 1988–1989.

Pontecorvo, Bruno. *Bruno Pontecorvo: Selected Scientific Works*. 2nd ed. Edited by S. M. Bilenky, T. D. Blokhintseva, I. G. Pokrovskaya, and M. G. Sapozhnikov. Bologna: Società Italiana di Fisica, 2013.

Popov, F. D. *The Atom Bomb and the KGB*. Moscow: 2003.

Reed, Nicholas. *My Father, the Man Who Never Was*. Folkestone, UK: Lilburne Press, 2011.

Rhodes, Richard. *Dark Sun: The Making of the Hydrogen Bomb*. New York: Simon and Schuster, 2005.

Rhodes, Richard. *The Making of the Atomic Bomb*. New York: Penguin, 1988.

Robotti, Nadia. "The Beginning of a Great Adventure: Bruno Pontecorvo in Rome and Paris." Presented at the Pontecorvo centenary conference, Rome, September 2013.

Robotti, Nadia and Francesco Guerra. "Bruno Pontecorvo in Italy." Presented at the Pontecorvo centenary conference, Rome, September 2013.

Segrè, Emilio. *A Mind Always in Motion: The Autobiography of Emilio Segrè*. Berkeley: University of California Press, 1993.

Smith, Michael. *The Spying Game: The Secret History of British Espionage*. London: Politico's, 2004.

Sudoplatov, Pavel, Anatoli Sudoplatov, Jerrold L. Schecter, and Leona P. Schecter. *Special Tasks: The Memoirs of an Unwanted Witness—A Soviet Spymaster*. New York: Little, Brown, 1994.

Turchetti, Simone. *The Pontecorvo Affair: A Cold War Defection and Nuclear Physics*. Chicago: University of Chicago Press, 2012.

Wallace, P. R. "Atomic Energy in Canada: Personal Recollections of the Wartime Years." *Physics in Canada* 56, no. 2 (March/April 2000): 123–131.

Weart, Spencer R. *Scientists in Power*. Cambridge, MA: Harvard University Press, 1979.

West, Nigel. *The Circus: MI5 Operations 1945–1972*. New York: Stein and Day, 1983.

Williams, Robert Chadwell. *Klaus Fuchs: Atom Spy*. Cambridge, MA: Harvard University Press, 1987.

Zheleznykh, Igor. "Early Years of High-Energy Neutrino Physics in Cosmic Rays and Neutrino Astronomy (1957–1962)." Proceedings of the ARENA workshop, May 2005.

Index